In-Situ Synthesis of Aluminum Matrix Composites

Yutao Zhao

In-Situ Synthesis of Aluminum Matrix Composites

Science Press
Beijing

Springer

Yutao Zhao
School of Materials Science
and Engineering
Jiangsu University
Zhenjiang, China

ISBN 978-981-16-9119-5 ISBN 978-981-16-9120-1 (eBook)
https://doi.org/10.1007/978-981-16-9120-1

Jointly published with Science Press
The print edition is not for sale in China (Mainland). Customers from China (Mainland) please order the print book from: Science Press.

Translation from the Chinese language edition: *Yuanwei Hecheng Lvji Fuhe Cailiao* by Yutao Zhao,

Responsible Editor: Jiajia Zeng

This Springer imprint is published by the registered company Springer Nature Singapore Pte Ltd.
The registered company address is: 152 Beach Road, #21-01/04 Gateway East, Singapore 189721, Singapore

Preface

With the rapid development of modern science and technology, the requirements for high-performance materials continue to increase and composite materials have become one of the inevitable trends in the development of new materials. Metal matrix composites are among the high-performance structural materials with great development potential in the twenty-first century, especially the particle-reinforced aluminum matrix composites that not only have high specific strength and specific stiffness, but also have the advantages of high wear resistance, good thermal conductivity, low coefficient of thermal expansion, and great versatility in designs. It has broad application prospects in high-tech fields such as aerospace, defense, and advanced manufacturing.

The in-situ aluminum matrix composite materials are new type of composites prepared by advanced in-situ synthesis technology. Because the reinforcement phase in these composites is a thermodynamically stable phase that nucleates and grows in-situ from the aluminum matrix, the surface of the reinforcement is free of pollution, reinforcement/matrix compatibility is good, high interface bonding strength, high specific strength, specific stiffness, good toughness, fatigue resistance, creep resistance, good damping performance, low thermal expansion coefficient. Thus, it has become a promising direction with great application potential of metal matrix composites and has received extensive attention from scientists and industry experts. It will have a major impact on the transformation and development of aluminum alloy industry and supply–demand chain in near future.

Although there are many developments in the field of in-situ synthesis of aluminum matrix composites at home and abroad, most of them are published in the form of journal and conference articles and have not yet formed into a systematic monograph. This monograph is based on the author and his research team's undertaking of more than 30 research projects, including the National Natural Science Foundation of China, the National "863" Project, the Key and Doctoral Funds of the Ministry of Education, the Natural Science Foundation of Jiangsu Province, and the Special Project of High Technology and Achievement Transformation.

This book is written by Professor Zhao Yutao, Chen Gang, Kai Xizhou, Li Guirong, Zhang Songli, Jiao Lei, Zhang Zhenya, etc., of Jiangsu University. Chapter-wise contribution of authors is as follows: Professor Zhao Yutao was responsible for the overall concept of this book and wrote Chaps. 1 and 2; Chap. 3 is written by Professor Li Guirong; Chap. 4 is written by Associate Professor Zhang Songli; Chaps. 5 and 6 are written by Associate Professor Kai Xizhou; Chap. 7 is written by Zhang Zhengya; Chap. 8 is written by Associate Professor Jiao Lei; Chap. 9 is written by Professor Chen Gang. The whole book is drafted by Zhao Yutao and Chen Gang. The book content is novel, comprehensive in structure, and practical. It is forward-looking in the research field of metal matrix composite materials. It can serve as a reference for professionals, technical experts and college teachers engaged in the research and application of aluminum matrix composite materials and can also be used as a textbook or reference book for postgraduate students in the field of materials science at higher education institutions.

This book adopts or quotes literature in the form of publications of relevant scholars and experts at home and abroad (see the references at the end of each chapter of this book), and I would like to express my sincere thanks.

Due to the vast research and limited writing experience, it is inevitable that there might be some errors in the book; readers are welcome to criticize and correct.

Zhenjiang, China
September 2021

Yutao Zhao

Contents

Chapter 1
Introduction

1.1 The Development History of Metal Matrix Composites

In today's rapidly developing world of science and technology, the field of materials science is tremendously changing; new theoretical concepts, ideas, and technologies continue to emerge, and preparation of composites is a new trend. Composites are new materials obtained by optimizing the combination of two or more materials of completely different properties. These materials have passed through various stages of development from natural materials to artificial materials, from simply structured materials to complex structured and functional materials; each stage is gradually entering into a higher, more refined, and faster development phase, thus incorporating new high-tech ideas and rendering rapid development pace to aerospace, transportation, information, biology, and other industries [1].

Metal matrix composites (MMCs) are composite materials that use metals or alloys as the matrix and fibers, whiskers, particles, or a combination of these as reinforcements. Because composite materials receive the properties of metal or alloy matrix (plasticity, toughness) and ceramic reinforcements (high strength, high stiffness), they have the advantages of higher strength and elastic modulus, good high-temperature performance, excellent wear resistance, etc., thus having important applications and broad prospects in the fields of aerospace, automotive, electronics, packaging, and sports industries. Due to the alarming conditions of energy and environment for human survival and for further development of transportation, aerospace, electronics, information technology, national defense, military industry, and sports industries, the demand for lightweight and high-strength multifunctional MMCs is growing [2].

The USA and Japan are currently the world's leading countries in the research and development of metal matrix composites. U. S. Department of Defense has recognized the metal matrix composite materials as the key technology and invested a lot of money, manpower, making it the world leader. The USA has been conducting research on metal matrix composites since the 1960s. It entered into the practical phase in the 1970s and began to widely use MMCs in the aerospace industry in the

Y. Zhao, *In-Situ Synthesis of Aluminum Matrix Composites*,
https://doi.org/10.1007/978-981-16-9120-1_1

1980s. For example, boron fiber-reinforced aluminum matrix composite materials were used for the cargo tank truss of the USS Columbia, launched in 1987. After that, the MMCs industry has developed rapidly in the USA, and in 2000, the total production value of advanced MMCs in the USA had reached 20 billion US dollars [3].

In Japan, the research and development of MMCs started later, and the investment in the research of MMCs only began in the early 1980s. However, the development rate was very fast, and it took only about 20 years for Japan to quickly occupy a very important position in the production and application research of metal matrix composites in the world. Japan successfully manufactured large-scale long fiber, whisker and other types of MMCs, and only two years later, Japan's Honda Motor Company first applied this technology to the cylinder block piston by utilizing Al_2O_3 short fibers as reinforcement in aluminum alloy composites. Japan achieved large-scale industrial production such that it quickly reached the forefront of MMCs in the world. At present, according to incomplete statistics, there are at least 40 companies in Japan active in research and development of MMCs [4, 5].

According to the latest business survey of the American Business Information Corporation, the total MMCs market in the world reached more than 10,000 tons in 2014. At present, there are hundreds of unique MMCs companies in the world that have exclusive technologies (such as DWA's powder metallurgy), or specialize in a certain material (such as Alcan's aluminum-based MMCs), or focus on a specific product type (such as CPS's thermal packaging substrate). It is predicted that the global MMCs market will maintain an 8% annual growth rate in 2016. According to different application fields, the MMCs market can be subdivided into five categories: ① land transportation, ② electronics/thermal control, ③ aerospace, ④ industry, and ⑤ consumer products. Among these, land transportation (including automobiles and rail vehicles) and high value-added heat dissipation components are still the dominant markets for MMCs, with consumption rates exceeding 60% and 30%, respectively.

1.2 In-Situ Reaction Synthesis Technology

1.2.1 Self-propagating High-Temperature Synthesis (SHS) Method

The self-propagating high-temperature synthesis (SHS)method was proposed by Soviet scientists Merzhanov et al. [6] while studying the combustion of Ti and C compact mixture. The principle is to mix the raw materials of reinforced phase with the metal powder, and a compact is formed. Preheating and ignition in a vacuum or inert gas cause an exothermic chemical reaction between the reaction components, and the released heat causes the unreacted adjacent components to continue to react until the reaction is completed. The reaction product is a reinforcement phase

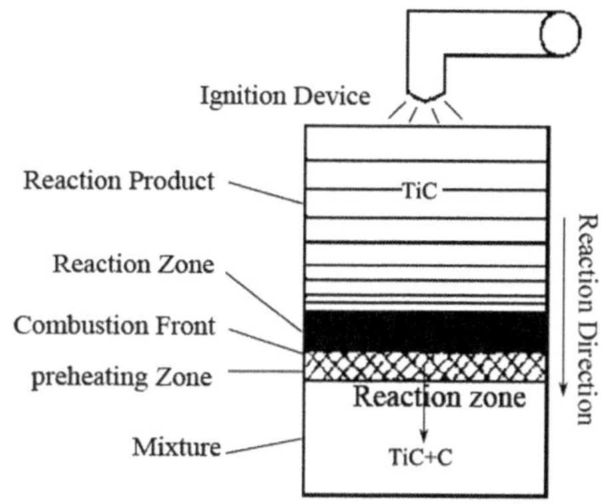

Fig. 1.1 Schematic diagram of conventional SHS method in reaction mode

which is dispersed in the matrix, and the particle size can reach submicron level. The schematic diagram of experimental method is shown in Fig. 1.1.

Compared with traditional methods, the SHS method has the following main advantages: ① simple process equipment, short process cycle, and high production efficiency; ② low energy and material consumption; ③ extremely high temperature during the synthesis can self-purify the product, and at the same time, very fast heating and cooling rates can result in products with non-equilibrium structure, so the product quality is good. The main disadvantages of this method are: ① high porosity, low density, and secondary processing are needed to obtain the final product; ② the reaction process is fast and difficult to control; ③ the process is prone to high defect concentration, and non-equilibrium transition phases are easy to appear in the product; ④ it is difficult to directly synthesize composite materials with low particle content.

1.2.2 Exothermic Dispersion (XD^{TM}) Method

The XD^{TM} (exothermic dispersion) method [7] was proposed by the American Martin Marietta Laboratory in 1983, and it was improved on the basis of SHS method. Its basic principle is: ① uniformly mixing the raw materials of reinforcement phase with the metal-based powder in specified proportions, cold-press or hot-press to form a block, ② preheating the sample at a suitable heating rate to a certain temperature (usually higher than the melting point of the matrix and lower than the melting point of the reinforced phase), ③ an exothermic chemical reaction taking place between the raw materials to generate the reinforcement phase which has small size and is dispersed within the matrix. The schematic diagram of XD^{TM} method for preparing metal matrix composites is shown in Fig. 1.2.

Compared with the SHS method, the XD^{TM} method has the following advantages: ① high density, because of the reaction carried out in a liquid/molten matrix; ② no ignition loss, equipment is simple to use, and the cost is low; ③ the melting point

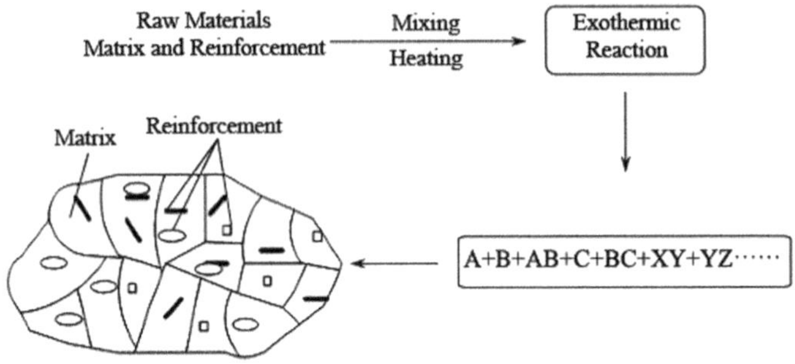

Fig. 1.2 Schematic diagram of metal matrix composites prepared by XD^{TM} method

of aluminum matrix is low (about 670 °C), so heating to more than 700 °C is fine. However, this method has also some shortcomings: ① the raw materials required for reactive synthesis are all powders, which are limited by the variety of powder supply; ② it requires many long process cycles, such as, ball milling, powder mixing, vacuum degassing, compact molding, reaction sintering; ③ it cannot cast complex shapes, and only some products with simple shapes can be produced.

1.2.3 Contact Reaction (CR) Method

The contact reaction (CR) method [8] was developed on the basis of SHS method and XD^{TM} method. Its basic principle is: to fully mix the raw elements or the compounds containing raw elements of reinforced phase to make a compact and then add it into the molten alloy matrix; the green compact has a direct contact with the high-temperature liquid alloy matrix to carry out a chemical reaction which generates a reinforcement phase. Mechanical stirring is carried out to further disperse the reinforcement phase, and then, it is poured into the mold. The process flow diagram is shown in Fig. 1.3.

At present, the CR method has successfully prepared TiC/Al, TiC/Al-Si, TiB_2/Al and other composite materials. Murthy and Rao [9] and Premkumar and Chu [10]

Fig. 1.3 Process flow chart of the CR method

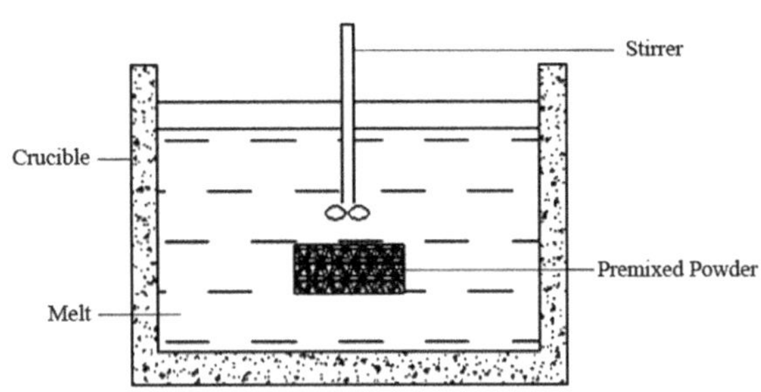

have prepared TiCp/Al composites by the CR method. The size of reinforcement phase in composites prepared by the former is only 0.3–0.7 μm, the distribution is uniform, and the interface is well intact. M. I. Johnson et al. used titanium powder and boron powder to prepare TiB_2/Al composite material by mixing and pressing it into the molten aluminum. It was found that the TiB_2 particles formed by the reaction could easily become the core of nucleating crystals in solidified microstructure, thereby refining the grains and improving the mechanical properties. H. Asanuma prepared the Al_3Ti/Al composite material by direct contact reaction of titanium powder and aluminum powder and found that the morphology of hard and brittle Al_3Ti particles changes with the cooling rate and the amount of titanium. When the cooling rate and the amount of titanium are reduced, its wear resistance is significantly improved and even becomes superior to that of the SiCp/Al composite with the same volume fraction.

1.2.4 Vapor Liquid Synthesis (VLS) Method

The vapor liquid synthesis (VLS) method is a patented technology invented by Koczak et al. [9]. The schematic diagram of the device used for this method is shown in Fig. 1.4. The principle is: carbon or nitrogen-containing inert gas is passed through high-temperature metal melt, carbon or nitrogen generated by the decomposition of gas react quickly with Ti in the alloy to produce thermodynamically stable fine TiC or TiN particles. The principle reactions can be written as follows:

$$CH_4 \rightarrow [C] + 2H_2(g) \tag{1.1}$$

$$M - X + [C] \rightarrow M + XC(s) \tag{1.2}$$

$$2NH_3(g) \rightarrow 2[N] + 3H_2(g) \tag{1.3}$$

$$M - X + [N] \rightarrow M + XN(s) \tag{1.4}$$

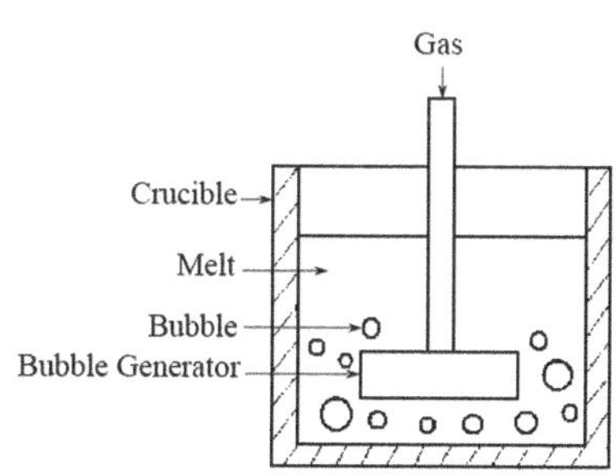

Fig. 1.4 Sketch of VLS device

where M is a metal; X is an alloy element; and M-X is a base alloy.

At present, this technology has been successfully used to prepare particle strengthened AlN/Al, TiN/Al, SiC/Al–Si and HfC/Al, TaC/Al, NbC/Al MMCs. The carrier inert gas used in this technology is Ar; carbon-containing gas used is generally CH_4, but C_2H_6 or CCl_4 can also be used; the source of nitrogen is generally N_2 or NH_3. Different gases require different decomposition temperatures, but they can be fully decomposed at 1200–1400 °C. American Lanxide company [10] passed N_2 (or NH_3) through molten aluminum–titanium alloy to produce AlN, TiN particle-reinforced aluminum matrix composite, and it was found that adding appropriate amounts of magnesium and lithium can reduce the surface energy of liquid aluminum and improve the reinforcement/matrix interface compatibility. The advantages of VLS method are: ① reinforcement particles' generation rate is fast, the surface is clean, and the particle size is fine (0.1–5 μm); ② the process continuity is good; ③ after reaction, the melt can be closer to net shape; ④ low cost. This method has also some shortcomings: ① the types of strengthening phases are limited; ② the particles' volume fraction is not high enough (generally less than 15%); ③ the required processing temperature is very high, generally between 1200 and 1400 °C.

1.2.5 Lanxide Method

The Lanxide method [11] developed by Lanxide company in the USA also utilizes the principle of above-mentioned gas–liquid reaction method, which combines the direct metal oxidation (DIMOX™) method and the metal pressureless infiltration (PRIMEX™) method. The schematic diagram of Lanxide method for preparing composite materials is shown in Fig. 1.5.

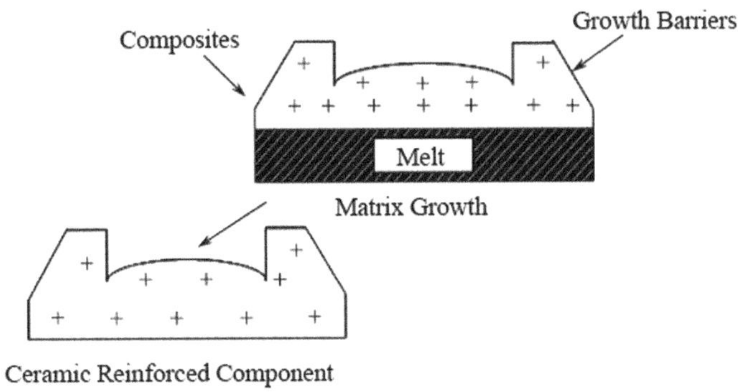

Fig. 1.5 Schematic diagram of composite materials prepared by Lanxide method

1.2.5.1 DIMOX™ Method

The principle of the DIMOX™ (direct metal oxidation) method [12] is to expose high-temperature liquid metal (Al, Ti, Zr, etc.) to air to first oxidize the surface of melt so as to form an oxide film (Al_2O_3, TiO_2, ZrO_2, etc.); the inner metal gradually diffuses to the surface through oxide layer and is oxidized after being exposed to the air. This process is repeated, and the metal oxide-enhanced MMCs or CMCs are finally formed. Murthy used this method to prepare Al_2O_3 reinforced Al–Mg-Si alloy matrix composite and studied the evolution of microstructure through crystal growth experiments. Dhandapani and Narciso used the DIMOX™ method to prepare Al_2O_3 and SiC-reinforced aluminum matrix composites. In order to ensure that the oxidation reaction of metal proceeds to completion, Newkirk et al. [13] studied the addition of a small amount of Mg, Si, and other alloying elements to Al. These elements were found to destroy the continuity of Al_2O_3 film on melt surface and kept the micro-channels between molten aluminum and Al_2O_3 open and reduced the surface energy of liquid aluminum alloy, thereby enhancing the compatibility of Al_2O_3 and the molten aluminum, so that the oxidation reaction continues to completion.

1.2.5.2 PRIMEX™ Method

The PRIMEX™ (pressureless metal infiltration) method [14] differs from the DIMOX™ method in that the atmosphere used is non-oxidizing. The principle of this method is that the base alloy is placed inside a heating furnace with controlled atmosphere and temperature is raised above the liquids temperature of base alloy. The reinforcement phase preform is immersed into the molten matrix. Two processes occur simultaneously: The first one is the penetration of liquid alloy into the ceramic preform under the action of ambient atmosphere; the second one is the reaction of liquid alloy with surrounding gas to generate new particles. Hunt [14] put aluminum ingots containing 3–10% Mg and Al_2O_3 ceramic preforms into an (N_2 + Ar)-mixed atmosphere furnace; when heated to above 900 °C and kept for some time, the above-mentioned two processes occurred simultaneously. After solidification, an aluminum matrix composite reinforced with in-situ formed AlN particles and original Al_2O_3 particles in the preform was obtained. The study found that the number density and the size of in-situ formed AlN particles mainly depend on the penetration rate of aluminum liquid, which is related to the N_2 partial pressure in the ambient atmosphere, the temperature and composition of the melt. Therefore, the microstructure and performance of composite material can be easily controlled by adjusting the composition of the melt, the partial pressure of N_2, and the processing temperature.

At present, the Lanxide method is mainly used to prepare aluminum matrix composites or ceramic matrix composites. The volume fraction of the strengthening phase can reach up to 60%. The types of strengthening phase particles usually include Al_2O_3, AlN, SiC, MgO, and other particles. The process is simple, and the cost of raw materials is low; in addition, the final product is nearly in net shape. Its products have been employed in automobiles, gas turbines, and heat exchangers.

1.2.6 Mixed Salt Reaction (LSM) Method

LSM method is the patented technology of London & Scandinavian Metallurgical (LSM) company [15]. Its basic principle is that after mixing of Ti and B containing salts (such as KBF_4 and K_2TiF_6), it is added into the melt at a high temperature. Under high temperature, Ti and B in the salts will be reduced by metal and react with metal melt to form TiB_2 reinforcement phase particles. After the removal of unnecessary by-products, casting, and solidification, an in-situ TiB_2 particles reinforced metal matrix composite is obtained. The schematic diagram of LSM method is shown in Fig. 1.6.

Wood et al. [16] added KBF_4 and K_2TiF_6 into the liquid Al-7%Si-0.3%Mg melt and obtained a uniform distribution of 4_8% in-situ TiB_2 particles in the matrix with a size of 0.5–2 μm. Compared to SiC/Al composite with the same reinforcement content, the obtained TiB_2/Al composite has superior mechanical properties and abrasion resistance. Research by Davies et al. [17] showed that the tensile strength, yield strength, and elastic modulus of 9% TiB_2/2021Al and 8%TiB_2/A356 composites prepared by LSM method are higher than the matrix, while the elongation is lower than the matrix. The main advantages of LSM method are as follows: ① simple process, short cycle, no need for vacuum and inert gas protection system, and no need of ball milling for powder mixing, compaction, and other processes; ② materials can be directly cast for molding, easy for mass production and promotion; ③ raw materials are salts, which have a wide range of sources and are low cost. This method also has some disadvantages: ① the generated TiB_2 particles are often covered with salt film, which weakens the strengthening effect of TiB_2; ② a large amount of gas escapes during the reaction process and a good ventilation device must be required; ③ the volume of generated particles is low, liquid slag is difficult to remove, and has a corrosive effect on the crucible and operating tools.

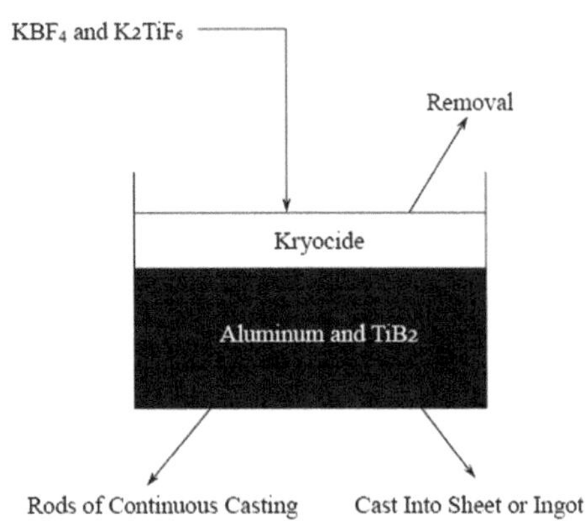

Fig. 1.6 Schematic diagram of mixed salt reaction [15]

1.2.7 Direct Melt Reaction (DMR) Method

The direct melt reaction (DMR) method [18], also known as the melt reaction method, is a new method developed by combining the characteristics of CR method and LSM method. The basic principle is that solid particles or powders, which form reinforcement phase, are added to the surface of the molten aluminum alloy at a certain temperature, and then, the melt is fully stirred to prepare the endogenous particles reinforced composite. Compared with other preparation methods, the characteristics of melt direct reaction process are: ① This process is based on the existing aluminum alloy smelting process, it directly generates reinforced phase particles in the melt, which can be cast into various shapes. So, the process is simple, the preparation time is short, the cost is low, and it is easy to promote; ② the size and distribution of reinforcement particles is easy to control, and the number density can be adjusted within a wide range; ③ The method can produce composite materials with high strength and high toughness simultaneously.

Nakata et al. [19] prepared TiC_p/Al composite material by DMR method. Al-Ti alloy ingot containing stable carbide forming elements and unstable carbide SiC or Al_4C_3 particles was used as raw materials. First, under the protection of argon, the Al–Ti alloy was melted in MgO crucible and superheated to 1200 °C, and then, the dried SiC or Al_4C_3 particles were added into the melt and fully stirred to produce TiC particles. Finally, the melt was poured into a metallic mold to prepare TiCp/Al composite. The TiC particles synthesized by this process are fine, with a size close to 1 μm, and the volume fraction can reach up to 10%. It was also found that adding Mg or Cu to the in-situ formed TiCp/Al melt can further improve the ultimate tensile strength, yield strength and elastic modulus of the composite.

1.2.8 Other Methods

Tjong et al. [20] have conducted a series of studies on the reaction hot pressing (RHP) method to prepare $(Al_3Ti + TiB_2 + Al_2O_3)/Al$, $(TiB_2 + Al_2O_3)/Al$ and other composite materials. The research has shown that the higher the content of Al_3Ti particles in the $(Al_3Ti + TiB_2 + Al_2O_3)/Al$ composite, the higher the tensile strength and the lower the elongation; while in $(TiB_2 + Al_2O_3)/Al$ composite which does not contain Al_3Ti particles, the tensile strength and elongation are higher, and low cycle fatigue life and high cycle fatigue life are higher than $(Al_3Ti + TiB_2 + Al_2O_3)/Al$ composite. Yang et al. [21] prepared TiC/7075 composite material by injection molding technology and studied the effect of Ti/C ratio on the reinforcement phase. H. Asgharzadeh et al. used reactive mechanical alloying (RMA) technology in the Ar/O_2 environment and prepared $Mg_2Si/6063Al$ composite material with high micro-hardness value and reinforcement particles of 40–100 nm size.

In summary, there are many technologies for in-situ reaction synthesis of composite materials, and each has its own advantages. Among these, the melt direct reaction method is based on the existing aluminum alloy smelting process, and the particle-reinforced phase is formed directly in the aluminum melt by in-situ reaction. It can be directly produced in castings of various shapes. The process is simple, the cost is low, and it is easy to promote. It has broad application prospects, however, the composite materials prepared by this method face the problems of coarse reinforced particle size and its non-uniform distribution within the matrix. Therefore, the application of different external fields (electromagnetic, ultrasonic, acousto-magnetic coupling field, etc.) has become the current research hotspot to optimize the solidification microstructure, achieve uniform distribution of reinforcement phase particles, and thus improve the mechanical properties of composite materials.

1.3 Current Status of In-Situ Aluminum Matrix Composites

Aluminum-based composite materials are known as one of the most competitive green engineering materials in the twenty-first century due to their excellent comprehensive properties such as low density, high specific strength, high specific modulus, and good wear resistance, especially in in-situ aluminum matrix composites prepared by reactive synthesis technology (i.e., in-situ reactive synthesis). Since the reinforced particles are grown from the matrix nuclei, they have the advantages of good thermal stability, small size, uniform distribution, and good compatibility with the matrix, which rectifies a lot of shortcomings such as the large size of reinforced particles and poor interface bonding faced in traditional synthesis methods. It has achieved a great attention of domestic and foreign researchers and has shown broad application prospects in high-tech fields such as aerospace, national defense, and advanced manufacturing. Industrially producing aluminum matrix composite has already appeared in foreign countries and successfully employed in the engine fan guide vanes of aircrafts such as F-16 fighter ventral fins, Boeing 777 and Airbus A380, as well as in compressor stator blades, high-speed train brake disks, light and efficient energy-saving automobile engine block, cylinder heads, wheel hub and as heat sink components of chips for high-performance computers. In China, the research work on aluminum matrix composite materials began in the early 1980s. At present, many milestones have been achieved in the basic theory, but the practical application is still in the experimental stage. The next 10 to 30 years will witness the China's large aircraft, high-speed rail, and light energy-saving vehicles. During the rapid development of manufacturing industry, the demand for high-performance, low-cost in-situ aluminum matrix composite is very urgent and huge. At present, the research on in-situ aluminum matrix composite is mainly focusing on the following aspects at home and abroad.

1.3.1 Design and Simulation of In-Situ Aluminum Matrix Composites

The system design and simulation of high-performance in-situ aluminum matrix composites involves the type, shape, size, content and distribution of particles in the matrix, and the structural characteristics of the particle/matrix interface. Since the in-situ aluminum matrix composites are prepared by means of the chemical reaction between the base aluminum alloy and reactant (gas phase, liquid phase or powder/solid phase) at an appropriate temperature; therefore, the reaction system determines the preparation method and final properties of in-situ composite materials. The level of preparation difficulty and the cost are important factors. The selection criteria of reaction systems include the following aspects: ① the performance of in-situ generated reinforcement; ② the difficulty of reinforcement morphology control; ③ nature of the interface between the reinforcement and the matrix; ④ the severity of reaction and the initial reaction temperature; ⑤ source and price of the reactants. The reinforcement phases produced by in-situ reaction synthesis are mainly oxides, nitrides, carbides, borides, and intermetallic compounds. The common ones are Al_2O_3, MgO, AlN, TiC, ZrC, TiB_2, ZrB_2, and Al_3Ti, Al_3Zr, etc. Aluminum-based in-situ synthesis reactions are mainly divided into four categories: direct synthesis reaction, oxide reduction reaction, inorganic salt reaction, and catalytic reaction. At present, the in-situ synthesis reaction systems for aluminum matrix composite materials are mainly Al–Ti–B, Al–Ti–C, Al–Ti–B–O, Al–Zr–B, Al–Zr–O, Al–Zr–B–O, etc. [22].

Since the spatial configuration of reinforcement particles in in-situ aluminum matrix composites and at reinforcement/matrix interface structure is multi-scale and multi-component, it is very complicated. At present, there is no designated theory and simulation method for in-situ aluminum matrix composites preparation. Existing studies have shown that the type, size, content, and distribution of reinforcements are the key factors in the design of in-situ aluminum matrix composites; reducing the size of the reinforcement or increasing its content is conducive to improving the yield strength of composite, and this mechanism can be explained by Ashby's strain gradient theory [23]. As the applications of aluminum matrix composite materials are extended to important load-bearing structures, the elongation and fracture toughness have become the focus of research work. During the loading process, the stress concentration near the interface causes the particles to crack, and then the cracks expand into the matrix and connect with each other, which becomes the key factor restricting the improvement of toughness of the composite. Song et al. used the Weibull distribution and the Eshelby model to establish particle size and content, elongation and fracture toughness models which can better describe the relationship between reinforcement particles and properties of the composites.

1.3.2 Preparation and Forming Technology of In-Situ Aluminum Matrix Composites

In the past two decades, there has been a lot of research on in-situ aluminum matrix composites at home and abroad, and in-situ synthesis technology has been developed rapidly. In summary, there are mainly the exothermic dispersion (XD™), vapor–liquid synthesis (VLS), self-propagating high-temperature synthesis (SHS), mixed salt reaction (LSM), reactiver mechanical alloying synthesis (RMA), and direct melt reaction (DMR) methods, etc. Among these, the direct melt reaction method is considered to be the most promising technology for industrial exploitation due to its simple process, low cost, short cycle, and easy large-scale production. However, this new technology is still imperfect, mainly in the following aspects: ① There are few reaction systems, mostly falling in Al–Ti–X (Al–Ti–O, Al–Ti–B, Al–Ti–C) system; ② the initial reaction temperature is high, often higher than 900 °C, or even more than 1000 °C, deteriorating the liquid aluminum; ③ The formation and growth mechanisms of particles are rarely engaged, and current research is mainly focusing on the phase identification and morphology observation of particles; ④ The study of interface structure between the reinforced particles and the matrix is not comprehensive. Most of the literature focuses on the observation of interface morphology and provides no information about whether or not there is a certain orientation relationship between the particles and the matrix. Little research is carried out on the interface modeling, simulation of interfacial atoms, and the quantitative analysis of dislocations near the interface area; ⑤ most of the studies have focused on room temperature tensile properties, and there are very few reports on the super-plasticity and high-temperature creep properties; ⑥ research on wear resistance is rarely reported.

In the synthesis of high-performance in-situ aluminum matrix nanocomposites, there are still problems that have not been resolved internationally: ① The morphology and size of generated reinforcement particles are not easy to control, the submicron level can be achieved, but the nanometer level is difficult to achieve; ② when volume fraction is more than 3%, the generated nanoparticles can easily agglomerate; ③ the reaction time is too long, it is not thorough, and there exist many intermediate phases. Therefore, the in-situ reaction synthesis technology needs to be further improved and innovated. In view of the existing problems in the preparation of in-situ aluminum matrix composites from traditional synthesis technology and the above-mentioned international issues, the author's research group seeks new ideas to control and improve the synthesis process of in-situ aluminum matrix composites by using alternating electromagnetic fields, high-energy ultrasonic fields, acoustic-magnetic fields coupling, and through in-depth discussion of the effects of magnetic or ultrasonic fields on in-situ reinforcements' nucleation and growth mechanisms, structure–property relationships of in-situ aluminum matrix composites. At the same time, the establishment of related models and dynamic equations has important engineering applications.

Moreover, the forming and processing of in-situ aluminum matrix composites is one of the key technologies determining their applications. In-situ aluminum matrix composites are composed of hard reinforced particles distributed in soft aluminum matrix and their plastic formability is poor. Components with complex shapes are difficult to form and process, which limits their applications to a large extent. In order to solve this problem, near-net shape forming method is used to produce in-situ aluminum matrix composite components, which not only saves raw materials but also greatly reduces costs. The key to near-net shape forming method is superplastic deformation, the strain rate during traditional superplastic forming is too low (in the range of 10^{-5}–10^{-3} s^{-1}), which is not conducive for mass production. The development of high-strain rate superplastic (HSRS) molding with a strain rate greater than 10^{-2} s^{-1} can greatly improve production efficiency. Therefore, high-strain rate ($>10^{-2}$ s^{-1}) superplastic molding is a solution to the plastic forming of aluminum matrix composites in recent years. It is the best route for processing and overcoming the shortcomings of low strain rate ($<10^{-3}$ s^{-1}) in traditional superplastic forming. What's more, it has become a research hotspot in high-tech fields of in-situ aluminum matrix composites and advanced manufacturing [24].

Due to the presence of reinforced particles, the solidification process of in-situ aluminum matrix composites makes the solidification of aluminum matrix more complicated. The existing metal solidification theory is obviously not applicable: the generation of reinforcements, the temperature distribution during the solidification process, thermodynamic and kinetic processes of crystal growth all change; the fluid flow behavior, solute redistribution behavior, and the structure of the solidified part also change accordingly. For endogenous particles reinforced aluminum matrix composites, reinforcement/matrix interactions and solid/liquid interface have always been the focus of researchers. On the basis of actual growth morphology of solid phase while in contact with the solidification interface, Nakae and Wu [25] proposed an interface energy model in combination with the wettability of solid/liquid particles. Uhlmann et al. [26] through theoretical analysis found that there is a critical value v_c for the solid/liquid interface advancing speed. Han and Hunt defined the distribution coefficient k of the particles during solidification of composite materials, and their model solved the problem of particles redistribution in the conventional solidification process.

1.3.3 Interface, Microstructure, and Performance Control of In-Situ Aluminum Matrix Composites

Because the interface structure and properties play a decisive role in the stress–strain distribution, thermal conductivity, electrical conductivity and thermal expansion, load transfer, and fracture of in-situ aluminum matrix composites, modern analysis methods are used to deeply study the interface microstructure and properties. The study of interface evolution principle on various properties of composite materials, the interface structure, performance optimization, control methods, and the stability of the interface structures have become important research directions. Using high-resolution transmission electron microscopy (HRTEM) and high-resolution scanning electron microscopy (HRSEM) analysis, intuitive images of the interface can be obtained, and then by combining with the information of elemental analysis (EDS and nano-probe), the atomic structure, chemical bonding, defect structure of the surface, the relationship between the geometric structure of the interface, energy of the interface can be estimated. By clarifying the relationship between the atomic structure of interface, the physical and chemical properties, and the performance of the material, a theoretical basis for extending the lifetime of composite material through property improvement can be established. Kostka et al. used HRTEM to study the TiC/Al and TiB_2/Al composite materials and found that both materials have a higher dislocation density. The dislocation density near the interface is greater, and the lattice constant changes as the distance from interface changes. Heinrich et al. [27] used large-angle convergent beam electron diffraction (LACBED) technology to study the interfacial strain fields of short fiber-reinforced aluminum matrix composites. Scudino et al. [28] studied aluminum matrix composites produced by powder metallurgy route and found that the second phase improves the mechanical properties of the material through dislocation pinning. Li et al. [29] used HRTEM and EELS to study the grain-boundary, phase-boundary interface characteristics and composition segregation in B_4C/50883Al ultrafine grain composites and found that interface characteristics are closely related to the overall strength of the composite. Study of orientation relationships between the reinforcement phase and the matrix and their atomic arrangement can further clarify the relationship between the atomic structures, physical and chemical properties, etc., of the interface and composite. Bai et al. [30] used TEM to study the interface structure of squeeze cast Al_2O_3/2024Al composite material and showed that there was a certain orientation relationship between the Al matrix and the Al_2O_3 particles. Dislocations were observed near the interface area of Al_2O_3 particles, but only the Frank partial edge dislocations with the Burgers vector $b = 1/3\,[111]$ appeared in the matrix, which reduced its plasticity to a certain extent.

In recent years, new analytical methods have been used to study metal matrix nanocomposites [31, 32]. Inkson et al. observed the three-dimensional morphology of individual grains in extruded FeAl nanocomposite using three-dimensional focused plasma beam (3DFIB) X-ray radiographic analysis. The microstructure of each 2DFIB intersection was imaged at several angles to locate the boundaries of the grains. The 3D morphology of individual submicron grains was determined by imaging and computer reconstruction of many parallel 2D slices through the grains. Xia et al. [31] used a nano-indenter combined with finite element analysis to calibrate the Al/Si interface strength in the A356 alloy, and the measured values were found to be of the same order of magnitude as the atomic model. In terms of performance and application of aluminum matrix nanocomposites, it is mainly the evolution of microstructure during synthesis and molding process, which determines the relationship and precise control of microstructure and the performance. The design of aluminum matrix nanocomposites for properties mainly includes static performance, functional properties (damping, thermal conductivity, etc.), service behavior and mechanisms taking place under industrial environments, microstructure evolution and performance reliability evaluation, etc. Studies have shown that in-situ nano-reinforcements can significantly improve the mechanical properties of aluminum matrix composites including strength, creep resistance, fatigue resistance, and damping, etc. It has broad application prospects in aerospace, national defense, military, transportation, electronic, information technology, precision instruments, and other fields [33].

1.3.4 Service Behavior and Damage Failure Mechanisms of In-Situ Aluminum Matrix Composites in Simulated Environment

The in-situ aluminum matrix composites are mostly used in aerospace, national defense, and other high-tech fields. These materials are directly tested under service conditions, not only the information and control factors of the evolution process cannot be obtained, but also the risk is high, the test cost is high, and it is difficult to accomplish. Carrying out research on service behavior and lifetime evaluations of in-situ aluminum-based composite materials in simulated service environment is an important subject to realize their engineering applications [34]. At present, since in-situ aluminum-based composite materials are emerging as high-performance metal matrix composite materials, the research on behavior and lifetime evaluation under simulated service environment has not been reported in the world; however, the reliability enhancement tests of aeronautical materials under simulated service environment can be used for reference.

Material damage and failure during their service is an important research direction. For in-situ particle-reinforced aluminum matrix composites, a key issue is how to determine the size and volume fraction of in-situ particles so that the reinforced composite material can achieve the best properties. On one hand, the in-situ particles greatly improve the material's Young's modulus, yield strength and other parameters; on the other hand, the in-situ particles act as cavity nucleation sites during the deformation process of composite. This greatly accelerates the damage evolution, thereby reducing the ductility and fracture toughness of the material. Therefore, researchers give great importance to the study of composite material's damage theory to improve the ductility and fracture toughness. However, so far, the model for quantitative description of damage evolution is not satisfactory. The main problem is that the stress-based or strain-based models cannot evaluate the micro-damage evolution well. The quantitative relationship between meso-damage and particle fracture is not established, and the meso-damage mechanism of micro-pore formation is disconnected from the macro damage criterion. Therefore, the establishment of fracture theory on the dynamics of meso-scale damage evolution is a hot topic for researchers. When studying damage evolution, it is necessary to consider the damage initiation at the meso-scale level and the reduction in material properties at the macro-scale level and combine the mesoscopic damage process with the macroscopic damage characteristics. The establishment of a reasonable, reliable, and practical constitutive model is of great importance for the development and applications of in-situ reinforced aluminum matrix composites [35].

References

1. Mohammad Sharifi E, Karimzadeh F, Enayati MH. Fabrication and evaluation of mechanical and tribological properties of boron carbide reinforced aluminum matrix nanocomposites. Mater Des; 2011;32(6):3263–71.
2. Li Y, Zhao YH, Ortalan V, et al. Investigation of aluminum-based nanocomposites with ultra-high strength. Mater Sci Eng, A. 2009;527(1–2):305–16.
3. Louzguine-Luzgin DV, Inoue A. Investigation of a rapidly solidified Al-based nanocomposite with extremely high number density of precipitates. Mater Sci Eng, A. 2007;449–451:1026–8.
4. Witkin DB, Lavernia EJ. Synthesis and mechanical behavior of nanostructured materials via cryo-milling. Prog Mater Sci. 2006;51(1):1–60.
5. Emamy M, Mahta M, Rasizadeh J. Formation of TiB_2 particles during dissolution of $TiAl_3$ in Al-TiB_2 metal matrix composite using an in-situ technique. Compos Sci Technol. 2006;66(7–8):1063–6.
6. Merzhanov AG, Shkiro VM, Borovinskaya I. Method of producing refractory carbides, borides, silicides, sulfides, andnitrides of metals of groups iv, v, and vi of the periodic system, US3726643A; 1973.
7. Christodoulou L, Nagle DC, Brupbacher JM. Process for forming metal-second phase composites and product thereof, US4751048A; 1988.
8. Mitra R, Chiou WA, Fine ME, et al. Interfaces in as-extruded XD Al/TiC and Al/TiB_2 metal matrix composites. J Mater Res. 1993;8(9):2380–92.
9. Murthy VSR, Rao BS. Microstructural development in the directed melt-oxidized (DIMOX) Al-Mg-Si alloys. J Mater Sci. 1995;30(12):3091–7.
10. Premkumar MK, Chu MG. Synthesis of TiC particulates and their segregation during solidification in in-situ processed Al-TiC composites. Metall Trans A. 1993;24(10):2358–62.
11. Aghajanian MK, MacMillan NH, Kennedy CR, et al. Properties and microstructure of Lanxide^{TM} Al_2O_3-Al ceramic composite materials. J Mater Sci. 1989;24(2):658–70.
12. Dhandapani SP, Jayaram V, Surappa MK. Growth and microstructure of Al_2O_3-SiC-Si(Al) composites prepared byreactive infiltration of silicon carbide preforms. Acta Metall Mater. 1994;42(3):649–56.
13. Newkirk MS, Urquhart AW, Zwicker HR, et al. Formation of Lanxide^{TM} ceramic composite materials. J Mater Res. 1986;1(1):81–9.
14. Hunt M. Automotive MMCs: better and cheaper. Mater Sci Eng A. 1989;10:45–8.
15. Davies P, Kellie JLF, Wood JV. UK Patent 2257985A; 1992.
16. Wood JV, Davies P, Kellie JLF. Properties of the reactively cast aluminium-TiB_2 alloys. Mater Sci Technol. 1993;9(10):833–40.
17. Davies P, Kellie JLF, Wood JV. Development of cast aluminium metal matrix composites. Key Eng Mater. 1993;77–78:357–62.
18. Koczak M. Reactive processing of MMC. In: Proceeding of TMS; 1993; p. 223–30.
19. Nakata H, Choh T, Kanetake N. Fabrication and mechanical properties of in-situ formed carbide particulate reinforced aluminium composite. J Mater Sci. 1995;30(7):1719–27.
20. Tjong SC, Wang GS, Mai YW. High cycle fatigue response of in-situ Al-based composites containing TiB_2 and Al_2O_3 submicron particles. Compos Sci Technol. 2005;65(10):1537–46.
21. Yang B, Wang F, Huang ZJ, et al. The development of spray deposition forming particle reinforced metal matrix composite preparation technology. Material Guide. 2005;15(3):4–9 (in Chinese).

22. Zhao YT, Zhang SL, Chen G, et al. In-situ (Al_2O_3+Al_3Zr) np/Al nanocomposites synthesized by magneto-chemical melt reaction. Compos Sci Technol. 2008;68(6):1463–70.
23. Ladeveze P, Nouy A, Loiseau O. A multiscale computational approach for contact problems. Comput Methods Appl Mech Eng. 2002;191(43):4869–91.
24. Louzguine G, Luzgin DV, Inoue A. Investigation of a rapidly solidified Al-based nanocomposite with extremely high number density of precipitates. Mater Sci Eng A. 2007;449(13):1026–1028.
25. Nakae H, Wu S. Engulfment of Al_2O_3 particles during solidification of aluminum matrix composites. Mater Sci Eng, A. 1998;252(2):232–8.
26. Uhlmann DR, Chalmers B, Jackson KA. Interaction between particles and a solid-liquid interface. J Appl Phys. 1964;35(10):2986–93.
27. Heinrich H, Vananti A, Kostorz G. Stress fields near interfaces of aluminium-based metal-matrix composites. Philosophical Magazine Part A. 2002;82(11):2269–85.
28. Scudino S, Liu G, Sakaliyska M, et al. Powder metallurgy of Al-based MMCs reinforced with β-Al_3Mg_2 intermetallic particles: analysis and modeling of mechanical properties. Acta Mater. 2009;57(15):4529–38.
29. Li Y, Zhang Z, Vogt R, et al. Boundaries and interfaces in ultrafine grain composites. Acta Mater. 2011;59(19):7206–18.
30. Bai PC, Dai XJ, Zhao CW, et al. The interface structure characteristics of Al_2O_3/Al composites. J Compos Mater. 2008;25(1):88–93 (in Chinese).
31. Xia SM, Qi Y, Perry T, et al. Strength characterization of Al/Si interfaces: a hybrid method of nano-indentation and finite element analysis. Acta Mater. 2009;57(3):695–707.
32. Ouyang QB, Zhang GD, Zhang D. Research and application progress of discontinuously reinforced aluminum matrix composites. China Materials Progress. 2010;29(4):36–40 (in Chinese).
33. Wang HM, Li GR, Zhao YT, et al. Microstructure, billet surface quality and tensile property of (Al_2O_3 + Al_3Zr)p/Al composites in-situ synthesized with electromagnetic field. J Alloy Compd. 2011;509(18):5696–700.
34. Onat A. Mechanical and dry sliding wear properties of silicon carbide particulate reinforced aluminium-copper alloy miatrix composites produced by direct squeeze casting method. J Alloy Compd. 2010;489(1):119–24.
35. Du SY. Advanced composite materials and aerospace. J Compos Mater. 2007;24(1):1–12 (in Chinese).

Chapter 2
Design and Development of In-Situ Reaction Systems

Since in-situ metal matrix composites are prepared by means of the chemical reaction between the base metal and a suitable reactant (gas, liquid, or solid powder) at an appropriate temperature, the reaction system is an important factor determining the performance, ease of preparation, and cost of in-situ composites. The criteria for the selection of reaction systems include the following aspects [1]: ① the performance of reinforcement phase; ② difficulty in morphology-control of the reinforcement phase; ③ nature of interface between the reinforcement and the matrix; ④ the severity and initial temperature of reaction; ⑤ the source and price of reactants. At present, there are three main types of endogenous reinforcement phases in in-situ aluminum matrix composites: ① ceramic phase reinforcement particles; ② intermetallic compound reinforcement phase; ③ ceramic-intermetallic compound hybrid reinforcements.

Currently, in-situ synthesized ceramic particle reinforcements mainly include TiC, TiB_2, ZrB_2, Al_2O_3, SiC, and AlN. Table 2.1 shows reaction systems of some common endogenous ceramic reinforcement phases and their properties. Table 2.2 shows reaction systems of some common intermetallic compound reinforcement phases and their properties.

2.1 Thermodynamics and Kinetics of Reaction Systems

In a chemical reaction system, the reaction thermodynamics, kinetics, and reaction mechanism are closely related. The study of thermodynamics tells about the possibility of a chemical reaction. In other words, thermodynamics can only predict the direction and extent of a chemical reaction under given conditions, and in order to realize it, it requires the knowledge of kinetics; i.e., kinetics tells how a reaction proceeds. It focuses on various factors influencing the speed and mechanism of the chemical reaction. In in-situ aluminum matrix composites, the reinforcement phase is formed by means of a chemical reaction in the aluminum melt. The reaction system, thermodynamics and kinetics, and mechanisms of a chemical reaction in the melt

Y. Zhao, *In-Situ Synthesis of Aluminum Matrix Composites*,
https://doi.org/10.1007/978-981-16-9120-1_2

Table 2.1 Reaction systems of some common endogenous ceramic reinforcement phases and their properties [2]

Reinforcement phase	In-situ synthesis system	Density/(g/cm^3)	Thermal expansion coefficient/($\times 10^{-5}$/°C)	Ultimate tensile strength σ_b/MPa	Elastic modulus E/GPa
TiC	Al–Ti–C Al–T–SiC	4.93	7.6	55(1100 °C)	269(1100 °C)
TiB_2	Al–K_2TiF_6–KBF_4	4.5	8.28	–	414(1100 °C)
ZrB_2	Al–Zr–B_4C Al–K_2ZrF_6–KBF_4	6.9	8.28	–	495
Al_2O_3	Al–CuO Al–SiO_2	3.98	7.92	221(1100 °C)	329(1100 °C)
SiC	Al–SiO_2–C	3.2	3.40	–	450(25 °C)
AlN	Al–N_2	3.26	3.30	–	345(25 °C)

Table 2.2 Reaction systems of some common intermetallic compound reinforcement phases and their properties [3]

Intermetallic compound	In-situ synthesis system	Density/(g/cm^3)	Thermal expansion coefficient/($\times 10^{-5}$/°C)	Melting point/K
Al_3Ti	Al–Ti–TiC Al–K_2TiF_6	3.36	166–230	1610
Al_3Ni	Al–Ni–Co	4.00	116–140	1160
Al_3Fe	Al–Fe	3.20	130	1420
Al_3Zr	Al–K_2ZrF_6	3.26	196–205	1850

are of significant importance to control the synthesis process, microstructure, and morphology of reinforcement phase in the composite, as well as to obtain an in-situ aluminum matrix composite of higher comprehensive performance [4].

From a thermodynamics perspective, in order to make the chemical reaction proceed forward and obtain the required reinforcement, it is necessary to ensure that the chemical reaction occurs at a temperature such that the free energy of formation $\Delta G < 0$, and the free energy of formation of reinforcement phases is lower than the other intermediate phases. In the synthesis of in-situ aluminum matrix composites, the oxide replacement reduction reaction is the main in-situ chemical reaction. Reaction kinetics affect the speed and progress of melt chemical reaction, thereby determining the shape, size, and distribution of products in the matrix which affects the degree of synthesis process and performance of the composite. Therefore, it is necessary to study the kinetics of in-situ synthesis reaction, especially the kinetics of in-situ melt reaction under the influence of physical fields.

During the synthesis of composite materials by in-situ reaction, in general, one phase of the reactants should form a liquid phase like in SHS, CR, XD™, DIMOX™, VLS, and other synthesis methods. Matrix aluminum in Al–Ti–B, Al–Ti–C, Al–Ti–B–C, or Al–Ti–B_4C systems transforms into a liquid state at a synthesis temperature. Through the principles of liquid mass transfer, capillary diffusion, solid_liquid adsorption, etc., liquid aluminum comes in direct contact with titanium powder, carbon powder, boron powder or B_4C powder, and chemical reaction between reactants produces a reinforcement phase or a transition phase which further transforms into a stable phase. Kanury [5] studied the reaction kinetics of TiC synthesis by SHS method and proposed a layer-by-layer reaction mechanism and established a mutual reaction kinetic model. Wang [6] and Zhang [7] established a kinetic model of Al–Ti–C reaction in molten metal on the basis of fundamental kinetic and diffusion theories and deeply analyzed the influence of various factors on the entire reaction.

At present, there are three main kinetic mechanisms of in-situ chemical reactions: diffusion mechanism, collapse mechanism, and reaction–diffusion mechanism. In diffusion mechanism [8], the chemical reaction proceeds by migration and exchange of atoms, and it is an important mechanism of chemical reactions and controls the growth of reaction products. In collapse mechanism [9], during the reaction of system, reaction products (reinforcements) do not cover the surface of reactant particles, and liquid aluminum always keeps in contact with reactants; as the reaction proceeds, the reactant particles continue to shrink and collapse, the reaction products (reinforcement particles) continue to nucleate and grow in molten aluminum. Reaction–diffusion mechanism [10] states that reaction rate is not only controlled by diffusion but also controlled by the combined reaction and diffusion mechanism. The results show that within a certain reaction time, the speed of reaction is proportional to the square root of time, and when the reaction time exceeds that certain value, the reaction speed is directly proportional to the time (i.e., linear relationship). Obviously, the diffusion mechanism alone cannot explain this phenomenon. In short, the process of in-situ melt chemical reaction to synthesize aluminum matrix composites is very complicated, especially multiphase reactions under the influence of the physical field. So in order to lay a strong theoretical foundation of the preparation method and synthesize higher performance aluminum matrix composites, the reaction kinetics and mechanisms need to be studied further.

2.2 Development of New Reaction Systems for In-Situ Aluminum Matrix Composites

At present, reaction systems for in-situ aluminum matrix composites mainly consist of Al–Ti–X (Al–Ti–O, Al–Ti–B, Al–Ti–C) system, and initial reaction temperature for composites synthesis is high, often higher than 900 °C, or even more than 1000

Table 2.3 Properties of Al_3Zr and Al_2O_3 [12, 13]

Particles	Density/(g/cm^3)	Melting point/K	Vickers hardness/MPa	Strength/MPa	Elastic modulus/GPa	Linear expansion parameter/($\times 10^{-6}$/°C)
Al_3Zr	4.10	1850	7400	–	–	12.10
Al_2O_3	3.98	2313	–	221 (1090 °C)	379 (1090 °C)	7.92

°C. It not only makes it difficult to control the morphology of the reinforcement phase but also severely deteriorates molten aluminum. It is worth noting that the reaction system containing Zr has not been reported in the literature before the publication [11] of the author's research group in 2001. The author chose Al–Zr–X (X = O, B, Ti, etc.) system as the research object, the reinforcements synthesized by in-situ melt reaction mainly include Al_3Zr, Al_2O_3, ZrB_2, etc., phase particles and have the following advantages:

(1) Al_3Zr and Al_2O_3 phases have a high melting point, high hardness, high elastic modulus, and similar density (specific gravity), as shown in Table 2.3.

(2) Al_3Zr and α-Al phases have coherent correspondence, while Al_2O_3 and Si phases have semi-coherent relationship. According to the literature [14], Al_3Zr and Al_3Ti both have square lattices which are similar to face-centered cubic lattices of Al and have a coherent relationship with Al, such as $(001)_{Al}//(001)_{Al3Ti}$, $[100]_{Al}//[100]_{Al3Ti}$, and lattice matching difference in the a- and b-directions is 5.2%; and $(001)_{Al}//(001)_{Al3Zr}$, $[100]_{Al}//[100]_{Al3Zr}$, the lattice matching difference in a- and b-directions is only 0.9%. It can be seen that the lattice matching difference of Al_3Zr is small, aluminum is more likely to have epitaxial growth on external nuclei of Al_3Zr, and the refinement effect is inevitably better than Al_3Ti. The literature [15] has reported that the refinement effect of Zr is not as good as that of Ti. The main reason is that the preparation of Al–Zr master alloy produces coarse Al_3Zr particles which have large specific gravity and thus easily settle down. They are difficult to dissolve at the actual aluminum melting temperature of 750–760 °C, and the amount of Zr used for refinement is small. The use of Zr salts and Zr oxides after reaction with molten Al scatters Zr atoms, which is conducive to the formation of a large number of dispersed Al_3Zr crystal nuclei. The literature [16] has reported that α-Al_2O_3 has a hexagonal structure and the degree of mismatch between $(001)_{Al_2O_3}$ surface and the $(110)_{Si}$ surface is $\delta = 5.08\% < 6\%$, which shows that the Al_2O_3 and Si phases basically meet the coherent relationship criteria, and Si can nucleate on the surface of Al_2O_3.

(3) The morphology of Al_3Zr and Al_2O_3 reinforcements is easy to control. The literature [17] has reported that Al_3Zr phase exists in a granular form below 1200 °C and appears in needle-shape form when the reaction temperature is greater than 1200 °C. Therefore, in usual smelting temperature range of

aluminum alloy, by selecting appropriate Zr-containing reactant and controlling the reaction temperature of Al below 1200 °C, the granularity of Al_3Zr can be ensured. Al_2O_3 is produced by the reaction, its shape is easier to control, and it is usually has a granular form.

(4) The Al–Zr–O system has a stable reaction, and the reaction starting temperature is low and easy to achieve. For Al–Zr–O system, the reaction can usually be carried out at 750 °C, and it is stable, generating fine and dispersed Al_3Zr and Al_2O_3 particles. This is advantageous from industrial applications' point of view. The literature [18] has used the solid–liquid reaction method to study CuO/Al system. The results showed that the reaction of CuO/Al system only occurs when the reaction temperature is greater than 1000 °C (1273 K), generating Al_2O_3 particles, but at the same time at 1273 K there are some shortcomings such as violent synthesis reaction, difficult control of particle addition process and coarsening of self-generated particles. Moreover, the addition of CuO in the Al melt is sometimes prone to explosion.

(5) Zr can increase the recrystallization temperature of Al, thereby improving its high-temperature strength. The literature [19] has reported that Zr can increase the recrystallization temperature of Al by 100 K. When Zr exists in the form of extremely fine Al_3Zr precipitates, it is most effective; when Zr is rapidly cooled and then precipitated at high temperature, a fine and even distribution of Al_3Zr can be obtained.

(6) The raw materials are in abundance, and the price is low.

In summary, the new Al–Zr–O in-situ reaction system has good prospects for the development of in-situ aluminum matrix composites and their applications.

2.2.1 Al–Zr–O System Development

2.2.1.1 Reaction Thermodynamics

The Al–Zr–O system reactants for synthesis of aluminum matrix composites mainly include $ZrOCl_2$, $ZrSiO_4$, and ZrO_2 powders. After preheating at 250 °C for 2 h, these reactants are added into the aluminum melt, and the following metallurgical reactions occur:

$$2ZrOCl_2 \rightarrow ZrCl_4 + ZrO_2 \tag{2.1}$$

$$3ZrCl_4 + 4Al(l) \rightarrow 3[Zr] + 4AlCl_3(g) \tag{2.2}$$

$$3ZrO_2 + 4Al(l) \rightarrow 3[Zr] + 2Al_2O_3 \tag{2.3}$$

$$3ZrSiO_4 + 8Al(l) \rightarrow 3[Zr] + 3[Si] + 4Al_2O_3 \quad (2.4)$$

$$Zr + 3Al(l) \rightarrow Al_3Zr \quad (2.5)$$

The formula for the calculation of standard free energy of formation is

$$\Delta G_T^{\ominus} = \Delta H^{\ominus} - T\Delta S_T^{\ominus} \quad (2.6)$$

$$\Delta H_T^{\ominus} = \Delta H_{298}^{\ominus} + \int_{298}^{T} \Delta C_p \mathrm{d}T \quad (2.7)$$

$$\Delta S_T^{\ominus} = \Delta S_{298}^{\ominus} + \int_{298}^{T} \frac{\Delta C_p}{T} \mathrm{d}T \quad (2.8)$$

where $\Delta H_T^{\ominus}$ is the enthalpy change in standard state; $\Delta S_T^{\ominus}$ is the entropy change in standard state; and ΔC_p is the specific heat capacity ($\Delta C_p = a + b \times 10^{-3} \cdot T - c \times 10^5\ T^{-2}$).

According to the data in the literature [20] and the above formulas, the standard free energy (J/(mol K)) of the reactants is

$$\Delta G_{ZrO_2}^{\ominus} = -1082151.4 + 198.52T \quad (2.9)$$

$$\Delta G_{Al_2O_3}^{\ominus} = -1675285.5 + 329.24T \quad (2.10)$$

$$\Delta G_{Si}^{\ominus} = -6918.2 + 3.88T \quad (2.11)$$

$$\Delta G_{Zr}^{\ominus} = -8308.6 - 11.23T \quad (2.12)$$

$$\Delta G_{Al}^{\ominus} = -5609.9 - 11.35T \quad (2.13)$$

$$\Delta G_{AlCl_3}^{O} = -684781.2 + 144.51T \quad (2.14)$$

And with the calculated data according to the literature [21], we have:

$$\Delta G_{ZrCl_4}^{\ominus} = -1003361.1 + 378.99T \quad (2.15)$$

$$\Delta G_{ZrSiO_4}^{\ominus} = -2030835.2 + 377.32T \quad (2.16)$$

The free energy of formation for reaction (2.2) is

$$\Delta G^{\ominus} = 3\Delta G^{\ominus}_{Zr} + 4\Delta G^{\ominus}_{AlCl_3} - 3\Delta G^{\ominus}_{ZrCl_4} - 4\Delta G^{\ominus}_{Al} = 268472.3 - 547.22T \quad (2.17)$$

Let $\Delta G^{\ominus} = 0$, the starting temperature for reaction (2.2) is 490.6 K. It can be seen that in molten aluminum ($\geq$ 933 K), the reaction (2.2) can proceed spontaneously. The free energy of formation for reaction (2.3) is

$$\Delta G^{\ominus} = 3\Delta G^{\ominus}_{Zr} + 2\Delta G^{\ominus}_{Al_2O_3} - 3\Delta G^{\ominus}_{ZrO_2} - 4\Delta G^{\ominus}_{Al} = -106602.8 + 74.64T \quad (2.18)$$

Let $\Delta G^{\ominus} = 0$, $T = 1428.4$ K for reaction (2.3), which shows that when the melting temperature of aluminum is lower than 1428 K (i.e., within 660–1150 °C), the reaction (2.3) can proceed spontaneously.

The free energy of formation for reaction (2.4) is

$$\begin{aligned}\Delta G^{\ominus} &= 3\Delta G^{\ominus}_{Zr} + 3\Delta G^{\ominus}_{Si} + 4\Delta G^{\ominus}_{Al_2O_3} - 3\Delta G^{\ominus}_{ZrSiO_4} - 8\Delta G^{\ominus}_{Al} \\ &= -609437.6 + 253.75T\end{aligned} \quad (2.19)$$

Let $\Delta G^{\ominus} = 0$, $T = 2401.7$ K for reaction (2.4), which shows that when the melting temperature of aluminum is lower than 2401 K (i.e., within 660–2128 °C), the reaction (2.4) can proceed spontaneously.

In summary, reactions (2.2), (2.3), and (2.4) are within the usual melting temperature range of aluminum (660–1000 °C), and all can take place spontaneously. Moreover, according to the data in the literature [22], the relevant parameters are listed in Table 2.4.

The enthalpy of chemical reaction is

$$\Delta H^{\ominus} = \Delta H^{\ominus}_{298} + \int_{298}^{T} \Delta C_p \mathrm{d}T \quad (2.20)$$

where $\Delta H^{\ominus}_{298} = \sum\left(\Delta H^{\ominus}_{298}\right)_{\text{Produtcs}} - \sum\left(\Delta H^{\ominus}_{298}\right)_{\text{Reactants}}$; and $\Delta C_p = \sum\left(C_p\right)_{\text{Products}} - \sum\left(C_p\right)_{\text{Reactants}}$.

Table 2.4 Enthalpy and isobaric heat capacity of several substances

Substance	$\Delta H^0_{298}/J$	C_p/(J/(mol K))
α-Al_2O_3	– 1672.0	$114.66 + 12.79 \times 10^{-3}\ T - 35.40 \times 105\ T^{-2}$
α-ZrO_2	– 1096.4	$59.56 + 7.52 \times 10^{-3}\ T - 14.04 \times 105\ T^{-2}$
AlC13 (g)	– 546.3	$81.89 + 0.63 \times 10^{-3}\ T - 9.91 \times 105\ T^{-2}$
ZrC14 (g)	– 865.3	$107.47 + 0.25 \times 10^{-3}\ T - 8.03 \times 105\ T^{-2}$
α-Zr	0	$21.95 + 11.62 \times 10^{-3}\ T$
Al(l)	0	31.77

Therefore, the enthalpy of reaction (2.2) is

$$\begin{aligned}\Delta H^{\ominus} &= \left(3\Delta H^{\ominus}_{\mathrm{Zr},298} + 4\Delta H^{\ominus}_{\mathrm{AlCl_3},298} - 3\Delta H^{\ominus}_{\mathrm{AlCl_4},298} - 4\Delta H^{\ominus}_{\mathrm{Al},298}\right) \\ &\quad + \int_{298}^{T} \left(\sum C_{p\mathrm{Products}} - \sum C_{p\mathrm{Reactants}}\right)\mathrm{d}T \\ &= 410.7 + \int_{298}^{T} \left(-56.08 + 36.63 \times 10^{-3}T - 15.55 \times 10^{5}T^{-2}\right)\mathrm{d}T \\ &= 18749.43 - 67.0T + 18.32 \times 10^{-3}T^{2} \end{aligned} \tag{2.21}$$

The enthalpy of reaction (2.3) is

$$\begin{aligned}\Delta H^{\ominus} &= -54.8 + \int_{298}^{T} \left(-40.59 + 37.88 \times 10^{-3}T\right)\mathrm{d}T \\ &= 13722.97 - 51.88T + 18.94 \times 10^{-3}T^{2} \end{aligned} \tag{2.22}$$

The graphical Gibbs free energy–temperature ($\Delta G^{\ominus} - T$) relationship for reactions (2.2), (2.3), and (2.4) is shown in Fig. 2.1, and the enthalpy–temperature ($\Delta H^{\ominus} - T$) relationship for reactions (2.2) and (2.3) is shown in Fig. 2.2.

It can be seen from Fig. 2.1 that the Gibbs free energy $\Delta G^{\ominus}$ of reactions (2.2), (2.3), and (2.4) is negative, indicating that it is thermodynamically feasible. And from Fig. 2.2, it is clear that the enthalpy $\Delta H^{\ominus}$ of reactions (2.2) and (2.3) is large, which means that a large amount of heat is released during the reaction. The above analysis is consistent with the experimental results. The X-ray diffraction pattern (XRD) of composite material synthesized from Al–Zr–O system is shown in Fig. 2.3. The results show that in addition to Al, two phases of Al_3Zr and Al_2O_3 are also present, which further confirms the in-situ reactions (2.1)–(2.5) of $ZrOCl_2$, $ZrSiO_4$, and ZrO_2 with molten aluminum.

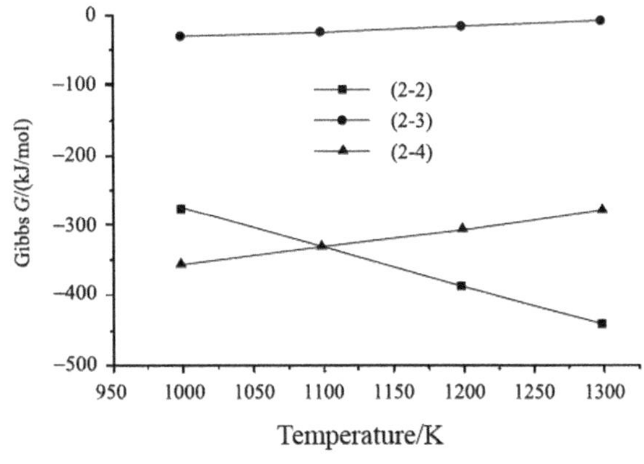

Fig. 2.1 Variation in Gibbs free energy ($\Delta G^{\ominus}$) of reactions (2.2), (2.3), and (2.4) with temperature (T)

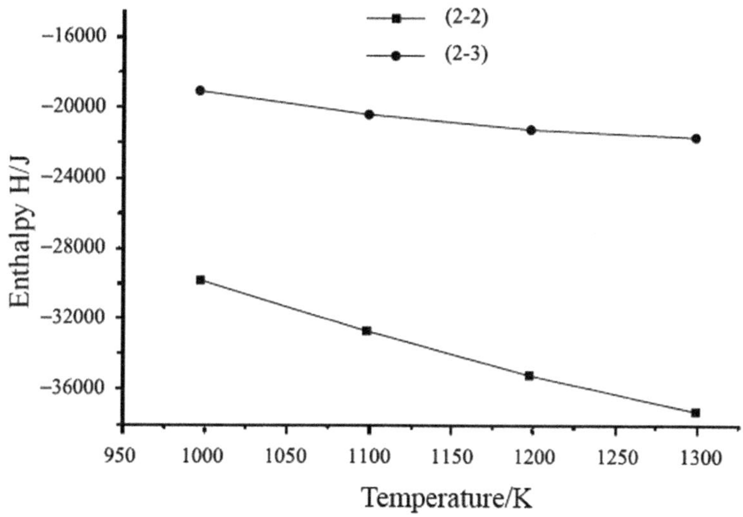

Fig. 2.2 Variation in enthalpy ($\Delta H^{\ominus}$) of reactions (2.2) and (2.3) with temperature (T)

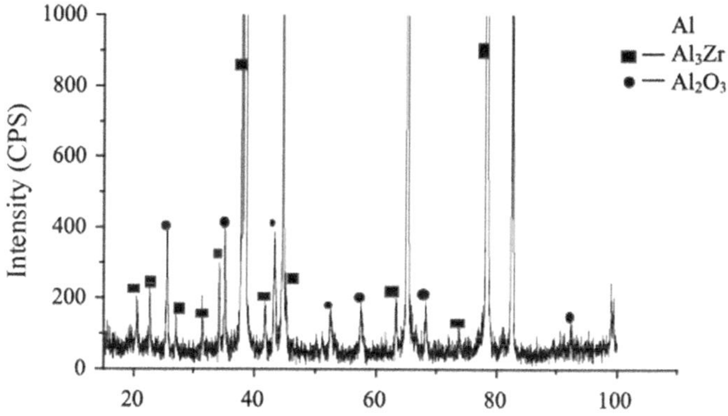

Fig. 2.3 XRD pattern of composite material produced from Al–Zr–O reaction system

2.2.1.2 Reaction Kinetics

The chemical reactions in Al–Zr–O system (Al–ZrO_2, Al–$ZrOCl_2$, $Zr(CO_3)_2$, and Al–$ZrSiO_4$) are mainly thermite reduction reactions. Although the reaction takes place between $ZrOCl_2$ and molten aluminum, there is an intermediate reaction between $ZrCl_4$ and Al, and the reaction rate is very fast (confirmed by the actual reaction observations). Therefore, the thermite reduction reaction is the key in the whole reaction of Al–Zr–O system. In the below, we take the analysis of the reaction between ZrO_2 and molten aluminum as an example.

Since the ZrO_2/Al reaction is carried out at a relatively high temperature (800–950 °C), and the products are only Al_3Zr and Al_2O_3 particulate phases, these particles exist as solid phases in the molten aluminum. During the reaction, liquid aluminum can easily come in contact with ZrO_2 and participate in the reaction. Based on this, it can be assumed that the ZrO_2/Al solid–liquid reaction is controlled by the chemical reaction rate, and its kinetic model is a hard-core model for which the following assumptions are made:

(1) ZrO_2 particles produced from $ZrOCl_2$ are spherical and are uniformly distributed in the liquid aluminum.
(2) The reaction takes place at the ZrO_2/Al interface.
(3) The concentration of Al in the melt is constant.

According to thermal stress calculation [23], when the initial temperature of a solid sphere suddenly rises from T_0 to T_1, the outer surface of the sphere will expand, resulting in a uniform distribution of tensile stress on the sphere. The tensile stress formula is expressed as

$$\sigma_r = \sigma_t = 0.771\frac{\alpha E}{2(1-\nu)}(T_1 - T_0) \tag{2.23}$$

where α is the thermal expansion coefficient; E is the elastic modulus; and ν is Poisson's ratio.

Under the general research conditions, taking $\alpha = 9.6 \times 10^{-6}/°C$, $E = 200$ GPa, $T_1 - T_0 = 200$ °C, and substituting these values into the Eq. (2.23), we have

$$\sigma_r = \sigma_t = \frac{148}{1-\nu}\text{MPa} > 148\ \text{MPa}$$

This tensile stress is greater than the rupture strength of ZrO_2, i.e., about 8 MPa, which causes the phenomenon of ZrO_2 particle breakage, increases the contact area between reactants and molten aluminum, and allows the products to diffuse out of the reaction layer more quickly, thus accelerating the reaction.

Figure 2.4 shows the XRD pattern of water-quenched sample after a reaction time of 15 min. As can be seen from the figure, in addition to generated Al_2O_3 and Al_3Zr-reinforcement particles during synthesis, unreacted ZrO_2 particles and active Zr atoms are also present that penetrated the reaction layer of molten aluminum, indicating that the reaction has not been completed yet. Thus, the ZrO_2 particles generated by decomposition of the zirconium-containing reactant and the Zr atoms produced by the substitution reaction of ZrO_2 with Al have not completed their reaction with Al.

According to the above analysis, the kinetic mechanism of in-situ generated particles in the Al–Zr–O system is "reaction–fracture–diffusion." Figures 2.5 and 2.6 show the kinetic model and the principle of in-situ particles generation in the Al–Zr–O system, respectively.

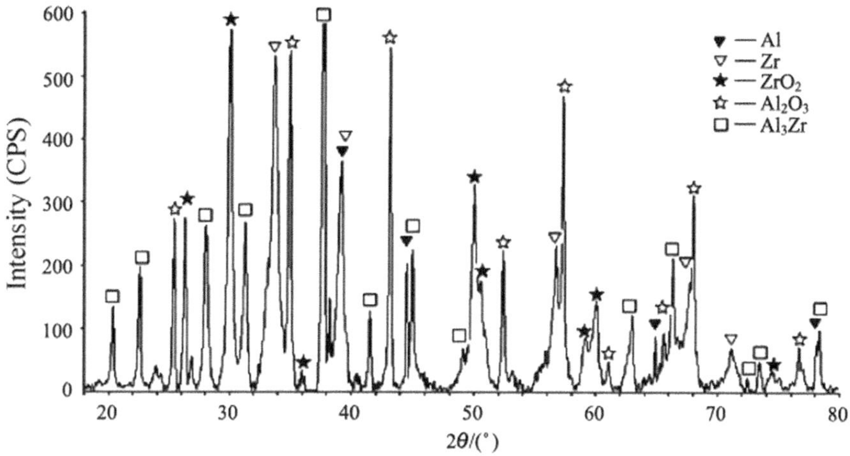

Fig. 2.4 XRD pattern of water-quenched sample after 15 min reaction time

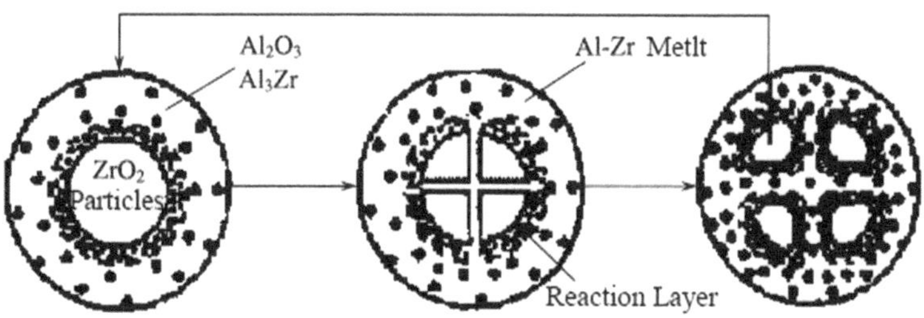

Fig. 2.5 Kinetic model of formation of endogenous particles in Al–Zr–O system

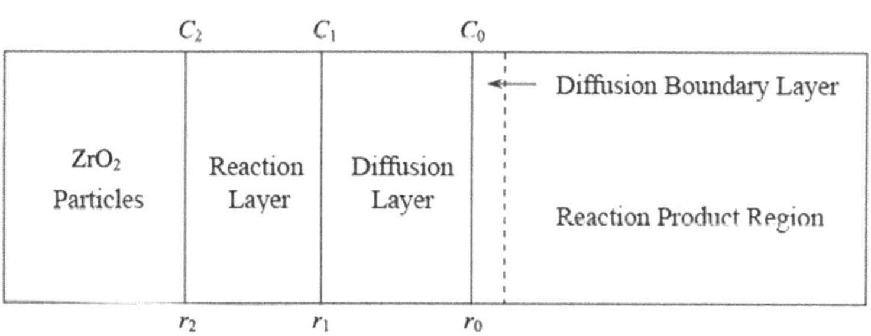

Fig. 2.6 Principle of in-situ reaction synthesis of Al–Zr–O system

In Fig. 2.6, C_0 is the concentration of Zr atoms in the outer boundary of reaction diffusion layer and the melt; C_1 is the concentration of Zr atoms at the outer boundary of the reaction layer and the inner boundary of reaction diffusion layer; and C_2 is the concentration of active Zr atoms in ZrO_2 particles at the inner boundary of reaction layer. r_0, r_1, and r_2 are the original radius of ZrO_2 at the outer boundary of reaction diffusion layer, the radius of the outer boundary of reaction layer and the inner boundary of the reaction diffusion layer, and the radius of the inner boundary of the particle reaction layer, respectively. In a chemical reaction, as the reaction proceeds, the total amount of reactants gradually decreases while the total amount of products gradually increases. Therefore, according to the change of Zr atoms concentration and assuming that the volume of the system does not change during the reaction process, the characteristic reaction kinetics equation of the system is established as follows:

The reaction rate in the reaction layer is:

$$N_{\mathrm{R}} = 4\pi r_1^2 k_{\mathrm{R}}(C_2 - C_1) \tag{2.24}$$

where k_{R} is the reaction coefficient.

The diffusion rate in the diffusion layer is:

$$N_D = 4\pi r_0^2 D\frac{\mathrm{d}C}{\mathrm{d}r} \tag{2.25}$$

where D is the diffusion coefficient.

Taking the average of these two, we have

$$N_{\mathrm{avg}} = 4\pi D r_1 r_0 \frac{C_1 - C_0}{r_0 - r_1} \tag{2.26}$$

Taking Al–$Zr(CO_3)_2$ as an example, in the Al–ZrO reaction system, the consumption rate of ZrO_2 is

$$N = -\frac{4}{3}\pi\delta_0\frac{\mathrm{d}r_1^3}{\mathrm{d}t} \tag{2.27}$$

where δ_0 is the molar concentration of ZrO_2 particles.

The degree of the reaction is

$$R = 1 - \frac{r_1^3}{r_0^3} = \frac{C_0}{C_*} \tag{2.28}$$

where C_* (mol/m^3) is the Zr atomic molar concentration obtained after the melt reaction is completed.

When in-situ chemical reaction enters the stable stage, the above two steps reflect the rate of synthesis, and the two equations are combined to obtain the total reaction rate:

$$N = -\frac{4\pi r_0^2}{\frac{1}{k_R}\cdot\frac{r_0^2}{r_1^2}+\frac{r_0}{D}\cdot\frac{r_0-r_1}{r_1}}(C_2 - C_0) \tag{2.29}$$

From Eqs. (2.27)–(2.29), we have

$$\frac{1}{C_*}\cdot\frac{\mathrm{d}C_0}{\mathrm{d}t} = -\frac{3(C_2 - C_0)}{\delta r_0\left\{\frac{1}{k_R}\left(1-\frac{C_0}{C_*}\right)^{-\frac{2}{3}}+\frac{r_0}{D}\left[\left(1-\frac{C_0}{C_*}\right)^{-\frac{1}{3}}-1\right]\right\}} \tag{2.30}$$

The reaction stops when the concentration of active Zr atoms is the same as the concentration of Zr atoms in the melt. At this time, $C_2 - C_* = 0$. Let $C = C_* - C_2$, then the Eq. (2.30) can be expressed as

$$C_* t = \frac{\delta_0 r_0}{2k_R}\left[\left(\frac{C}{C_*}\right)^{-\frac{2}{3}} - 1\right] + \frac{\delta_0 r_0^2}{3D}\left[\ln\frac{C}{C_*} + 3\left(\frac{C}{C_*}\right)^{-\frac{1}{3}} - 3\right] \tag{2.31}$$

Assuming that all the reactants are consumed to form Al_3Zr and Al_2O_3 particles, then $C_* = \delta_0$, thus Eq. (2.32) can be obtained

$$t = \frac{r_0}{2k_R}\left[\left(\frac{C}{C_*}\right)^{-\frac{2}{3}} - 1\right] + \frac{r_0^2}{3D}\left[\ln\frac{C}{C_*} + 3\left(\frac{C}{C_*}\right)^{-\frac{1}{3}} - 3\right] \tag{2.32}$$

The first and second terms on the right-hand side of Eq. (2.32) represent the reaction times occupied by chemical reaction and diffusion resistance in the reaction process, respectively.

Equation (2.32) can be simplified and expressed as $t = t_1 + t_2$, where

$$t_1 = \frac{r_0}{2k_R}\left[\left(\frac{C}{C_*}\right)^{-\frac{2}{3}} - 1\right] \tag{2.33}$$

$$t_2 = \frac{r_0^2}{3D}\left[\ln\frac{C}{C_*} + 3\left(\frac{C}{C_*}\right)^{-\frac{1}{3}} - 3\right] \tag{2.34}$$

Under certain experimental conditions, we take $r_0 = 120\ \mu\text{m}$, $D = 1.2 \times 10^{-11}$ m^2/s, $k_R = 3 \times 10^{-6}$ m/s and define $F_1 = \frac{t_1}{t_1+t_2}$, $F_2 = \frac{t_2}{t_1+t_2}$ to represent the time fraction occupied by chemical reaction and diffusion resistance, respectively, in the course of the reaction. Energy dispersive X-ray (EDX) analysis is used on water-quenched samples prepared at different times to analyze the change of Zr content with time, and then the curve is fitted according to the Eq. (2.32). Figure 2.7 shows the effect of chemical reaction and diffusion resistance on the reaction process. It can be seen from Fig. 2.7 that before the reaction approaches 20% of its completion, the time elapsed by chemical reaction plays a major role, and then the time elapsed

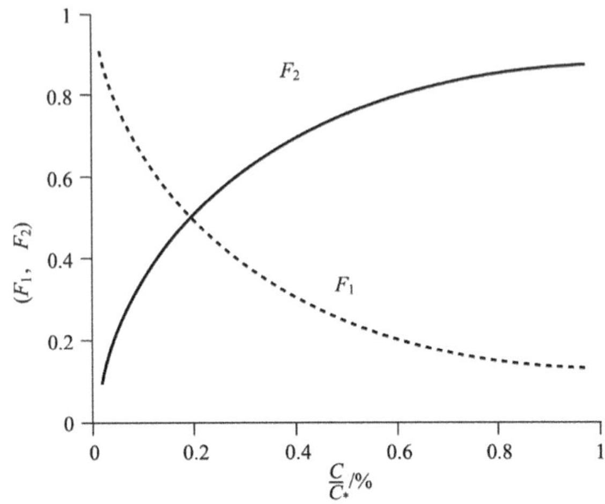

Fig. 2.7 Effect of chemical reaction and diffusion resistance on the reaction process

by diffusion resistance gradually increases. The main reason is that as the reaction proceeds further, the reaction generates a great amount of products. These products accumulate on the outer surface of the particles, forming a viscous layer of diffusion boundary, which greatly increases the resistance to diffusion, so that the remaining time is mainly elapsed by the diffusion process.

Figure 2.8 shows the fitted curve and EDX analysis of Zr content varying with the reaction time. It can be seen from the figure that the fitted curve and the EDX results have a good agreement. At the beginning of the reaction, the Zr content rises sharply, and after about 600 s, the Zr content reaches 90%. This is consistent with the effects of chemical reaction and diffusion resistance on the chemical reaction as demonstrated in Fig. 2.7.

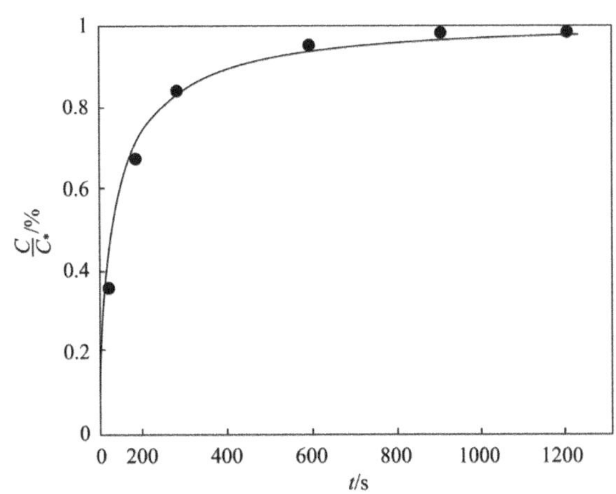

Fig. 2.8 Fitting curve showing variation in Zr content with reaction time

2.2.2 Al–Zr–B System Development

In Al–Zr–B system-based composites, Zr is mainly introduced by industrial-grade zirconium carbonate ($Zr(CO_3)_2$), potassium fluor zirconate (K_2ZrF_6), or other powders, and B is mainly introduced by potassium fluoroborate (KBF_4), borax ($Na_2B_4O_7 \cdot 10H_2O$), or boric anhydride (B_2O_3) powders. By studying multiple in-situ reaction systems, reaction efficiency, and particle yield for Al–K_2ZrF_6–KBF_4 system, it is found that the synthesis temperature is less than 900 °C. In addition, the yield of reaction salt is the highest which can be more than 90%, and the amount of by-products is small and easily removable. Secondly, borax ($Na_2B_4O_7 \cdot 10H_2O$)-type boride and K_2ZrF_6-type fluoride powder are reactive salts. The direct melt reaction method is used to synthesize particle-reinforced aluminum matrix composites, realizing control over the size of Al_2O_3 reinforcement particles at a nanoscale. It is a suitable reaction system for low-temperature synthesis of high-performance in-situ nanoparticles-reinforced aluminum matrix composites. Figure 2.9 shows the morphology of reactants, potassium fluoroborate (KBF_4), and potassium fluor zirconate (K_2ZrF_6).

The synthesis process of Al–Zr–B system-based composites is as follows: First, the mixed salts (K_2ZrF_6 and KBF_4 with a certain ratio, a small amount of Na_3AlF_6 flux, and a small amount of magnesium powder) are baked at 250 °C for 3 h; Then, industrial-grade pure aluminum ingot is put in graphite crucible for smelting in a 30 kW electric furnace; after the alloy is melted, it is heated to a certain temperature, the weighed reactants are pressed into the melt with a bell jar or powder sprayed into the melt and temperature is controlled at a set value. After the reaction is complete, the flux is used to degas the melt and remove slag, then it is poured into the copper mold at 720 °C; finally, samples are cut and polished to observe the microstructure of the composite.

Main factors affecting the solidification microstructure of Al–Zr–B system-based composites are the initial melt temperature, reaction time, and the amount of reactants

(a) KBF4 (b) K2ZrF6

Fig. 2.9 Morphology of raw reactant materials

Table 2.5 Orthogonal design experiment of Al–Zr–B system-based composites and properties of the reinforcement

Sample number	Influencing parameter			
	Initial melt temperature/°C	Reaction time/min	Amount of reactants (volume fraction)/%	Microstructure evaluation grade
1	800	10	1	9
2	800	15	2	5
3	800	20	3	7
4	850	10	2	1
5	850	15	3	3
6	850	20	1	2
7	900	10	3	8
8	900	15	1	6
9	900	20	2	4

added. The synthesis experiment is designed according to the orthogonal experiment principle ($L_9(3^4)$), and the content of design experiment is shown in Table 2.5.

The bases for microstructure evaluation of composite material are: ① the absolute amount of particles and the difference between the actual number density and theoretical number density of particles; ② particle size (submicron and nanometer as standard); ③ particle size distribution (measured according to the composite material standard evaluation method) to comprehensively assess the performance level "microstructure evaluation grade" is ranked from 1 to 9, and the smaller the number, the higher the level.

According to the data processing method of "orthogonal test design," the synthesis process of Al–Zr–B system-based composite material is finally optimized. The initial temperature of melt is 850 °C, the reaction time is 15 min, and the theoretical volume fraction of particles is 2%. Figure 2.10 shows the XRD pattern of composite, which confirms that the particles in the composite are ZrB_2 particles.

2.2.3 Al–Zr–B–O System

2.2.3.1 Al–Zr–B_2O_3 System Development

Zr and B_2O_3 powders are preheated at 250 °C for 2 h and added into the molten aluminum, and then the following in-situ chemical reactions take place:

$$Zr + 3Al \rightarrow Al_3Zr \quad (2.35)$$

$$B_2O_3 + 3Al \rightarrow AlB_2 + Al_2O_3 \quad (2.36)$$

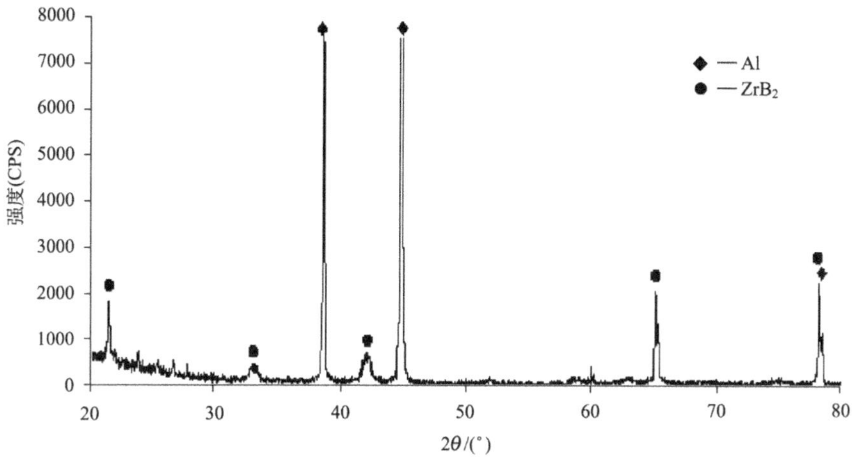

Fig. 2.10 XRD pattern of composite material synthesized by Al–Zr–B system

$$AlB_2 + Al_3Zr \rightarrow ZrB_2 + 4Al \tag{2.37}$$

According to the data in the literature [24] and calculation of standard free energy of formation, the standard free energy of substances participating in the reaction (J/(mol K)) is

$$\Delta G^{\ominus}_{Zr} = -8308.6 - 11.23T \tag{2.38}$$

$$\Delta G^{\ominus}_{Al} = -5609.9 - 11.35T \tag{2.39}$$

$$\Delta G^{\ominus}_{Al_3Zr} = -370858.3 + 123.02T \tag{2.40}$$

$$\Delta G^{\ominus}_{B_2O_3} = -71980.5 - 5.1729T \tag{2.41}$$

$$\Delta G^{\ominus}_{AlB_2} = -48072{+}36.363T \tag{2.42}$$

$$\Delta G^{\ominus}_{Al_2O_3} = -1675285.5 + 329.24T \tag{2.43}$$

$$\Delta G^{\ominus}_{ZrB_2} = -242568.7 - 38.617T \tag{2.44}$$

According to thermodynamic equation, the free energy calculated for reaction (2.35) is

$$\Delta G^{\ominus} = \Delta G^{\ominus}_{Al_3Zr} - 3\Delta G^{\ominus}_{Al} - \Delta G^{\ominus}_{Zr} = -345720 + 168.3T \tag{2.45}$$

Let $\Delta G^{\ominus} = 0$, we get $T = 2054.1$ K, it can be seen that when the aluminum melt temperature is lower than 2054.1 K(i.e., 660–1781.1 °C), the reaction (2.35) can proceed spontaneously.

The free energy calculated for reaction (2.36) is

$$\Delta G^{\ominus} = \Delta G^{\ominus}_{AlB_2} + \Delta G^{\ominus}_{Al_2O_3} - \Delta G^{\ominus}_{B_2O_3} - 3\Delta G^{\ominus}_{Al} = -1634547.3 + 404.8259T \quad (2.46)$$

Let $\Delta G^{\ominus} = 0$, we get $T = 4037.65$K (i.e., 660–3464.65 °C), the reaction (2.36) can proceed spontaneously.

The free energy calculated for reaction (2.37) is

$$\Delta G^{\ominus} = \Delta G^{\ominus}_{ZrB_2} + 4\Delta G^{\ominus}_{Al} - \Delta G^{\ominus}_{AlB_2} - \Delta G^{\ominus}_{Al_3Zr} = 153922 - 243.4T \quad (2.47)$$

Let $\Delta G^{\ominus} = 0$, the initial temperature of reaction is 632.4K (359.4°C), which means it can proceed spontaneously in molten aluminum ($\geq$ 660 °C).

Adding reactions (2.35), (2.36), and (2.37), the overall reaction for Al–Zr–B_2O_3 system is

$$2Zr + B_2O_3 + 5Al \rightarrow Al_3Zr + ZrB_2 + Al_2O_3 \quad (2.48)$$

Figure 2.11 shows the DSC curve obtained by reaction of Al–Zr–B_2O_3 system at a heating rate of 10 °C/min. It can be seen from the figure that at 542.7 °C an obvious exothermic peak appears corresponding to the decomposition reaction of flux in Zr-containing agent. The endothermic peak appearing at 662.7 °C is the melting temperature of aluminum powder. After the aluminum powder is melted, the first exothermic peak appears at 717.7 °C. At this time, a new phase Al_3Zr is produced in the melt. The melt temperature again rises to 857.7 °C, and an exothermic

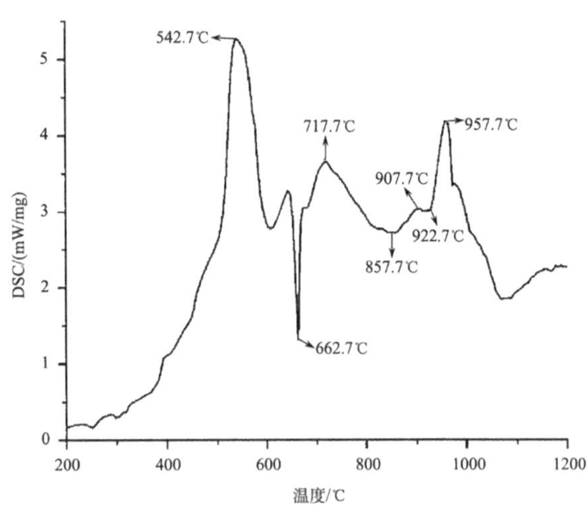

Fig. 2.11 DSC curve of aluminum and 20% (Zr agent + B_2O_3) powder under argon protection

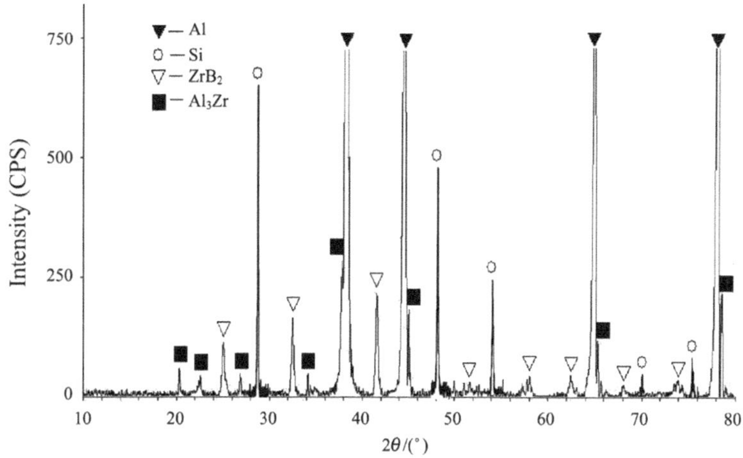

Fig. 2.12 XRD pattern of the composite prepared at a reaction temperature of 900 °C

reaction occurs corresponding to the formation of new Al_2O_3 and ZrB_2 phases in the melt. At this time, after B_2O_3 is considered to participating in the reaction, Al_2O_3 reinforcement phase particles are formed. However, the intermediate AlB_2 phase only slightly takes part in further reaction to form ZrB_2, and the reaction exothermic peak is small. After this, the temperature begins to rise sharply to 922.7 °C. At this point, the remaining AlB_2 and Al_3Zr completely react to form ZrB_2 phase. The highest temperature of the system reaches 957.7 °C. Figure 2.11 shows that aluminum powder reacts with mixed salt and new phases Al_3Zr, Al_2O_3, and ZrB_2 are formed. When ZrB_2 begins to form, the starting reaction temperature is rising to 857.7 °C, and a large amount of ZrB_2 phase particles are formed only when the temperature is further increased.

Figure 2.12 shows the XRD pattern of composite prepared from A356–Zr–B_2O_3 system. The figure shows that in addition to Al and Si peaks, ZrB_2 and Al_3Zr peaks are also observed, showing that ZrB_2 and Al_3Zr are formed in the reaction. Al_2O_3 peaks are not shown in the XRD pattern because its content is relatively small. Moreover, Fig. 2.13 shows the XRD pattern of slag produced during the chemical reaction. In addition to the presence of C, $KAlF_4$, K_3AlF_6, and KAl_4F_{13}, the solidified slag still contains B_2O_3, which shows that B_2O_3 reaction is not completed and the reaction process is difficult to control. The presence of Al_2O_3 in the slag shows that Al_2O_3 reinforcement particles are formed during the reaction.

2.2.3.2 Al–Zr(CO_3)$_2$–KBF_4 System Development

After analyzing in-situ reactions using the classical theory of thermodynamics, it is found that Al–Zr(CO_3)$_2$–KBF_4 system will undergo the following metallurgical reactions in aluminum melt:

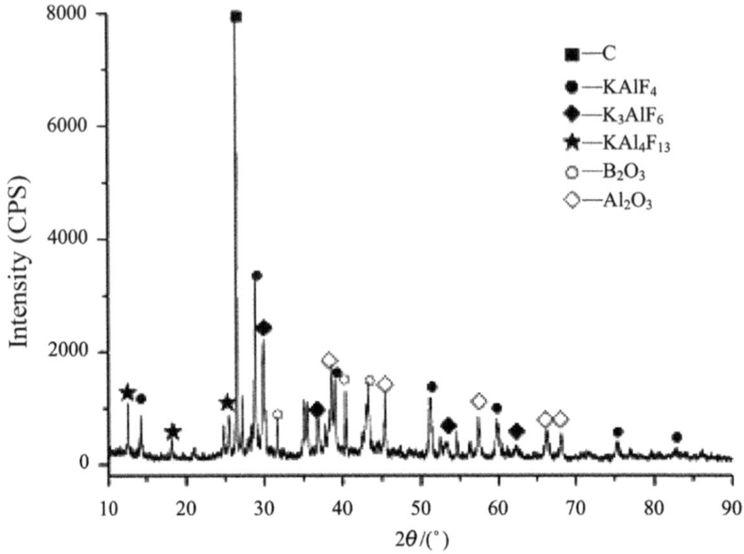

Fig. 2.13 XRD pattern of the slag of composite prepared at 900 °C

$$Zr(CO_3)_2 \rightarrow ZrO_2 + 2CO_2 \uparrow \quad (2.49)$$

$$2KBF_4 + 3Al \rightarrow AlB_2 + 2KAlF_4 \quad (2.50)$$

$$3ZrO_2 + 13Al \rightarrow 3Al_3Zr + 2Al_2O_3 \quad (2.51)$$

$$AlB_2 + Al_3Zr \rightarrow ZrB_2 + 4Al \quad (2.52)$$

In reaction (2.50), $\Delta G^{\ominus}_{AlB_2} = 38730.4 - 69.80T$, $\Delta G^{\ominus}_{ZrB_2} = -3182236 + 19.706T$. According to the thermodynamics equation, the free energy calculated for reaction (2.51) is

$$\Delta G^{\ominus}_{Al_2O_3} = -354267.7 + 343.8T \quad (2.53)$$

Let $\Delta G^{\ominus} = 0$, we get $T = 1030.4$ K(757.3 °C), it shows that when aluminum melt temperature is lower than 1030.4 K (i.e., 660–757.3 °C), the reaction (2.51) can proceed spontaneously.

By adding reactions (2.49), (2.50), (2.51), and (2.52), we get the overall reaction as:

$$2KBF_4 + 3Zr(CO_3)_2 + 12Al \rightarrow 2KAlF_4 + 6CO_2 \uparrow + 2Al_2O_3 + 2Al_3Zr + ZrB_2 \quad (2.54)$$

Table 2.6 shows the standard Gibbs free energy of all the substances involved in Al–$Zr(CO_3)_2$–KBF_4 system. From a thermodynamics point of view, the free energies of formation of various reinforcements should be lower than those of the intermediate phases, so that it is possible to obtain the required strengthening phase in the final product. Utilizing the data in Table 2.6, free energy of three strengthening phases can be obtained:

$$\Delta_r G^{\ominus}_{Al_2O_3} = -1313\,\text{kJ},\ \Delta_r G^{\ominus}_{Al_3Zr} = -390\,\text{kJ},\ \Delta_r G^{\ominus}_{ZrB_2} = -296\,\text{kJ}$$

Therefore, the order of stability in molten aluminum is $Al_2O_3 > Al_3Zr > ZrB_2$.

Figure 2.14 shows the DSC curve of Al–$Zr(CO_3)_2$ reaction system obtained with a heating rate of 10 °C/min. It can be seen from the figure that during the entire heating process, the endothermic peak at 647.3 °C corresponds to the melting of aluminum. The enthalpy of Al–$Zr(CO_3)_2$ reaction system increases significantly at 850 °C and an exothermic reaction occurs. At this point, a new phase is formed in the melt and the temperature of the Al–$Zr(CO_3)_2$ reaction system reaches to the highest point, i.e., 1006.7 °C, as shown in the DSC curve. It is clear that the Al–$Zr(CO_3)_2$ system synthesizes particle-reinforced aluminum matrix composites through an in-situ reaction. The obvious feature of this reaction is that after the system reaches a certain initial temperature, the exothermic effect increases significantly, and the melt temperature reaches to the highest point. During the experiment, the liquid level in the melt fluctuates violently at this point and the reaction proceeds in full swing.

Table 2.6 Expressions of standard Gibbs free energy of formation for various elements and compounds in Al–$Zr(CO_3)_2$–KBF_4 system

Substance	Expressions of $\Delta G_i^{\ominus}$ for elemental substances and $\Delta_r G_i^{\ominus}$ for compounds	Applicable temperature range/k	$\Delta_r G_i^{\ominus}$ at 1143 K/kJ
Al_2O_3	$\Delta G^{\ominus}_{Al_2O_3} -$ $-1682927 + 323.24T$	933–2315	– 1313
B_2O_3	$\Delta G^{\ominus}_{B_2O_3} =$ $-1228800 + 210.04T$	723–2316	– 989
ZrO_2	$\Delta_r G^{\ominus}_{ZrO_2} =$ $-1092000 + 183.7T$	298–2123	– 882
Al_3Zr	$\Delta_r G^{\ominus}_{Al_3Zr} =$ $-25138.3 \quad 319.4T$	298 2500	– 390
ZrB_2	$\Delta_r G^{\ominus}_{ZrB_2} =$ $-318236 + 19.706T$	350–2450	– 296
AlB_2	$\Delta_r G^{\ominus}_{AlB_2} =$ $38730.4 - 69.8T$	298–2500	– 41
Zr	$\Delta G^{\ominus}_{Zr} = -8308.6 - 11.23T$	300–2350	– 21
Al	$\Delta G^{\ominus}_{Al} = -5609.9 - 11.35T$	298–2500	– 19

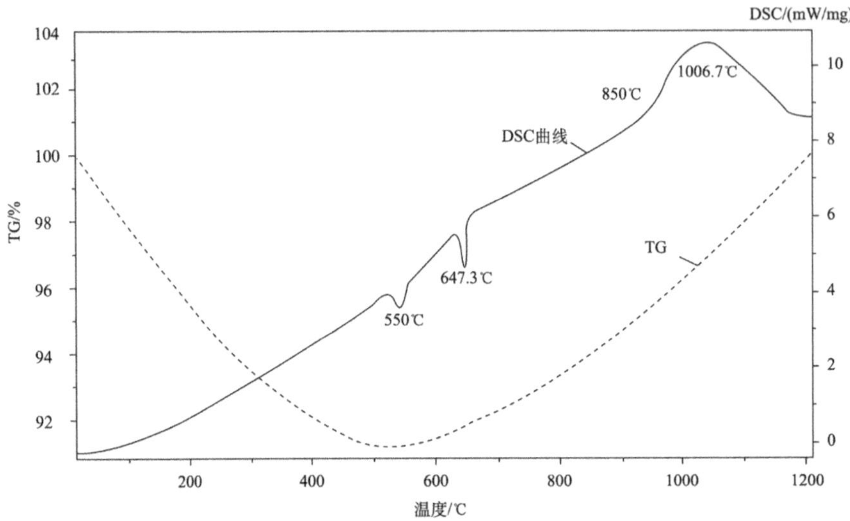

Fig. 2.14 DSC curve of Al–$Zr(CO_3)_2$ reaction system

In addition, the real-time temperature measurements show the same pattern as the thermal analysis curve shows.

Figure 2.15 shows the DSC curve of composite synthesized from Al–$Zr(CO_3)_2$–KBF_4 reaction system with a heating rate of 10 °C/min. It can be seen that when the system temperature is higher than 870 °C, there is an obvious exothermic process, and the maximum temperature reaches 974.8 °C. In the whole heating process, the endothermic reaction at 560.4 °C corresponds to the decomposition of $Zr(CO_3)_2$, and the endothermic reaction at 667.2 °C corresponds to the melting of aluminum.

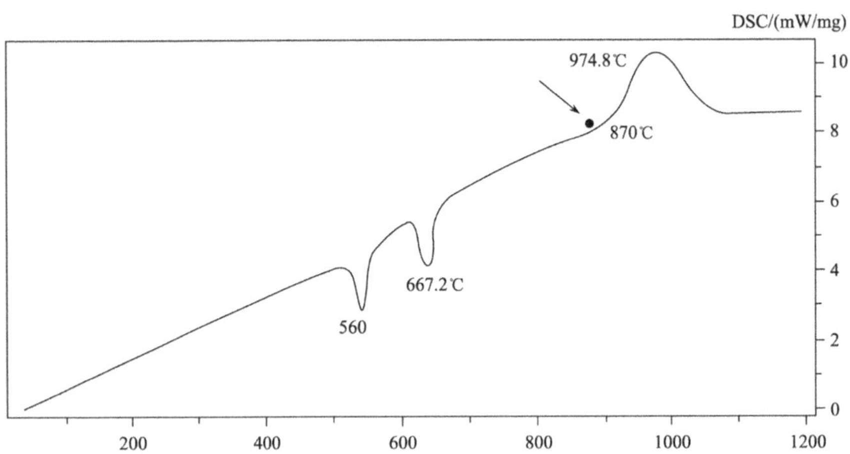

Fig. 2.15 DSC curve of composite synthesized from Al–$Zr(CO_3)_2$–KBF_4 reaction system

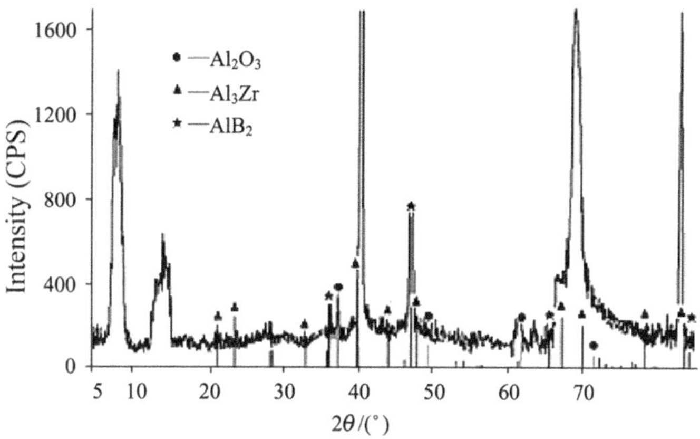

Fig. 2.16 XRD pattern of water-quenched composite sample synthesized by in-situ reaction of $Al–Zr(CO_3)_2–KBF_4$ system

An exothermic reaction is observed when the reaction system is around 850 °C. Followed by the exothermic reaction, a new phase is formed in the melt, and the maximum temperature reaches 974.8 °C. These features of the DSC curve show that after the system reaches a certain initial temperature, the exothermic effect increases significantly and melt attains the highest temperature, which reflects the general characteristics of in-situ synthesis of composite materials by melt reaction.

Figure 2.16 shows the XRD pattern of water-quenched composite melt sample during in-situ synthesis from $Al–Zr(CO_3)_2–KBF_4$ system. Composite samples are extracted from the melt after a predetermined synthesis time (i.e., 5, 10 $_{min}$). The results show that the XRD patterns of synthesized composites are similar, and a large number of AlB_2 intermediate phase particles are found in the sample (shown in Fig. 2.16), verifying that the reaction (2.50) can take place. However, AlB_2 phase is not found in the final synthetized composite. It is believed that AlB_2 appearing in the $Al–Zr(CO_3)_2–KBF_4$ reaction system is only an intermediate phase of in-situ synthesis process and not the ultimate strengthening phase. It can be observed from Fig. 2.16 that the produced final reinforcement phases are ZrB_2, Al_2O_3, and Al_3Zr.

2.2.3.3 $Al–K_2ZrF_6–KBF_4–Na_2B_4O_7$ System Development

The $Al–K_2ZrF_6–KBF_4–Na_2B_4O_7$ system will have the following metallurgical reactions in aluminum melt:

$$2KBF_4 + 3Al \rightarrow AlB_2 + 2KAlF_4 \tag{2.55}$$

$$3K_2ZrF_6 + 13Al \rightarrow 3Al_3Zr + K_3AlF_6 + 3KAlF_4 \tag{2.56}$$

$$Na_2B_4O_7 + 6Al \rightarrow Na_2O + 2Al_2O_3 + 2AlB_2 \tag{2.57}$$

$$4K_2ZrF_6 + 4Al + 2Na_2O \rightarrow 4K_2NaAlF_6 + 2ZrO_2 \tag{2.58}$$

$$3ZrO_2 + 13Al \rightarrow 2Al_2O_3 + 3Al_3Zr \tag{2.59}$$

Among these, the reactions (2.55) and (2.57) generate AlB_2 phase; and reactions (2.56) and (2.59) generate partial Al_3Zr phase:

$$AlB_2 + Al_3Zr = ZrB_2 + 3Al \tag{2.60}$$

Figure 2.17 shows the XRD pattern of composite prepared by in-situ reaction of A356–K_2ZrF_6–KBF_4–$Na_2B_4O_7$ system. As can be observed from the figure that in addition to Al and Si peaks of A356 matrix alloy, there are three main diffraction peaks of ZrB_2, Al_3Zr, and Al_2O_3 phases. It illustrates that the system has produced three new phases of ZrB_2, Al_3Zr, and Al_2O_3 after in-situ melt reaction. In addition, the slag removed after in-situ reaction is analyzed by X-ray diffraction, as shown in Fig. 2.18. The slag is found to be composed of K_2NaAlF_6, $KAlF_4$, and K_3AlF_6 substances.

Since no AlB_2 phase is observed in Fig. 2.18, it is confirmed that the reaction has been completed. Finally, three types of reinforcement particles (ZrB_2, Al_2O_3, and Al_3Zr) are obtained in the composite.

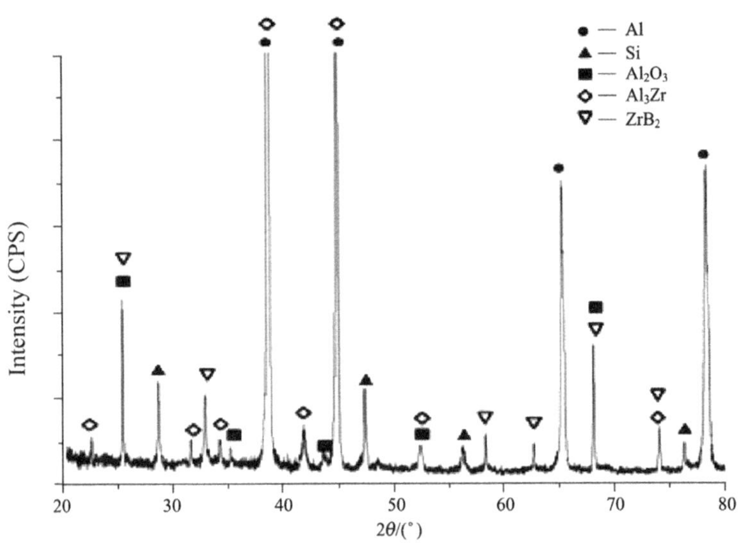

Fig. 2.17 XRD pattern of composite prepared by in-situ reaction of A356–K_2ZrF_6–KBF_4–$Na_2B_4O_7$ system

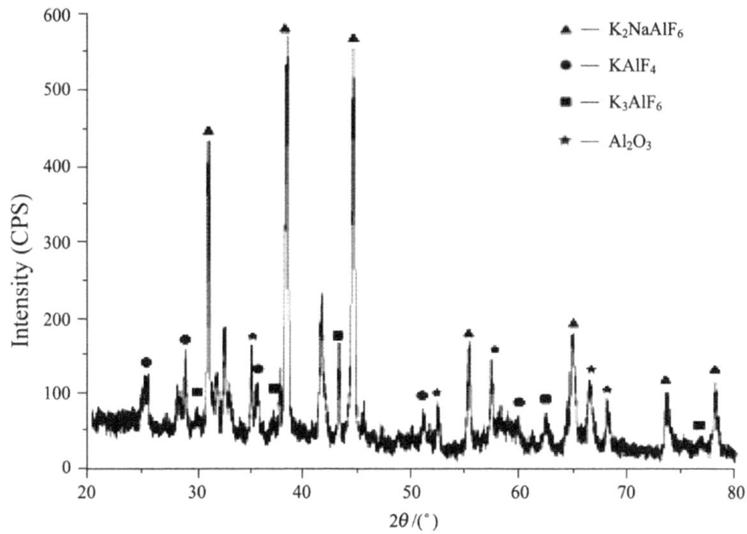

Fig. 2.18 XRD pattern of slag produced after in-situ reaction of A356–K_2ZrF_6–KBF_4–$Na_2B_4O_7$ system

References

1. Koczak MJ, Premkumar MK. Emerging technology for the in-situ production of MMCs. JOM. 1993;45(1):44–8.
2. Song IH, Kim DK, Hahn YD, et al. Investigation of Ti_3AlC_2 in the in-situ TiC-Al composite prepared by the exothermic reaction process in liquid aluminum. Mater Lett. 2004;58(5):593–7.
3. Zhao YT. Study on the microstructure and properties of particle reinforced composites formed by the reaction of Al-Zr-O system. Nanjing: Southeast University; 2001. (in Chinese).
4. Ma ZY, Tjong SC. In situ ceramic particle-reinforced aluminum matrix composites fabricated by reaction pressing in the TiO_2 (Ti) Al B (B_2O_3) systems. Metall Mater Trans A. 1997;28(9):1931–42.
5. Kanury AM. A kinetic model for metal + nonmetal reactions. Metall Trans A. 1992;23(9):2349–56.
6. Wang ZD. Microstructure and mechanical properties of 2024/TiC composite prepared by contact reaction method. Harbin: Harbin Institute of Technology; 1994. (in Chinese)
7. Zhang EL. Study on the thermodynamic and kinetic processes of Al/TiCp composites. Harbin: Harbin Institute of Technology; 1996. (in Chinese)
8. Gao Y, Jia J, Loehman RE, et al. Microstructure and composition of Al-Al_2O_3 composites made by reactive metal penetration. J Mater Sci. 1996;31(15):4025–32.
9. Chen ZY. Microstructure and properties of endogenous Al-4.5Cu/TiB_2 composite through melt reaction. Harbin: Harbin Institute of Technology; 1998. (in Chinese)
10. Breslin MC, Ringnalda J, Xu L, et al. Processing, microstructure, and properties of co-continuous alumina-aluminum composites. Mater Sci Eng A. 1995;195(94):113–9.
11. Zhao YT, Sun GX. In-situ synthesis of novel composites in the system Al-Zr-O. J Mater Sci Lett. 2001;20(20):1859–61.
12. Mondolfo LF. The microstructure and properties of aluminum alloy. In: Wang ZT, et al. (eds.) Beijing: Metallurgical Industry Press; 1998. (in Chinese)
13. Zhang GD, Zhao CZ. Metal matrix composites. Shanghai: Shanghai Jiaotong University Press; 1996. (in Chinese).

14. Doyle WM. Aluminum alloys: structure and properties. Metal Sci. 1976;35(11):408.
15. Napakov B. Refining treatment of deformed aluminum alloy. In: Wang YH, Zhang FB (eds.) Beijing: Metallurgical Industry Press; 1992. (in Chinese)
16. Cao G. (Al_2-$x$$SO_2$) ZL109 composite material interface structure and performance research. Nanjing: Southeast University; 1999. (in Chinese).
17. Lee MS, Terry BS. Effects of processing parameters on aluminide morphology in aluminium grain refining master alloys. Mater Sci Technol. 1991;7(7):608–12.
18. Chen G, Sun GX. Research on Al_2O_3(p)/Al-Cu composites formed by reaction. Special Casting Non-ferrous Alloys. 1997;3:1–4 (in Chinese).
19. Yu K, Li SG, Li WX, et al. The effect of trace Sc and Zr on the recrystallization behavior of 2618 aluminum alloy. Chin J Nonferrous Metals. 1999;12:709–13 (in Chinese).
20. Barin I, Knacke O, Kubaschewski O. Thermochemical properties of inorganic substances. New York: Springer-Verlag; 1973. p. 792–802.
21. Lin CX, Bai ZH, Zhang ZR. Manual of thermodynamic data of minerals and related compounds. Beijing: Science Press; 1985. (in Chinese).
22. Ye DL. Practical inorganic thermodynamic data handbook. Beijing: Metallurgical Industry Press; 1981. (in Chinese).
23. Dunmead SD, Readey DW, Semler CE, et al. Kinetics of combustion synthesis in the Ti-C and Ti-C-Ni systems. J Am Ceram Soc. 1989;72(12):2318–24.
24. Subrahmanyam J, Vijayakumar M. Self-propagating high-temperature synthesis. J Mater Sci. 1992;27(23):6249–73.

Chapter 3
Synthesis of In-Situ Aluminum Matrix Composites by Electromagnetic Method

The electromagnetic field belongs to a contactless field; that is, the electromagnetic force can transfer energy into a material fabricating system without having physical contact with it. It can exhibit the comprehensive advantages as wide adjusting range of parameters, high operating flexibility, and superior operating efficiency. On the basis of available reports on the effects of electromagnetic field on the metallic melt and chemical reaction, in this chapter, the electromagnetic field is introduced into the synthesis process of in-situ aluminum matrix composites, i.e., fabricating the in-situ aluminum matrix composites by electromagnetic method. The influence of electromagnetic field parameters and synthesizing parameters on the microstructure evolution is investigated together with the mechanism of in-situ aluminum matrix composites synthesis by electromagnetic method.

3.1 Effect of Electromagnetic Field on Melt and Chemical Reaction

The effect of electromagnetic field on the metallic melt, at first, lies in the distribution of magnetic induction intensity (B) and electromagnetic force (F) in the electromagnetic field. The influence of electromagnetic field on the chemical reaction is connected with its effect on reaction particles than on the reaction thermodynamics and kinetics conditions.

3.1.1 Distribution of B and F

To describe quantitatively the distribution rules and movement states of composites' melt in the electromagnetic field, the necessary parameters concerning magnetic field, electric field, and velocity field should be analyzed and calculated in detail. In

Y. Zhao, *In-Situ Synthesis of Aluminum Matrix Composites*,
https://doi.org/10.1007/978-981-16-9120-1_3

the quantitative calculation, the Maxwell equation set and Ohm's law are demanded, among which the Maxwell equation set is the basic equations of magnetic fluid as follows:

$$\nabla \times H \cdot J + \frac{\partial D}{\partial t} \quad \text{(Ampere circuital theorem)} \tag{3.1}$$

$$\nabla \times E = -\frac{\partial B}{\partial t} \quad \text{(Law of electromagnetic induction)} \tag{3.2}$$

$$\nabla \cdot D = \rho_0 \quad \text{(Gauss flux theorem)} \tag{3.3}$$

$$\nabla \cdot B = 0 \quad \text{(Flux continuity theorem)} \tag{3.4}$$

where ∇ is rotation; H is magnetic intensity, unit: A/m; J is the induction current, unit: A; D is electric displacement, unit: C/m^2; t is time, unit: s; E is the induced voltage, unit: V; B is the magnetic induction intensity, unit: T; ρ_0 is the electric charge density, unit: C/m.

The corresponding Ohm's law is:

$$J = \sigma(E + u \times \mathrm{B}) \tag{3.5}$$

where u is the movement velocity of electric charge, unit: m/s; σ is the conductivity of melt, unit: S/m.

Suppose that the length along height direction z of melt is infinite and the melt is incompressible, and a small unit is extracted in the aluminum melt in the presence of electromagnetic field. The stress of this unit along radial direction r and tangential direction θ is investigated. Figure 3.1 shows the stress analysis of liquid aluminum mass.

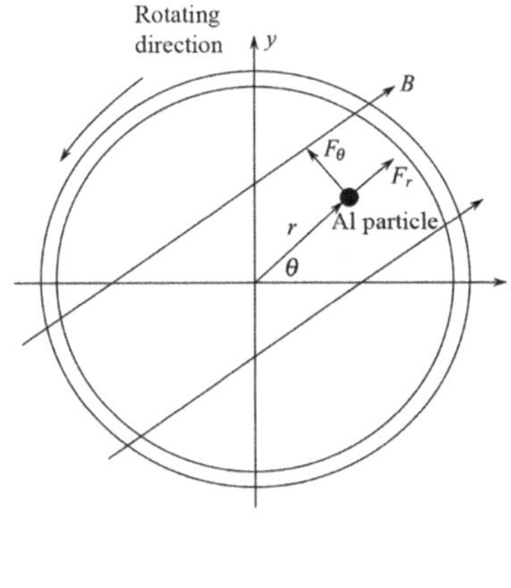

Fig. 3.1 Stress on Al liquid particles in electromagnetic field

In the presence of alternative magnetic field *B*, the induced current *J* will be generated in the aluminum melt; therefore, the aluminum melt turns into current-carrying melt. The interaction between the *J* and *B* will generate electromagnetic force and drive melt motion.

$$F = J \times B \tag{3.6}$$

where the unit of *F* is N/m^3. When compared to the gravitational field, the dimension equals to ρg (N/m^3). The ρ stands for the melt density and equals to a force density; therefore, the *F* is called as electromagnetic volume force, also known as Lorentz force. For the microzone with V_L volume, the electromagnetic force is:

$$F_L = J \times B \times V_L \tag{3.7}$$

where the induced current *J*, magnetic induction intensity *B* and electromagnetic force *F* are vectors, and their relationship is multiplication cross; furthermore, the *J* is orthogonal to the *B*, so the electromagnetic force can be calculated by the dot product of *J* and *B*. To simplify [1], in the following equations, the formula is expressed by scalar.

On the other hand, the continuity equation for incompressible fluid can be expressed by:

$$\nabla \times V = 0 \tag{3.8}$$

where the *V* stands for the flow velocity, unit: m/s.

The (3.9) is used to describe the Navier–Stokes equation:

$$\rho\left[\frac{\partial V}{\partial t} + (\nabla V)V\right] = -\nabla P + \mu \cdot \nabla^2 V + \rho g \tag{3.9}$$

where $(\nabla V)V = V_x\frac{\partial V_x}{\partial x} + V_y\frac{\partial V_y}{\partial y} + V_z\frac{\partial V_z}{\partial z}$, $\nabla^2 V = \frac{\partial^2 V_x}{\partial x^2} + \frac{\partial^2 V_y}{\partial y^2} + \frac{\partial^2 V_z}{\partial z^2}$, *P* and μ are pressure and viscosity coefficient. The first term on the right is the pressure term, the second term is the viscous momentum term, and the third term is the gravity term.

In the presence of magnetic field, in the conductive flow the velocity, the electric and magnetic fields appear simultaneously and with interact each other [2]. In common sense, the electromagnetic field will affect the velocity field, in turn, the velocity field will influence the electromagnetic field to some extent. In view of the Reynolds number $\text{Rm} = \mu\sigma UL << 1$, among which μ is the permeability of the medium (unit: H/m), *U* is the characteristic voltage, *L* is the characteristic length. In view that the Reynolds number is so small, the effect of velocity field on the electromagnetic field is often ignored.

The rotation ∇ is introduced into the equation $\nabla \times B = \mu B$, on the basis of Eqs. (3.2) and (3.5), when eliminating the electric parameter E and electric density J, Eq. (3.10) can be acquired:

$$\begin{aligned}\nabla \times \nabla \times B &= \mu \nabla \times J = \mu \nabla \times \sigma (E + V \times B) \\ &= -\mu\sigma \frac{\partial B}{\partial t} + \mu\sigma \nabla \times (V \times B)\end{aligned} \tag{3.10}$$

Using the two vector identities as $\nabla \times \nabla \times B = \nabla(\nabla \cdot B) - \nabla^2 B$ and $\nabla \cdot \mathrm{B} = 0$, Eq. (3.11) can be obtained:

$$\nabla^2 B + \mu\sigma \nabla \times (V \times B) = \mu\sigma \frac{\partial B}{\partial t} \tag{3.11}$$

Further, because $R_m = \mu\sigma UL << 1$, the item $V \times B$ can be ignored, Eq. (3.12) can be derived.

$$\nabla^2 B = \mu\sigma \frac{\partial B}{\partial t} \tag{3.12}$$

As shown in Fig. 3.2, the metal liquid surface is defined as x–y face and z-axis is vertical to the surface and directed to the interior (on the basis of right-hand rule). When the magnetic induction intensity along x-direction is defined as B_x, the magnetic field in the melt can be expressed by Eq. (3.12). Its component along x-direction can be calculated by Eq. (3.13).

$$\frac{\partial B_x}{\partial t} = \eta \frac{\partial^2 B_x}{\partial z^2} \tag{3.13}$$

Then, assigning $B_x = \mu H_x(z)e^{j\omega t}$($H$ is the magnetic field intensity, ω is the angular frequency), Eq. (3.14) is obtained:

$$j\omega h_x = \eta \frac{\mathrm{d}^2 H_x}{\mathrm{d}z^2} \tag{3.14}$$

The boundary conditions are:

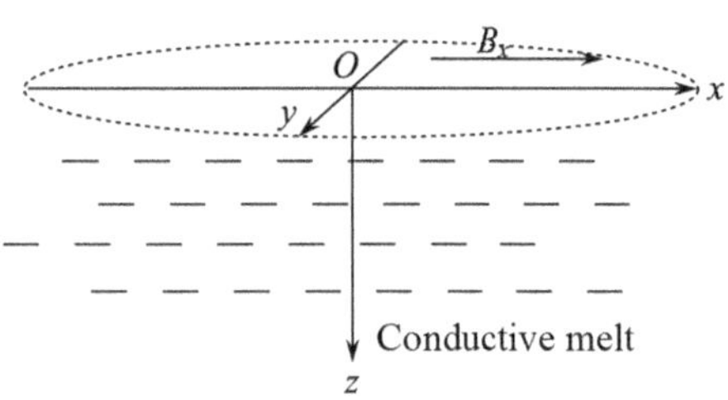

Fig. 3.2 Semi-infinite one-dimensional diffusion model in alternating electromagnetic field

$$\text{When} \quad x = 0,\, H_x = H_0; \quad \text{when} \quad x = \infty,\, H_x = 0 \tag{3.15}$$

When substituting into (3.14), Eq. (3.16) is obtained:

$$B_x = \mu H_0 \exp\left(-z\sqrt{\frac{\omega}{2\eta}}\right) \cdot \exp\left[j\left(-z\sqrt{\frac{\omega}{2\eta}} + \omega t\right)\right] \tag{3.16}$$

Based on $\nabla \times B = \mu J$ and Eq. (3.16), Eq. (3.17) is obtained:

$$J_x = -\sqrt{\frac{\omega}{2\eta}}(1 - j)H_0 \exp\left(-z\sqrt{\frac{\omega}{2\eta}}\right) \cdot \exp j\left(-z\sqrt{\frac{\omega}{2\eta}} + \omega t\right) \tag{3.17}$$

When the time is fixed, the B and J along z-direction will change periodically with the wavelength of $2\pi\sqrt{2\eta/\omega}$. With the x increase, the B and J will show exponential decay. In a certain radical distance, with the x increases, the B decreases rapidly. The certain distance is defined as skin depth δ. The result shows that in the electromagnetic field the distribution of H and B is non-uniform, which directly leads to the non-uniformity of force in melt.

On substitution of $J = (0, J_y, 0), B = (B_x, 0, 0)$ into Eq. (3.7), the electromagnetic force along z-direction can be obtained.

$$F_z = (0, 0, -\mu J_y H_x) \tag{3.18}$$

Further, substituting Eqs. (3.16) and (3.17) into (3.18), and then taking the average of time, Eq. (3.19) can be got:

$$F_z = \frac{1}{T}\int_0^T -\mu J_y H_x dt = -\frac{\eta H_0^2}{2J_y}\exp\left(-\frac{2z}{\delta}\right) \tag{3.19}$$

Here the expression of period T is: $T = 2\pi/\omega$.

When integrating Eq. (3.19), the electromagnetic pressure P_m can be calculated.

$$P_m = \int_0^\infty F_z \mathrm{d}z = \frac{\mu H_0^2}{4} \tag{3.20}$$

Further, substituting (3.20) into (3.16), Eq. (3.21) is obtained.

$$F = J \times B = \left(\frac{1}{\mu}\right) \cdot (\nabla \times B) \times B = \frac{(B \cdot \nabla)B}{\mu} - \frac{\nabla B^2}{2\mu} \tag{3.21}$$

where $\frac{(B\cdot\nabla)B}{\mu} \neq 0$, it will cause the melt to rotate and is defined as $F_{\theta..}$ It takes the effect of promoting the uniform distribution of temperature and elements, which is the main dynamic factor in the process of composite fabricating [3]. In Eq. (3.21), the second item is a radical force, which prompt the melt free surface to form a falcate shape. It is vertical to the melt lateral surface and point to the liquid core then balance with the static pressure of liquid. With the periodic change of alternating frequency, it will shrink and expand periodically, which constitutes the other dynamic factor in the process of composite fabricating.

According to the Maxwell equation, the relationship between F_r and F_θ *is:*

$$\left|\frac{F_r}{F_\theta}\right| \approx \frac{L}{\delta} \tag{3.22}$$

With the magnetic frequency f decreases, the δ increases. When the f is less than 10 Hz, $\delta >> L$, $F_\theta >> F_r$. On this condition, the electromagnetic force is expressed as tangential rotation force. When f is less than 10 Hz, $\delta << L$, $F_\theta << F_r$, the electromagnetic force is expressed as radical force. Therefore, whether in the low or high electromagnetic field, the electromagnetic force will exist and influence the composites fabricating process.

3.1.2 *Temperature Distribution in the Electromagnetic Field*

For the metal conduct in the electromagnetic field, the received unit active power can be expressed as:

$$P_e = 2 \times 10^{-4}\mathrm{k}(\mathrm{J\ N})^2\sqrt{\rho_r \mu_r f} \tag{3.23}$$

where P_e is the received magnetic field power (Unit: J); k is the correction factor and often selected as 0.5; J is the electric current (Unit: A); N is coils turns in 1 cm induct; ρ_r is the electrical resistivity (for Al it is 0.028 Ω mm^2/m). μ_r is the relative permeability. Due to the paramagnetism of Al, so $\mu_r \approx 1$; f is the frequency of alternating current (Unit: Hz).

Therefore, for the conductive melt, it will receive magnetic field power that will lead to the increase in temperature. According to Eq. (3.23), the power of high-frequency magnetic field is higher than others, so in the presence of high-frequency magnetic field, the effect of f on the temperature increase must be considered.

3.1.3 *Effect of Electromagnetic Field on the Melt*

While discussing the electromagnetic field on the metallic melt, both the force and heat effects should be considered comprehensively. Figure 3.3 reveals the relationship among the fluid, heat, and electromagnetic field. In the presence of electromagnetic field, the melt characteristic, electromagnetic force, and electromagnetic heat will interact with each other and take the effect of increasing melt quality.

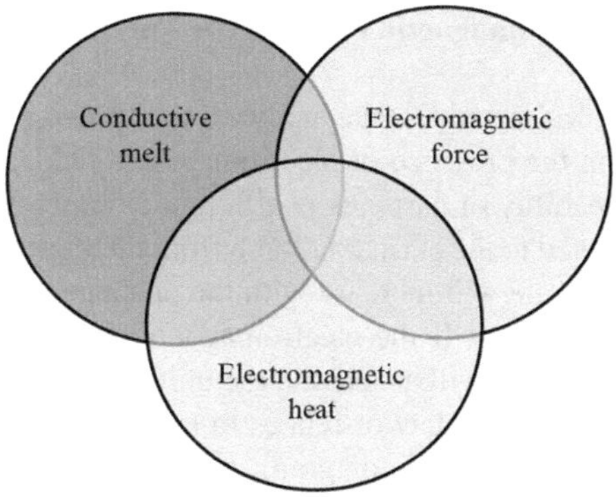

Fig. 3.3 Relationship among the conductive melt, electromagnetic force, and electromagnetic heat

3.1.3.1 Stirring Effect of Electromagnetic Field on Melt

Figure 3.4 shows the stirring apparatus for the aluminum melt in the presence of traveling magnetic field. When the low-frequency current is input into the electromagnetic inductor, the traveling magnetic field will generate. The interaction between the induced current and electromagnetic field will generate electromagnetic force and take the effect of melt stirring, whose direction and intensity can be adjusted through modifying the direction and intensity of traveling magnetic field.

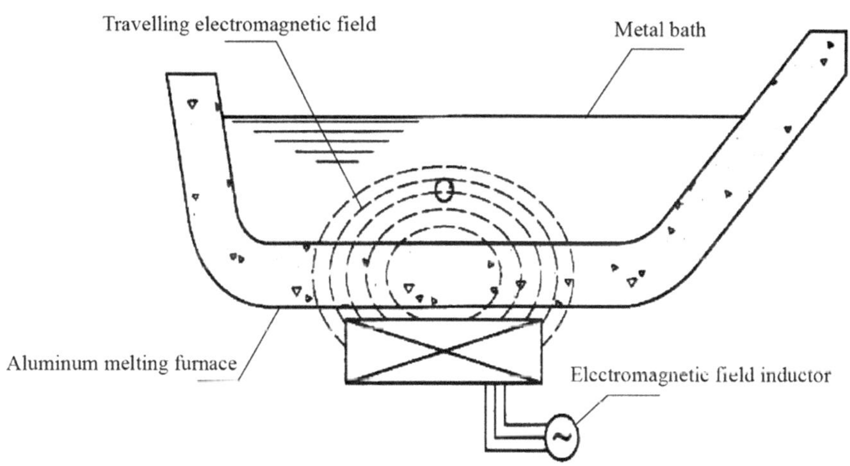

Fig. 3.4 Stirring apparatus for aluminum melt in traveling magnetic field

3.1.3.2 Effect of Electromagnetic Field on the Particles in Melt

In 1954, American A. Kolin and D. Pate analyzed the spherical particles behavior in the conducting flow in the presence of electromagnetic field. It is found that if the conductivity or permeability of particles is different from the surrounding flow, the particles will be subjected to the extrusion force from the flow, then the particles will migrate. The migration rate will increase with the enhancement of electromagnetic force of the surrounding flow. If the electromagnetic force is non-directional, the random migration of particle will accelerate the uniform distribution of particles.

Furthermore, the particles behavior relates to the uniformity of electromagnetic field. Figure 3.5 illustrates the force diagram in the uniform and non-uniform electromagnetic field. When considering force only, the particles in the non-uniform electromagnetic field are easy to migrate. Based on previous analysis, the distribution of magnetic induction intensity and magnetic lines of force varies with different locations along radical direction. The non-uniform distribution of magnetic field is beneficial for the particle's movement.

On the common condition, the electric conductivity of non-organic particles is much inferior to that of molten melt. The induced electric current in the particles is rather small; therefore, the subjected electromagnetic force is far below to that of the molten melt. The particles will be subjected to a force that is opposite to the magnetic force in the melt, which is called as the electromagnetic extrusion force. The particles will move along the direction that is opposite to the magnetic force. The moving velocity relates to the electromagnetic field kinds and electromagnetic parameters. The kinds of magnetic fields include low frequency, high frequency, traveling electromagnetic field, and so on. Furthermore, the particles movement relates

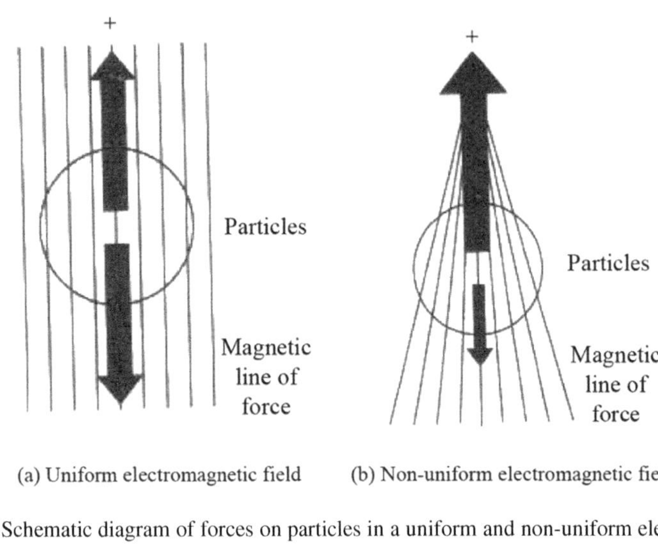

(a) Uniform electromagnetic field (b) Non-uniform electromagnetic field

Fig. 3.5 Schematic diagram of forces on particles in a uniform and non-uniform electromagnetic field

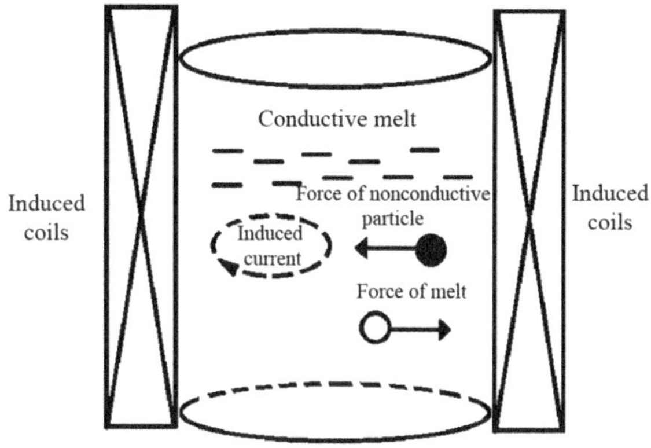

Fig. 3.6 Forces on the particles and their motion in conductive melt under electromagnetic field

to its morphology, if it exhibits an apparent shape singularity, the size difference from different directions will take a large effect on the electromagnetic repulsion [4].

If the electromagnetic force is fixed, the electromagnetic extrusion force will toward to a fixed direction. The directional migration of particles results in the particles' separation from melt. If the particles belong to an inclusion, it can be cleaned via electromagnetic force. The particle migration and separation effect relate to the electromagnetic field kind and parameters, such as in the presence of high-frequency electromagnetic field, a higher frequency, and magnetic induction intensity are beneficial to the inclusion separation. Figure 3.6 illustrates the particle's force and migration in the conducing melt.

Combined with the effects of electromagnetic field on the melt, electromagnetic force can play the role in stirring melt and driving particle movement, which plays a positive role in the synthesis of in-situ aluminum matrix composites.

3.1.4 Effects of Electromagnetic Fields on Chemical Reactions

In the early twentieth century, Faraday proposed that there was a close relationship between magnetic field and chemistry, and the combination of magnetic field and chemistry, namely the study of magnetic field's influence on chemical reactions, was also one of the important achievements of physical chemistry in the late twentieth century. It is well known that the elements involved in chemical reactions usually need to obtain certain energy in thermal motion, and only after effective collisions between the elements can the reaction be promoted. Thermal energy, light energy, and electric energy can be used as common means to stimulate chemical reaction, but they all need to reach a certain threshold before they become effective. Studies have found that the energy value generated by magnetic field interacting with molecules alone is extremely small. For example, the energy generated by 1 T magnetic field interacting with molecules is only 11.2 J/mol, which cannot reach the reaction activation energy (10–100 kJ/mol) required by common chemical reactions, so that the phenomenon of magnetic field affecting chemical reactions cannot be properly explained for a long time. It was not until the 1960s, due to the rapid development of quantum chemistry, that the internal connection between magnetic field and chemical reaction was gradually explained scientifically and reasonably, and a theoretical system with free radical pair theory as the core was gradually formed. It is gradually recognized that magnetic fields can change not only the rate but also the yield of chemical reactions.

According to the transition state theory of chemical reaction, the reaction rate k is:

$$k = \frac{k_B T}{hc} e^{\frac{\Delta S}{R}} e^{\frac{-\Delta H}{RT}} \tag{3.24}$$

where R is thermodynamic constant, T is temperature, h, c and k_B are constants in the transition state theory. ΔH is the activation enthalpy of the system and ΔS is the activation entropy of the system.

From the above equation, it can be seen that the reaction rate constant is mainly affected by three factors, including activation enthalpy and activation entropy, in addition to temperature T.

In the view of energy, the external magnetic field and the systematic mole free energy ΔGM can be expressed as:

$$\Delta G_M = -\frac{1}{2} \Delta \chi_M B^2 \tag{3.25}$$

where ΔX_M is the changes in molar susceptibility, B is the magnetic induction intensity.

Because the magnetic susceptibility is on the order of 10^{-7}, and the industrial B value is tens of Tesla (T). From this equation, it can be known that the magnetic field has little influence on the free energy value of the reaction. Therefore, the effect of magnetic field on chemical reaction is mainly to affect the arrangement of molecules and atoms as well as the spin mode and spin orientation of electrons, that is, to affect the entropy of reaction system, thus affecting the reaction rate and material structure [5].

The mechanism of magnetic field affecting the internal chemical reaction is very complex, which involves the interaction between magnetic field and various molecules, atoms and electrons in the reaction system, and inevitably involves quantum theory. Only from the perspective of quantum analysis can we essentially explore the mechanism of magnetic field that influences on the internal chemical reactions of materials. According to the analysis, the influence of magnetic field on the internal chemical reaction of materials is mainly based on the free radical reaction process. The magnetic field can change the time of the existence of free radicals, thus affecting the kinetic process and chemical yield of the reaction.

When the external magnetic field acts, the free radical and the single electron pair will show a strong magnetic sensitivity, so the magnetic field has a more obvious influence on the process involving these particles. As for the influence of magnetic field on the internal chemical reaction of materials, it is believed that the effect is mainly based on free radical reaction. This theory is based on the spin orientation of electrons and nuclei in chemical reactions [6]. It is a theory that systematically tracks and describes the influence of magnetic field on reaction kinetics and chemical yield by changing the existence time of free radicals.

According to quantum theory, atoms at different energy levels have different energy states. An electron in the same energy level has multiple states of motion, and the energy level with two or more states is called the degenerate energy level, which is correspondingly called the degenerate state. The energy levels of one state are called singlet states, and the energy levels of three states are called triplet states. Free radicals produced in free radical reactions can have different orientations in space and thus different states. Free radicals usually exist as singlet or triplet states. Under certain conditions, there can be mutual transformation between them, which is called intersystem transition. In a free radical reaction, it is considered that the reaction is terminated when two singleton free radicals collide and bond to a dimer. According to the viewpoint of free radical pair mechanism, there are two kinds of internal reactions in materials: transition reaction between free radicals with quantum states and free radical escape reaction. Under the action of magnetic field, when the free radical pairs transit from singlet state to triplet state, Zeeman splitting occurs, which causes magnetic field effect. It also reduces the binding of free radicals during transport and increases the free radical escape reaction. It is the magnetic field that changes the relationship between the two reactions, which in turn affects what happens inside the material.

The free radical pair theory discusses the theoretical feasibility of magnetic field affecting chemical reaction from the quantum scale and becomes a guiding theory for improving synthetic conditions of chemical reaction under magnetic field. At present, there are many types of chemical reactions affected by magnetic field, including photochemical reactions, polymerization reactions, electrochemical reactions, isotope enrichment reactions, etc., but there are few reports on inorganic chemical reactions affected by magnetic field.

3.2 Law of Electromagnetic Synthesis of Aluminum Matrix Composites

The in-situ aluminum matrix composites were prepared by in-situ reaction to produce reinforced particles, which belongs to inorganic chemical reaction. Taking Al–$Zr(CO_3)_2$ system as the object, the law of electromagnetic synthesis of aluminum-based composites is explored; that is, the effects of magnetic induction intensity, magnetic field processing time, additive amount of reactants, and initial reaction temperature on the microstructure of composites are studied.

3.2.1 Effect of Magnetic Induction Intensity

Figure 3.7 shows the solidification structure of Al-20 wt.%$Zr(CO_3)_2$ composite system under different magnetic induction intensities (B = 0 T, 0.03 T, 0.05 T, 0.07 T). Other synthetic conditions are as follows: The initial reaction temperature is 850 °C, and the reaction time is 20 min. It can be seen that under the condition of no electromagnetic field, fewer, coarser, and fewer uniform particles are generated. With

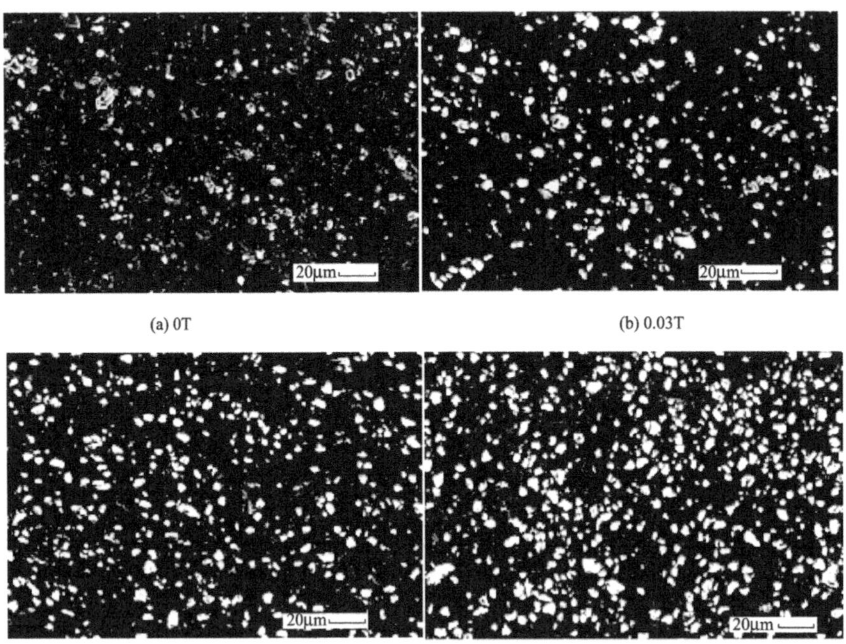

(a) 0T (b) 0.03T

(c) 0.05T (d) 0.07T

Fig. 3.7 Microstructure of in-situ Al matrix composite synthesized from Al-20 wt% $Zr(CO_3)_2$ system with different magnetic induction intensity the magnetic induction intensity increasing from 0 to 0.03 T, 0.05 T to 0.07 T in turn, the number of reinforced particles is increased, the size is refined, the morphology tends to be round, and the distribution is improved. When the magnetic induction intensity is 0.07 T, the reinforced particles of the composite are fine and more evenly distributed in the matrix, and the surface of the particles is smooth, round, and blunt.

Firstly, from the view of force, since the melt under the action of magnetic field is a cylinder, which is symmetric relative to the center of magnetic field. The electromagnetic force of electromagnetic field on the melt is mainly the force on the cylinder and is expressed as:

$$P = -\frac{B^2}{2\mu_0} \tag{3.26}$$

where "–" means the pressure, μ_0 is the vacuum permeability of melt.

Because B is an alternating magnetic field with high frequency, within the length range of the solenoid coil, the melt does reciprocate radial motion along the direction perpendicular to the symmetry axis, especially at the end of the coil, the melt moves up and down accompanied by turbulent motion and vibration. As the magnetic pressure P changes violently and its intensity is far greater than the dynamic pressure of the metal melt, the melt will oscillate strongly, which causes the melt to produce violent forced convection, then promotes the migration and exchange of atoms in the in-situ chemical reaction of the melt and accelerates the chemical reaction. Therefore, the higher the magnetic field strength is, the more violent oscillation will occur in the melt, which is more conducive to the migration and exchange of atoms in the chemical reaction, which inhibits the growth of grains, and finally makes the particles in the solidified microstructure finer and more evenly distributed, and the surface of particles smooth, round, and blunt.

From the perspective of the thermal effect by alternating magnetic field, according to the phase transition dynamics theory [7], the formation of crystal particles in aluminum melt mainly includes two processes of nucleation and growth, among which the nucleation rate of grains is:

$$I = I_0 \exp\left(-\frac{\Delta G_n^*}{kT}\right) \exp\left(-\frac{u}{kT}\right) \tag{3.27}$$

where I_0 is the coefficient of nucleation; ΔG_n^* is the thermodynamic barrier of critical nucleation; u is the activation energy of atomic migration, that is, the nuclear kinetic barrier; k is the Boltzmann constant, and T is the melt temperature.

When the high-frequency magnetic field is applied, the melt temperature increases. According to Eq. (3.27), nucleation rate I increases, thus accelerating the in-situ chemical reaction.

3.2.2 *Effect of Processing Time of Magnetic Field*

Figure 3.8 shows the microstructure of composite fabricated in Al-20 wt.%$Zr(CO_3)_2$ system under different magnetic field processing time (5, 10, 15, and 20 min). Other synthesis conditions are as follows: The initial reaction temperature is 850 °C, and the magnetic induction intensity is 0.07 T. As can be seen from Fig. 3.8a, when the magnetic field processing time is 5 min, the number of particles is small and the particles are large, and some white layer substances appear around the particles, indicating that the added reactants may not have completely reacted. With the increase of magnetic field processing time, the number and size of endogenous particles gradually increase, and the circular passivation effect becomes more obvious and the distribution becomes more uniform, as shown in Fig. 3.8b–d.

The main reason is that the reaction mechanism of Al–$Zr(CO_3)_2$ system is reaction–rupture–diffusion mechanism. When a magnetic field is applied at the initial stage of the reaction, the generated particles do not have enough time to disperse from the reaction layer, which affects the in-situ chemical reaction and results in incomplete reaction of some reactants. With the increase of magnetic field processing time,

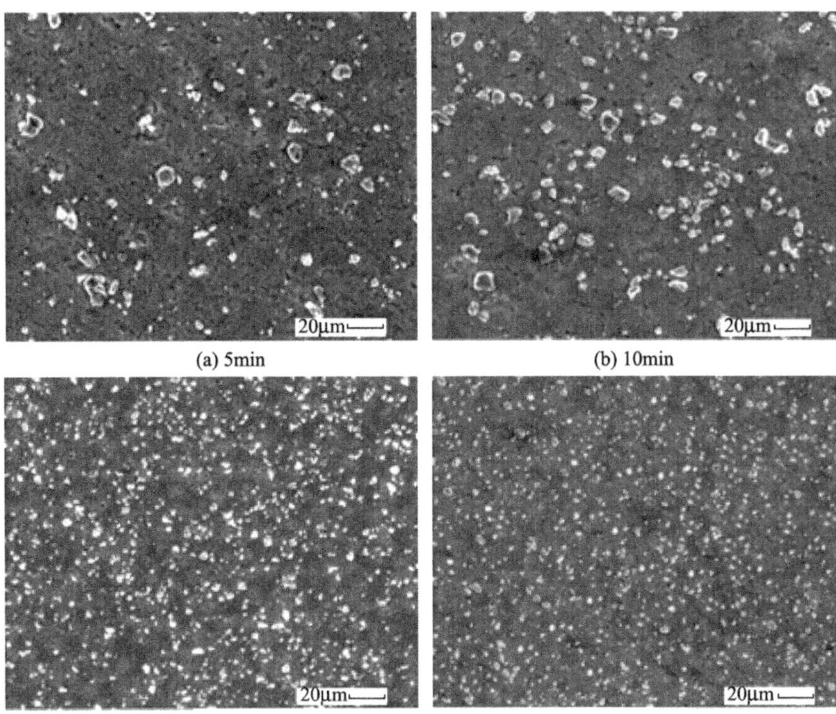

Fig. 3.8 Microstructure of in-situ Al matrix composites synthesized from Al-20 wt%t $Zr(CO_3)_2$ system with different processing time of magnetic field

the temperature of aluminum melt increases, prompting more particles to break and participate in the reaction. At the same time, the oscillation and forced convection of the aluminum melt accelerate the reaction diffusion and promote the decomposition of the bulk phase, which is beneficial to the reaction. Especially in the presence of magnetic field, the entire reaction process in the common drive of both magnetic field force and thermal stress and the generated Al_2O_3 particles and Zr atom due to the decomposition of ZrO_2 are rapidly spread out of conversion zone, speeding up the melt reaction, resulting in more practical preparation of Zr and Al atoms to form Al_3Zr reinforcing particle, together with its uniform distribution in the melt.

3.2.3 *The Effect of the Additive Mount of Reactants*

Figure 3.9 shows the XRD patterns of aluminum matrix composites synthesized by the melt reaction of Al-x%$Zr(CO_3)_2$ ($x = 5, 10, 15, 20, 25$) system at the initial reaction temperature of 850 °C, reaction time of 20 min, and magnetic induction intensity of 0.07 T. In addition to the diffraction peak of matrix Al phase, the diffraction peaks of Al_2O_3 and Al_3Zr phases also appear in the composites. At the same time, with the increase of the amount of $Zr(CO_3)_2$ powder x, the relative strength of the diffraction peak of Al_2O_3 and Al_3Zr phases also changes: When $x = 5$, the relative strength of the diffraction peak of Al_2O_3 and Al_3Zr phases is relatively low. With the increase

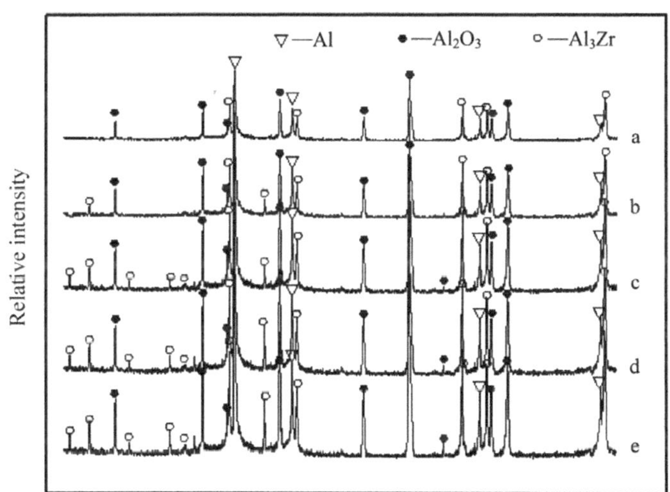

Fig. 3.9 XRD patterns of $(Al_2O_3 + Al_3Zr)_p$/Al in-situ aluminum matrix composites synthesized by electromagnetic method with different reactants additive amount: 5 wt%; **b** 10 wt%; **c** 15 wt%; **d** 20 wt%; **e** 25 wt%

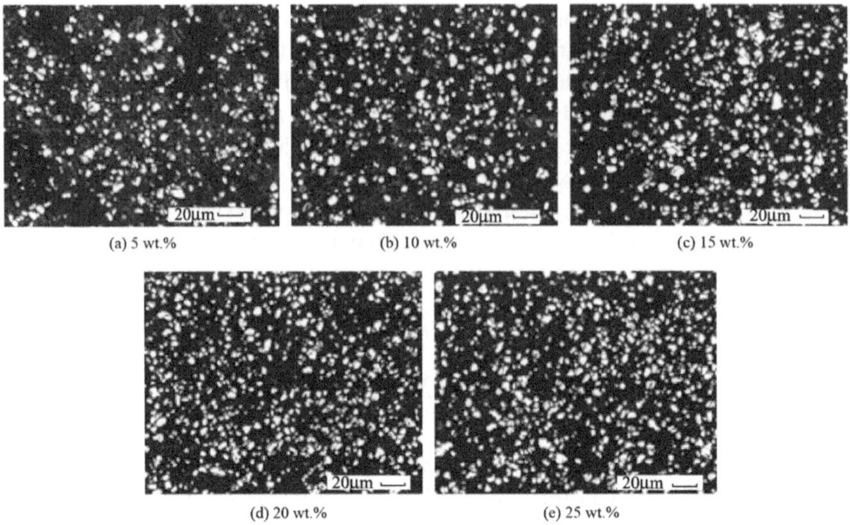

Fig. 3.10 Microstructure of in-situ Al matrix composite synthesized by electromagnetic method with different additive amount of reactants

of x, the relative strength gradually increases, especially when $x = 20$ and 25, strong diffraction peaks of Al_2O_3 and Al_3Zr phases appear.

Figure 3.10 shows the SEM figure of $(Al_2O_3 + Al_3Zr)_p/Al$ in-situ aluminum matrix composite synthesized by electromagnetic method with different amounts of $Zr(CO_3)_2$ reactants. When the additive amount is 5 wt.% and 10 wt.%, less particles are generated. With the x increase, the number of particles increases and the volume percentage increases. Especially when the additive amount of reactants is 20 wt.%, the size of the reinforced phase particles is small and the number is large, and they are uniformly distributed in the matrix. However, when the additive amount is 25 wt.%, the particles exhibit as agglomeration, and the particle size tends to increase. Thus, in the range of $x = 0$–20, with the increase of reactants amount, the volume fraction of the particles generated by the reaction gradually increases, and the size of the particles is smaller and the distribution is more uniform.

The reason is that when the amount of reactants is small, the ZrO_2 particles formed by $Zr(CO_3)_2$ are small, and the reaction can only take place in local areas of the melt. With the addition of more reactants, more ZrO_2 particles generate, that increase the contact area of ZrO_2 and liquid aluminum, at the same time, under the effect of electromagnetic field, owing to the strong stirring and oscillation of melt, makes the Al_2O_3 and Al_3Zr spread out from the reaction layer rapidly, thus increases the nucleation rate of Al_2O_3 and Al_3Zr. However, when the additive amount is too large, up to 25%, the wetting and diffusion of the reaction powder in the liquid aluminum become difficult, resulting in the phenomenon of "agglomeration" of particles. It is mainly due to the excessive addition of reactants, resulting in the decrease of melt mobility and the increase of viscosity. According to the modified fluid viscosity

Einstein calculation equation, the melt viscosity η_s is [8]:

$$\eta_s = \eta_0(1 + 2.5\phi + 10.52\phi^2) \quad (3.28)$$

where η_0 is the viscosity of aluminum matrix liquid and φ is the volume fraction of particles ($\varphi < 0.25$).

According to Eq. (3.28), it can be known that when the reactants amount increases by 5%, the melt viscosity will be increased by 10.95%, so the additive amount of reactants powder cannot increase infinitely. Experiments show that when additive amount is higher than 20 wt.%, the liquidity of melt will be reduced and viscosity be enhanced, resulting in the decrease of solute elements, causing the reactants to form a block. On this condition, the powder cannot contact with the liquid aluminum fully, to prevent the further reaction, reducing the generation of the reinforcing particles rate. Therefore, the microstructure of the composite is best when the percentage of reactants account for 20% of the liquid mass of aluminum.

3.2.4 *Effect of Initial Reaction Temperature*

Figure 3.11 shows the microstructure of in-situ composite synthesized from Al-20 wt.%$Zr(CO_3)_2$ system under electromagnetic field when the reaction time is 20 min, the magnetic induction intensity is 0.07 T, and the initial reaction temperatures are 800 °C, 850 °C and 900 °C. As can be seen from the figure, when the initial reaction temperature increases from 800 °C to 850 °C, the decomposition reaction of $Zr(CO_3)_2$ is more violent, the reaction between ZrO_2 and aluminum solution is more complete; therefore, the volume fraction of particles gradually increases, the size of particles is smaller, and the distribution is also more uniform. However, when the initial reaction temperature reaches 900 °C, particles agglomerates and grow up, as shown in Fig. 3.11c. This is mainly due to the excessive initial reaction temperature of the melt, and the melt temperature is higher due to the magnetic field thermal effect. Under this condition, the endogenous particle Al_3Zr will grow into sheets or large blocks. Therefore, the initial reaction temperature of Al-20 wt.%$Zr(CO_3)_2$ system should not be too high, and the range of 850 ± 5 °C is better.

3.3 Mechanism of Electromagnetic Synthesis of Composites

Compared with the composites without electromagnetic field, the microstructure of the composites synthesized by electromagnetic method is as follows: particle volume fraction increases, particle distribution uniformity improves, and particle roundness becomes better.

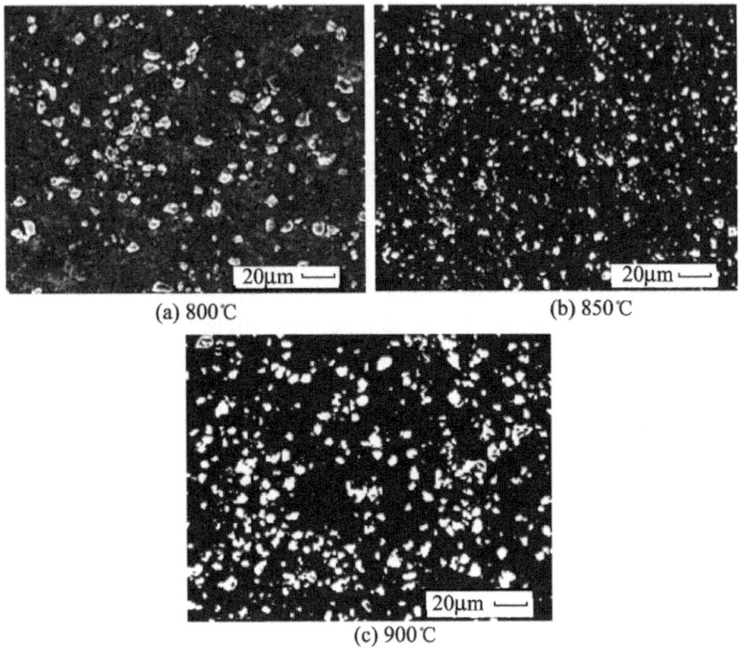

Fig. 3.11 Microstructure of in-situ $(Al_2O_3 + Al_3Zr)_p/Al$ composite synthesized by electromagnetic method with different initial reaction temperature

It is concluded that the increase of particle volume fraction is mainly due to the improvement of the conditions for the reactant powder particles to enter the melt, which is equivalent to the increase of reactant quantity under the condition of constant additive amount, which is helpful to increase the volume fraction of the generated particles. Furthermore, the electromagnetic field improves the thermodynamic and kinetic conditions of synthesis, so that the in-situ synthesis can be carried out under more favorable thermodynamic conditions, which helps to improve the reaction yield, the improvement of kinetic conditions helps to improve the reaction rate, and the reaction is more enough at the same time, which helps to increase the number of generated particles. The good dynamic conditions increase the diffusion ability of particles and improve the uniformity of particle distribution. During the process of formation and movement of particles, they are always in the disturbance and oscillating state, which helps to promote the roundness of their shape.

3.3.1 The Condition Under Which the Reactants Enter the Melt

In conventional synthesis process, the reactant particles enter the liquid aluminum completely depending on the graphite bell; that is, a mechanical approach, the entering condition for solid particles into the liquid aluminum conditions is poor, the electromagnetic stirring force due to electromagnetic field can force the reactant particles into the liquid aluminum and promote its uniform distribution. Firstly, the thermodynamic process of reactant particles entering the melt is discussed. Currently, the contact between solid particles and liquid aluminum is mostly on the melt surface. Because the surface heat dissipation condition is good, it is considered that the temperature condition for the in-situ reaction cannot be reached, and there is no interface reaction at the contact point. Figure 3.12 shows a schematic diagram of the entering process of a single reactant powder particle into the melt.

where γ_{gs} is the gas–solid interfacial tension, γ_{gl} is the gas–liquid interfacial tension, γ_{ls} is the liquid–solid interfacial tension, θ is the wetting angle.

When the particles enter the melt, the total free energy changes are: $\Delta E = \sum \Delta E_i$. ΔE_i includes surface energy change, kinetic energy change, thermal energy change, and potential energy change, et al. Since the particle velocity changes little from the surface to the melt and does not participate in chemical reaction, the kinetic energy and thermal energy change are ignored and only the surface energy E_{surf} and potential energy E_{pot} are considered.

Assume that the particle size of reactants is R, radius angle is ω, the entering process for particle corresponds to the 0°–180° change of ω.

In the process of particles entering the aluminum liquid, the surface energy and potential energy change are:

$$\Delta E_{surf} = -R^2\gamma_{lg}[2(1-\cos\omega)\cos\theta + \sin^2\omega] \tag{3.29}$$

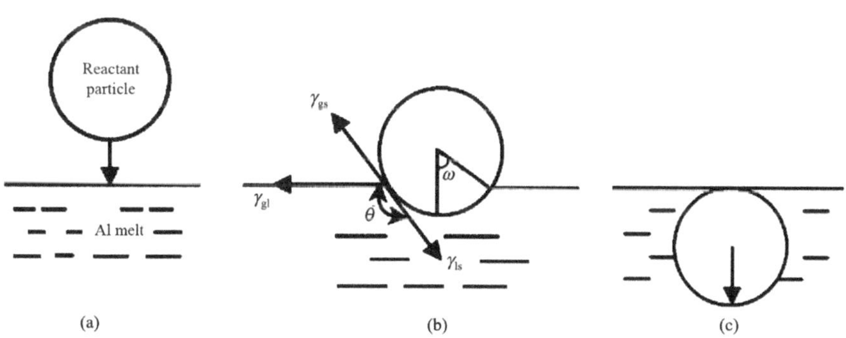

Fig. 3.12 Schematic diagram of the entering process for one reactant particles into aluminum melt under electromagnetic field, **a** Before entering the liquid aluminum; **b** on the surface of the liquid aluminum; **c** completely into the liquid aluminum

$$\Delta E_{\text{pot}} = -\frac{4}{3}\pi R^4 \rho_P g(1 - \cos\theta) \quad (3.30)$$

$$\Delta E_{\text{suyo}} = -\pi \frac{R^4}{48}\rho_l g(-3\cos 4\omega + 8\cos 3\omega + 12\cos 2\omega - 72\cos\omega + 55) \quad (3.31)$$

For the particles with $\theta > 90°$, $\Delta E > 0$, particles cannot spontaneously enter the melt. For the particles with $\theta < 90°$, particles can be spontaneously into the melt, mutual wetting between the particles and melt, the total free energy change $\Delta E < 0$; it is full and necessary condition for particles into the melt. On this condition, some measures can be used to improve the wetting effect, such as adding Mg element [9], or from some external powers, such as increasing mechanical stirring, electromagnetic stirring, and ultrasonic processing method.

When the in-situ reaction is carried out under electromagnetic field, the reactant particles are more likely to enter into the liquid aluminum under the eddy current caused by electromagnetic force. The number of participants in the reaction increases and the loss decreases, which will increase the yield of the product particles to a certain extent. When the output current is constant and the output frequency is constant, the electromagnetic stirring force generated by the alternating magnetic field is constant, and it is mainly manifested as shear force. Under the shear force, the composite melt is pushed to carry the solid reactant particles in the crucible for circular motion with constant angular velocity, forming a forced eddy current, as shown in Fig. 3.13.

The core of vortex is formed in the center, and the melt outside the core does free vortex motion. The real melt is a combined vortex formed by forced vortex and free vortex. In the region of forced vortex, the tangential velocity u is proportional to r; and their relationship is shown in (3.32).

$$\mu = kr \quad (3.32)$$

where k is the proportional constant, and the tangential velocity u of the free vortex part is inversely proportional to the radius r, i.e.,

$$\mu \cdot r^n = \text{constant} \quad (3.33)$$

In general, the index $n = 0.5–0.9$. When $n = 1$, the total melt moves as free vortex. When $n =$ -1, the total melt moves as forced vortex motion. As a fact, the fluid motion in the rotating magnetic field is between the forced vortex and the free vortex. Figure 3.13 is the pressure and velocity distribution diagram of the flow inside and outside the combined vortex under the electromagnetic field.

Assuming that the radius of vortex core is r_0, and the tangential velocity and pressure of vortex boundary are u_0 and $P_{0,}$ respectively; the pressure characteristics of free vortices can be determined by the Bernoulli equation. When the motion of the same plane is studied, Bernoulli's equation can be expressed as:

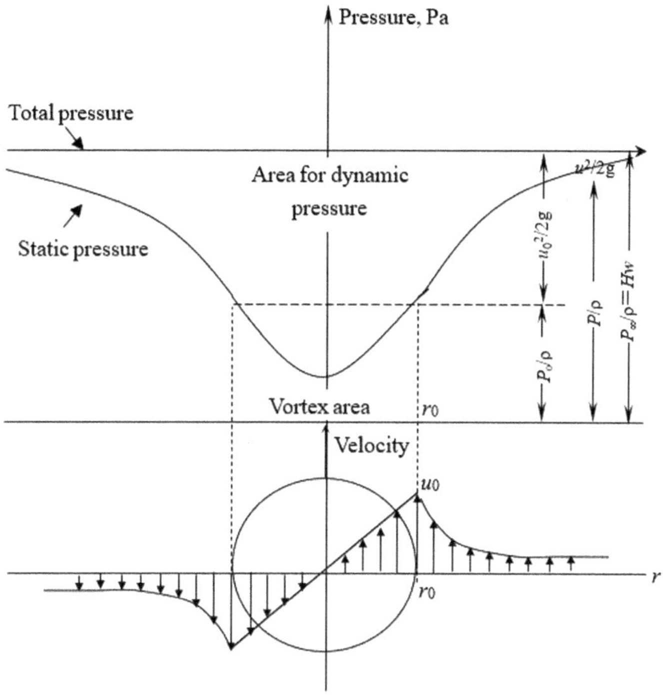

Fig. 3.13 Pressure and velocity distribution of the flow inside and outside the combined vortex under the electromagnetic field

$$\frac{P}{\rho}+\frac{u^2}{2g}=\frac{P_0}{\rho}+\frac{u_0^2}{2g}=\frac{P_\infty}{\rho}=H_w=\text{constant} \tag{3.34}$$

where P is the pressure at any point, H_w is the total water head, and P_0, u_0 are the pressure and tangent velocity at the eddy current boundary, respectively.

For the forced vortex, the radial pressure change inside the vortex core can be expressed as:

$$\frac{\mathrm{d}P}{\mathrm{d}r}=\frac{\rho}{g}\cdot\frac{u^2}{r} \tag{3.35}$$

Because the composite melt inside the vortex is a rigid melt, when substituting $u = \omega r$ and the eddy current boundary conditions into Eq. (3.35), Eq. (3.36) can be derived:

$$\frac{P}{\rho}=\frac{P_\infty}{\rho}+\frac{u^2}{2g}-\frac{u_0^2}{g} \tag{3.6}$$

The pressure distribution characteristics of the combined vortex are summarized as follows:

$$\frac{P}{\rho}+\frac{u^2}{2g}=\frac{P_\infty}{\rho}\quad(r>r_0),\ \frac{P}{\rho}=\frac{P_\infty}{\rho}+\frac{u^2}{2g}-\frac{u_0^2}{g}\quad(r\leq r_0) \tag{3.37}$$

Therefore, the pressure difference between any point on the vortex and the vortex core is:

$$\frac{\Delta P}{\rho}\cdot\frac{P_\infty-P}{\rho}=\frac{u^2}{2g}\quad(r>r_0) \tag{3.38}$$

$$\frac{\Delta P}{\rho}\cdot\frac{P_\infty-P}{\rho}=\frac{u_0^2}{g}-\frac{u^2}{2g}\quad(r\leq r_0) \tag{3.39}$$

Among these, the pressure difference $\frac{\Delta P}{\rho}$ is always greater than zero, which is the pressure for the particles to enter the melt. When the electromagnetic force increases, u_0 will increase, the pressure difference $\frac{\Delta P}{\rho}$ will also increase, and the particles will enter the melt more easily. From this point of view, the larger the electromagnetic force is, the higher the u_0 is, the better the conditions for reactants to enter the liquid aluminum. When the additive amount of is unchanged, the corresponding volume fraction of particles will increase, which is helpful to improve the yield of reactants.

3.3.2 *Thermodynamic Conditions by Electromagnetic Method*

The nature of a chemical reaction is the recombination of atoms or groups, that is, the breaking of old bonds in the reactants and the formation of new ones in the products. The atoms in a substance are bound together by chemical bonds. To break the bonds in the reactants, energy is absorbed; to form the bonds in the products, energy is released. The breaking and formation of chemical bonds are the main reasons for the change of energy in the process of chemical reaction. Whether a reaction absorbs or releases energy after completion depends on the relative amount of the total energy of the reactants and the total energy of the products [10]. It can be inferred from the nature of chemical reactions that changes in energy in the system also have a great influence on the breaking and formation of chemical bonds. A certain intensity of electromagnetic field can improve the initial state free energy of reactive atoms, relatively reduce the nucleation potential barrier of substances, and increase the number of nucleation [11]. In this section, the effect mechanism of external magnetic field on the thermodynamic conditions of in-situ reaction is studied from the perspective of the effect mechanism of magnetic field on entropy and energy of the reaction system.

3.3.2.1 The Entropy of the Reaction Under a Magnetic Field

According to the theory of thermodynamics, all possible macroscopic processes (including chemical changes and phase changes) proceed in the direction of increasing the entropy of the system, and the entropy value reaches the maximum when equilibrium is reached. Explanation of the principle about the entropy increase in statistical thermodynamics is: according to the statistical meaning of entropy, all possible macro-process is the microscopic state of the system total Ω (is the combination of all microparticle movement inside the system) increases, the process will continue until equilibrium Ω reaches maximum value.

At constant temperature T and with each component in its standard state, suppose a reaction is:

$$aA(g) + bB(g) = lL(g) + mM(g) \tag{3.40}$$

Its change in entropy is expressed as:

$$\Delta_r S^{\Theta} = lS_L^{\Theta}(g, T) + mS_M^{\Theta}(g, T) - aS_A^{\Theta}(g, T) - bS_B^{\Theta}(g, T) \tag{3.41}$$

According to the thermodynamic parameters table, the $\Delta_r S^{\Theta}$ is computable.

When a magnetic field is applied to the reaction system, it will induce the entropy change for A, B, L an M substance separately. They are assumed as ΔS_A^m, ΔS_B^m, ΔS_L^m, and ΔS_M^m, among which the superscript "m" to the right is the magnetic field.

Then the entropy of the reaction in a magnetic field becomes:

$$\begin{aligned}\Delta_r S^{\Theta,m} &= l[S_L^{\Theta}(g, T) + \Delta S_L^m] + m[S_M^{\Theta}(g, T) + \Delta S_M^m] \\ &\quad - a[S_A^{\Theta}(g, T) + \Delta S_A^m] - b[S_B^{\Theta}(g, T) + \Delta S_B^m]\end{aligned} \tag{3.42}$$

That is,

$$\Delta_r S^{\Theta,m} = \sum \mu_i s_i^{\Theta}(g, T) + \sum \mu_i \Delta S_i^m \tag{3.43}$$

where μ_i is the stoichiometric number of reactions; $S_i^{\Theta}(g, T)$ means the molar reaction entropy in standard state and T temperature; ΔS_i^{m} represents the molar entropy change of substance i in the magnetic field, which is mainly determined by the property of the substance itself (set this parameter as I) and the magnetic induction intensity B. It means that ΔS_i^m is the function of I and B and is marked as $\Delta S_i^m(I, B)$. Therefore, to quantitatively calculate the influence of magnetic field on the entropy change of the reaction system, it is necessary to figure out the magnitude of the entropy change ΔS_i^m caused by magnetic field.

According to Boltzmann's theorem [12],

$$S = f(\Omega) = k \ln \Omega \tag{3.44}$$

According to the principle of summation increase of statistical entropy [13], the entropy of the system is the sum of the contributions of various independent motion forms of particles to the entropy, namely

$$S = S_t + S_r + S_v + S_e + S_n \tag{3.45}$$

where S_t, S_r, S_v, $S_{e,}$ and S_n correspond to the spatial translation, rotation, vibration near the equilibrium, electron motion, and nuclear motion inside the atom, respectively. Usually under the common condition, the movement of electrons and core inside the atoms is in the ground state, namely in the general physical and chemical processing, the contributions of the electronic movement and nuclear movement to the entropy are same; in other words, the ΔS consists of S_t, S_r, S_v. When a magnetic field is applied to the system, the internal electrons will be excited and in an excited state. At this point, the contribution of electron motion and nuclear motion to entropy cannot be ignored.

When magneto-hydrodynamics describes the essence of magnetism, it is believed that the magnetism of matter comes from the atomic magnetic moment, which is composed of the nuclear magnetic moment and the electron magnetic moment. But both practice and theory show that the nuclear magnetic moment is small; only a few thousandths of the electron magnetic moment, and its contribution to the atomic magnetic moment is usually ignored. According to this conclusion, when studying the effect of magnetic field on chemical reaction, the entropy change caused by nuclear spin can also be ignored, i.e.,

$$\Delta S_i^m(I, B) \cong \Delta Se \tag{3.46}$$

In a system with a particle number greater than 10^{24}, the total microstate number is very large, and the microstate number and distribution probability of various distributions are also different. According to the law of equal probability (i.e., under the condition that the number of particles, energy and volume of the system are determined, the probability of the occurrence of each microstate in the system is equal), the distribution with the largest number of microstates must be the most likely to occur. Although systematic microstate changes all the time, the distribution W_B with the largest probability can replace the systematic distribution, namely $\Omega = W_{B,}$ the W_B is referred as the most probable distribution system, so there is:

$$S = k \ln \Omega = k \ln W_B \tag{3.47}$$

The magnetic field will affect the spin mode of the unpaired electrons in the reactive molecules, resulting in the change of entropy of the system. Under the condition of external alternating magnetic field, the microscopic motion state and distribution probability of chemical particles change, the unpaired electron spin mode of reactants becomes orderly, and entropy decreases. For the in-situ reactions to form the reinforced particle as $3ZrO_2 + 13Al = 3Al_3Zr + 2Al_2O_3$, the Gibbs free energy for the reaction is expressed as $\Delta G^{\ominus} = -1000065.4 + 756T$. So when entropy goes down, $\Delta G^{\ominus}$ tends to go down, which, from a thermodynamic point of view, helps in the in-situ reaction.

It can be seen from the table that, in the process of different systems, the expression of work should be the product of the corresponding intensive property and extensive property. Therefore, when pressure P, external force F, and magnetic field strength H simultaneously act on the system with volume V, length L, and magnetization intensity J, the work term should be written as:

$$\delta W = -P\mathrm{d}V + F\mathrm{d}l + H\mathrm{d}J \tag{3.48}$$

According to the first and second laws of thermodynamics:

$$\mathrm{d}U = \delta Q + \delta W \tag{3.49}$$

After combination, Eq. (3.50) can be obtained.

$$\mathrm{d}U - T\mathrm{d}S \leq \delta W \tag{3.50}$$

Substitute (3.47) into (3.49) and get Formula (3.51),

$$\mathrm{d}U - T\mathrm{d}S \leq -P\mathrm{d}V + F\mathrm{d}l + H\mathrm{d}J \tag{3.51}$$

If temperature T, pressure P, external force F, and magnetic field strength H are constant, then:

$$\mathrm{d}(U - TS + PV - Fl - HJ) \leq 0 \tag{3.52}$$

where "$<$" means irreversible process, and "$=$" means reversible process, so the Gibbs free energy G and enthalpy H of the system can be defined as:

$$G \equiv U - TS + PV - Fl - HJ \tag{3.53}$$

enthalpy $$H = G + TS \equiv U + PV - Fl - HJ \tag{3.54}$$

So the generalized G and H defined here are, respectively: $G \equiv U - TS - \sum XY, H \equiv U - \sum XY$, X and Y represent the intensive and extensive properties shown in Table 3.1, respectively. If the difference in the number of electrons with opposite spin is n in the presence of applied magnetic field, then the magnetic moment along the magnetic field direction is $n \cdot \mu_B$ (μ_B is the Bohr magneton). On this condition, the entropy consists of the part S_m related to the spin orientation and the part S_o independent of the spin orientation; that is, $S = S_m + S_o$. Then the Gibbs free energy for the system is expressed as:

Table 3.1 Forms of work in different processes and the expressions of the first law of thermodynamics

Kinds of external field	Process	Intensive property X	Extensive property Y	Work, J	expressions of the first law
No electromagnetic field	Expansion	Pressure P/Pa	Volume V/m^{-3}	$-\int P\mathrm{d}V$[a]	$\delta Q = \mathrm{d}U - P\mathrm{d}V$
	Force	Foreign force F/N	Length l/m	$\int F\mathrm{d}l$	$\delta Q = \mathrm{d}U + F\mathrm{d}l$
With EMF	Magnetic system	Magnetic intensity H/A m^{-1}	Magnetization intensity J/(A m	$\int H\mathrm{d}J$	$\delta Q = \mathrm{d}U + H\mathrm{d}J$

[a] The minus sign means work is done to the outside world; Q is the total energy of the system, and U is the internal energy

$$G = U - Hn\mu_{\mathrm{B}} - T(S_{\mathrm{m}} + S_{\mathrm{o}}) \quad (3.55)$$

where the part $U - TS_0$ is independent of n, the part dependent of n is:

$$G(n) = -Hn\mu_{\mathrm{B}} - TS_{\mathrm{m}} \quad (3.56)$$

So, in a magnetic field, the Gibbs free energy of the system changes. The study shows that the $G(n)$ value of paramagnetic substance decreases and tends to be stable, while the $G(n)$ value of diamagnetic substance increases and tends to be unstable, as shown in Fig. 3.14.

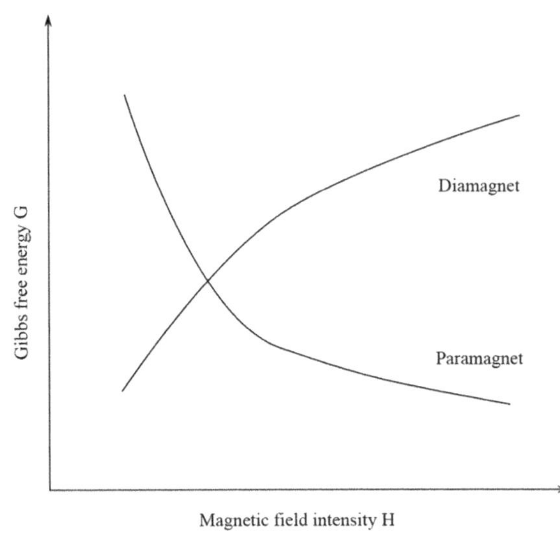

Fig. 3.14 Relationship between Gibbs free energy and magnetic field strength for non-ferromagnetic material

3.3.2.2 Thermodynamic Conditions for the Electromagnetic Synthesis

According to the transition state theory, for a reactant molecule, to become a product, it must first accumulate enough energy to form an intermediate substance with high potential energy. When the thermodynamic and kinetic conditions of the system are satisfied, equilibrium can be reached through chemical reaction and the chemical change process can be completed. Therefore, in the process of magnetochemical reaction, the reaction particle first absorbs magnetic field energy and transitions, generating a kind of high-energy particle with unstable thermodynamic state, which is defined as magnetic activated molecule. In the in-situ reaction process of preparing $(Al_2O_{3+}Al_3Zr)_p/Al$ composites, the core reaction is:

$$3ZrO_2 + 4Al = 2Al_2O_3 + 3[Zr] \quad (3.57)$$

When a magnetic field is applied to the in-situ reaction system, both ZrO_2 molecules and Al atoms absorb magnetic field energy and transform themselves into magnetically activated particles $(ZrO_2)^*$ and Al^*(Superscript * denotes magnetic activation). Compared with ordinary particles, these particles are in high-energy unstable state, and the chemical bond between Zr and O atoms in ZrO_2 is polarized and elongated, which is easy to break, which completes the primary process of magnetic chemical reaction. The new reaction equation is:

$$ZrO_2 + Al + E_m \rightarrow (ZrO_2)^* + Al^* \quad (3.58)$$

where E_m means the magnetic field energy.

When the energy of magnetically activated particles is enough to break the chemical bonds in ZrO_2, that is, the particle energy exceeds the threshold value of chemical reaction, the Zr atom in the $(ZrO_2)^*$ will be replaced, Al and O form a new chemical bond as Al_2O_3, complete the secondary process of magnetic chemical reaction, namely

$$(ZrO_2)^* + Al \rightarrow Al_2O_3 + Zr^* \quad (3.59)$$

The degree of reaction is related to the formation free energy of ZrO_2 and Al_2O_3 oxides. At the same temperature, the greater the difference in the standard free energy of formation between ZrO_2 and Al_2O_3, the easier the reaction is to proceed. Because both the ZrO_2 and Al_2O_3 are diamagnetic substances, under the action of an external magnetic field, the Gibbs free energy of both will rise and tend to be unstable. The magnetic susceptibility of Al^{3+} ion is -2×10^{-6}, and the data of the magnetic susceptibility of Zr^{4+} ion is -10×10^{-6}. The magnetic susceptibility of Al_2O_3 is less than that of ZrO_2, so the rise of Gibbs free energy of Al_2O_3 is less than that of ZrO_2. Figure 3.15 shows the free energy–temperature graph generated by Al_2O_3 and ZrO_2 with or without an applied magnetic field. It can be seen from the figure that at the same temperature, the difference of standard free energy generated by Al_2O_3 and

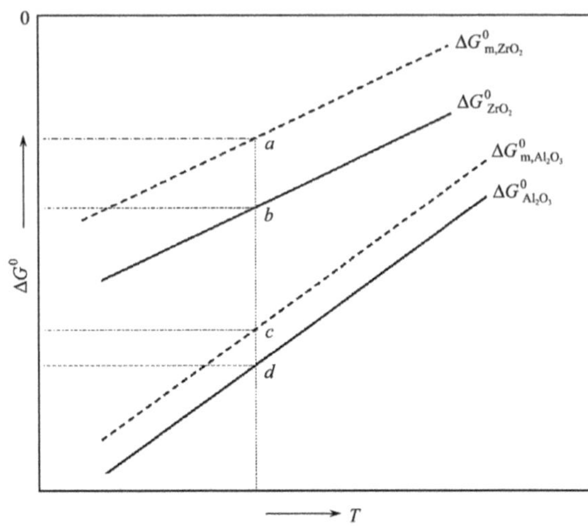

Fig. 3.15 Standard free energy–temperature curves of Al_2O_3 and ZrO_2 formation without (solid line) or with (dotted line) electromagnetic fields

ZrO_2 under the influence of external magnetic field is greater than that of without magnetic field. Just as the line segment ac > bd in the figure, it is more conducive to the in-situ reaction.

3.3.3 *Kinetic Conditions for the Electromagnetic Synthesis of Composites*

3.3.3.1 Microscopic Analysis of Dynamic Processes

There are two main reaction mechanisms for the in-situ synthesis of aluminum matrix composites, namely diffusion mechanism and reaction–diffusion mechanism. With the deepening of the research, the latter is easier to explain the formation process of in-situ reinforcement particles. Reaction–diffusion mechanism considers that the velocity of in-situ reaction is not only controlled by diffusion, but by both chemical reaction and diffusion. Therefore, it is of the same importance to study the influence of magnetic field on the synthesis kinetics as to study the thermodynamic conditions.

When the reactant particles enter the melt, two processes will occur. One is the macroscopic mixing, namely the crushing process. Under the action of shear force, the reactant powder will be separated in the form of granules to prevent the phenomenon of local sintering at high temperature. The second is microscopic mixing, that is, diffusion process, which eliminates the difference in concentration between adjacent areas of the mixed melt. Due to the convection effect in the melt caused by stirring force, reactant particles are uniformly distributed in the melt, increasing the contact area with the aluminum liquid, improving the dynamic conditions of in-situ reaction, and promoting the occurrence of the reaction.

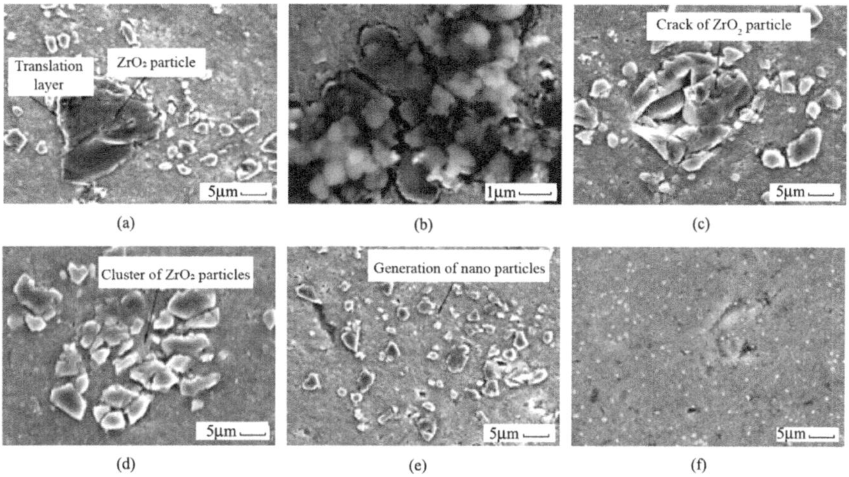

Fig. 3.16 SEM microstructure of water-quenched samples with different reaction time: **a** Initial state, ZrO_2 begin to decompose; **b** Amplified microstructure of the transformation layer in (**a**); **c** Crack of the ZrO_2; **d** Fine particle cluster formation; **e** Generation of nanometer particle; **f** As-cast microstructure

Figure 3.16 shows the SEM micrographs of Al-20 wt.%$Zr(CO_3)_2$ system obtained by water quenching and rapid cooling with different reaction time under the action of 0.07 T magnetic field. As can be seen, when the reaction time is 3 min, ZrO_2 particles appear in the form of near-circular polygons. A bright white transition layer appears around the ZrO_2 particles, indicating that the surface layer of the aluminum melt and the ZrO_2 particles begins to moisten and a trace chemical reaction occurs. Figure 3.16b shows the high-power SEM structure of the conversion layer. At the edge of the reactant particles, the reaction generates fine granular particles in a agglomerated state, and obvious cracks are generated between the aluminum matrix and the generated particles. This is mainly because in the early stage of the reaction, microzones of high-temperature metallurgy are formed at the reaction interface, and there is not enough time for enough heat exchange between the high-temperature metallurgical area and the surrounding environment. Therefore, great temperature difference and thermal stress will be generated between the reactive layer and the unreactive layer, leading to the generation of cracks between the aluminum matrix and particles. As the chemical reaction goes on, the melt reaction temperature increases further (>850 °C), and when the thermal stress is high enough, the rupture of the reaction particles will be caused (as shown in Fig. 3.16c, d), and the in-situ endogenous particles will form (as shown in Fig. 3.16e. When the reaction time is 20 min, the in-situ chemical reaction is over, and the in-situ endogenous particles are uniformly distributed in the aluminum matrix (as shown in Fig. 3.16f).

According to Grunberg's thermal stress calculation [14], when a solid sphere suddenly rises from the initial temperature T_0 to T_1, the outer surface of the sphere will expand and generate the tensile stress on the sphere. The tensile stress formula can be expressed as follows:

$$\sigma_r = \sigma_t = 0.771\frac{aE}{2(1-\nu)}(T_1 - T_0) \tag{3.60}$$

where a is the coefficient of thermal expansion, E is the modulus of elasticity, ν is Poisson's ratio. In this experimental condition, according to the references [15], $a = 9.6 \times 10^{-6}/°C$, $E = 200GPa$, $T_1 - T_0 = 200\ °C$.

When they are substituted into (3.60), it can be derived as:

$$\sigma_r = \sigma_t = \frac{148}{1-\upsilon}MPa > 148\ MPa \tag{3.61}$$

The tensile stress is far greater than the about 8 MPa fracture strength of ZrO_2, resulting in the crushing phenomenon of ZrO_2 particles. The contact area between the reactants and the aluminum solution is increased, and the reaction products are diffused from the reaction layer more rapidly, which accelerates the reaction.

Figure 3.17 shows the XRD diagrams for water-quenched specimen with the reaction time is 15 min. In addition to the generation of Al_2O_3 and Al_3Zr reinforcing particles, there are unreacted ZrO_2 particles as well as the active Zr atom, this shows that in the middle of the reaction, the ZrO_2 produced by decomposition of $Zr(CO_3)_2$ and Zr atom from the replacement reaction of Al-ZrO_2 have not fully reacted with Al.

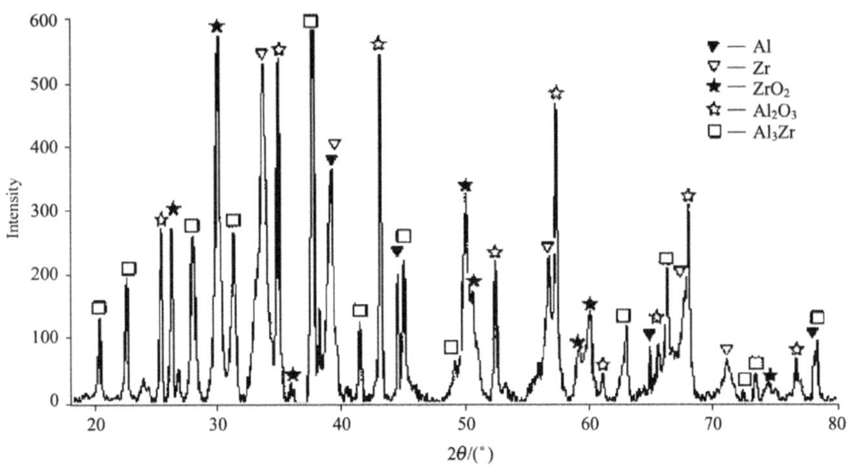

Fig. 3.17 XRD patterns of the water-quenched samples with reaction time of 15 min

3.3.3.2 The Influence of Dynamic Conditions on the Nucleation

The melt flow state will change under the conditions of electromagnetic field, the influence of the forced flow directly affects the reaction system in heat transfer, mass transfer process, as well as the products of nucleation and growth process. Furthermore, it will take significant and direct effects on the actual volume fraction, size of particles. Figure 3.18 shows a schematic diagram of the dynamic process of in-situ synthesis of Al_2O_3 and Al_3Zr reinforced phases.

When $Zr(CO_3)_2$ is added to molten aluminum, it is decomposed into ZrO_2 and CO_2. The Zr atoms in ZrO_2 are reduced by Al then combine with liquid Al atoms to form Al_3Zr. The formation of Al_2O_3 is formed on the surface of ZrO_2 after Al replaces Zr atoms in ZrO_2. Multiple Al_2O_3 particles can be generated on the surface of a ZrO_2. The generated Al_2O_3 grows attaching to ZrO_2, and its growth is controlled by O diffusion, resulting in a small space for aggregation growth.

Without electromagnetic field, the mass transfer of reactants and products mainly depends on the internal concentration difference, and the speed is slow. When the electromagnetic force is applied, the melt is in a strong mixed convective motion under Lorentz force, and the heat and mass transfer rate in the reaction system is significantly increased. At the initial stage of the reaction, the contact opportunities between aluminum liquid and solid ZrO_2 particles increase, the reaction interface expands, and the nucleation number of Al_2O_3 and Al_3Zr particles increased. It can be seen that the magnetic field improves the kinetics conditions, which in turn improves the particle distribution in the melt.

In addition, the forced movement of the melt promotes the temperature uniformity and the dissipation of the overheat, and convective heat transfer plays an important role in heat transfer. Temperature homogenization can prevent temperature differences in the reaction system, especially the low local temperature after the addition of reactants will reduce the reaction rate. Electromagnetic force not only improves the kinetic conditions of in-situ reaction, but also prevents the occurrence of local low temperature, which is conducive to the in-situ reaction.

According to the theory of "energy fluctuation" and "structure fluctuation" proposed based on thermodynamics condition of nucleation, the connection between atoms is three-dimensional, the vibration directions are chaotic, and they often collide with each other and transfer energy to each other. Not only is the energy of each atom

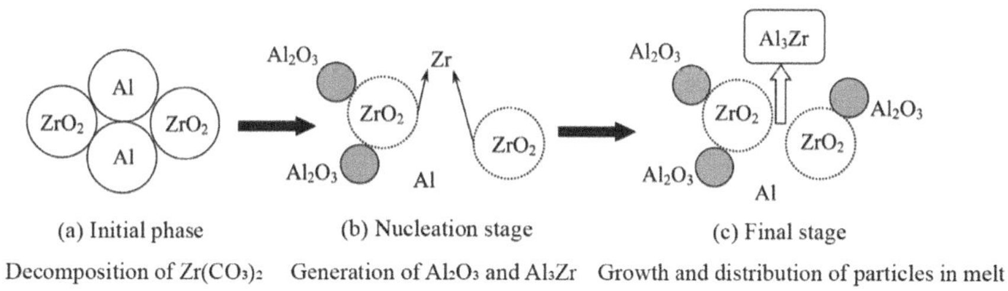

Fig. 3.18 Schematic for the dynamic process of in-situ synthesis of Al_2O_3 and Al_3Zr particles

changing in real time, but the thermal energy of each atom is not equal and varies greatly. In a collision, each atom and each atomic group has different energy at any time. As a result, some atoms always have energy greater than the average energy of the atoms, while others have energy less or far less than the average energy, which is called "energy fluctuation." Due to the energy fluctuation effect in liquid metals, the existing medium- and short-range ordered atomic groups will become smaller or larger, which is the so-called structural fluctuation effect. When electromagnetic field acts on the melt in the form of electromagnetic wave, it can effectively increase the energy fluctuation and structural fluctuation inside the melt, reduce the critical nucleation work of particles to a certain extent, and increase the nucleation probability and nucleation number of particles [16].

3.3.3.3 Effect of Dynamic Conditions on Dispersion Uniformity

According to particle dispersion techniques, particles in a dispersed system (a heterogeneous system in which one or more phases are distributed as particles in a continuous phase) are subjected to both gravity and diffusion forces.

If the density of the particles is greater than that of the medium, the particles will settle due to gravity. In the static condition, for particles with particle size below micron level, their settlement follows Stokes law [17], and the settlement velocity of particles can be expressed as:

$$v_0 = 54.5D^2\frac{\rho_P - \rho}{\mu_f} \tag{3.62}$$

where ρ and ρ_P are densities of particle and matrix (kg/m^3), respectively; D is the diameter of the spherical particle with an average size of 2 μm; μ_f is the viscosity of the matrix solution, which is tested as 0.28 Pa s. Table 3.2 shows the calculation results of settlement velocity for different particles in the composite material.

The sedimentation rate of particles is affected by particle size, density, melt density, and viscosity. It is found that the settling velocity of the submicron phase particles in the aluminum solution is very small, even negligible, and the stability of the composite melt will not be affected.

The second major force of the particles in the dispersed system is the diffusion force. After the molecules are heated, they will collide disorderly with each other, which will cause the diffusion displacement of the particles. After time t, the diffusion displacement x of the particles from the original position along a certain direction can be obtained from the Einstein equation, as shown in Eq. (3.63).

Table 3.2 Settling velocity of different particles in Al–Zr–O system composite

Particle	Al_2O_3	Al_3Zr	ZrB_2
Settling velocity, m/s	1×10^{-6}	1.09×10^{-6}	2.58×10^{-6}

$$\chi = \sqrt{\frac{RTt}{3\pi\mu DN_A}} \tag{3.63}$$

where $R = 8.314$ J /(mol K), N_A is Avogadro's constant, and the value is 6.022×10^{23}/mol.

It reveals that the diffusion displacement has nothing to do with particle density, but with melt viscosity and particle size and increases with the decrease of particle size. When the time is assumed to be 60 s and the melt temperature is 850 °C (1123 K), the diffusion displacement x value of the particles with particle size of 2 μm is 4.2×10^{-10} m. When the time is assumed to be 600 s, the diffusion displacement x value is 1.32×10^{-9} m. It shows that the value of diffusion displacement is very small, and agglomeration and segregation will not occur due to the thermal movement of particles in the composite melt.

From the above derivation, it is found that the submicron particles in the composite melt will not agglomerate due to sedimentation and thermal movement, but because the particle size is small, the surface area is large, the surface energy is high, it is easy to produce spontaneous coagulation and reduce its distribution dispersion in the matrix.

Increasing the wettability of particles and melts can promote the dispersion of particles in the matrix. The interfacial energy between alloy and particle is reduced by adding surfactant into the melt. The wettability of the matrix and the particles can be improved by adding Mg into the aluminum alloy, so A356 containing Mg element is favorable for the dispersion distribution of the particles when chosen as the matrix of the composite. For the in-situ chemical reaction, the reactants react with the aluminum matrix liquid to generate the particles, which are formed in the aluminum matrix liquid, and the infiltration effect between the particles and the matrix is better.

The external supply of energy contributes to the dispersion distribution of particles, which is usually produced by mechanical agitation, electromagnetic agitation, or ultrasonic magnetic field, on this condition the flow state will change. In the in-situ synthesis of composites, the melt flow state can be divided into two types as natural convection and forced convection. Natural convection is spontaneous convection caused by local concentration and temperature gradient changes in the melt. When no electromagnetic field is applied in the synthesis process, the mass transfer and diffusion of the system substance depend on its own concentration gradient, and the migration and flow of the substance inside the melt is natural convection. However, forced flow of melt occurs under the action of external field, and external forces will promote agglomeration particles to break up and suspend, which can effectively prevent agglomeration of particles. Especially after the in-situ reaction, the generated Al_2O_3 and Al_3Zr particles will disperse uniformly in the matrix along with the moving melt.

In addition, from the perspective of melt viscosity, under the action of electromagnetic shear force, the melt viscosity decreases compared with the melt viscosity without stirring [18]. The viscosity resistance of particles in the diffusion process decreases, which is conducive to the uniform diffusion of particles in the matrix.

Furthermore, CO_2 generated by the decomposition of $Zr(CO_3)_2$ acts as a gas agitator for the aluminum melt, and the vibration of the melt under the influence of exothermic reaction and gas promotion can be observed in the in-situ synthesis process.

3.3.3.4 Effect of Kinetic Conditions on Particle Shape

Firstly, electromagnetic force inhibited the selective growth of particle phase. The synthesized Al_3Zr particles are intermetallic compounds, that is, the appearance of facet growth. The growth of Al_3Zr particles has strong anisotropy. Under the action of systematic heat and electromagnetic heat, the local temperature rises, and Al_3Zr shows an obvious trend of preferential growth, which is not conducive to spheroidization. However, under the effect of strong electromagnetic force, the trend of selective growth is inhibited, which promotes the formation of more twin growth. At the same time, the electromagnetic force makes the temperature field of aluminum liquid tend to be uniform, which is conducive to the side deposition and step growth in the preferred growth direction, increasing its thickness, and inhibiting the preferred growth of particle phase.

As the second aspect, the alternating magnetic field promotes atomic diffusion. The surface curvature of the particle phase is different, so is the concentration of the solute around it. Where the curvature of the particle phase is large, the equilibrium solute concentration of the liquid phase is higher, while where the curvature of the particle phase is small, the equilibrium solute concentration of the liquid phase is low. According to the principle of crystal growth dynamics, solute diffuses from the place with large curvature to that with small curvature, which makes the sharp angle of particle phase passivate and is conducive to the roundness of particle phase. Meanwhile, the violent oscillation of pulsed magnetic field accelerates the diffusion of atoms and promotes the roundness process of particle phase.

The third is mechanism of mechanical action. The generated Al_3Zr particles occur scouring and collision in the violent turbulent melt due to electromagnetic field. The scouring force and collision force make the originally agglomerate or glued particles separate, and even fracture at the defective places of the particles, resulting in fining effect. At the same time, the scour force and the collision force also cause the collision and friction between the particles, so that the sharp angle of the particles can be passivated, resulting in the rounding effect.

References

1. Paesea E, Geier M, Homricha RP. A coupled electric–magnetic numerical procedure for determining the electromagnetic force from the interaction of thin metal sheets and spiral coils in the electromagnetic forming process. Appl Math Model. 2015;39:309–21.
2. Wang HM, Li GR, Zhao YT. Microstructure, billet surface quality and tensile property of $(Al_2O_3 + Al_3Zr)p/Al$ composites in situ synthesized with electromagnetic field. J Alloy Compd. 2011;509(18):5696–700.
3. Li GR, Zhao YT, Dai QX. In-situ fabrication of particulate reinforced aluminum matrix composites under high- frequency pulsed electromagnetic field. J Univ Sci Technol Beijing. 2007;14(5):460–3.
4. Bekir SU, Enver A. Tribological properties of journal bearings manufactured from particle reinforced Al composites. Mater Design. 2009;30(4):1381–5.
5. Li GR, Wang HM, Li PS. Mechanism of dislocation kinetics under magnetoplastic effect. Acta Physica Sinica 2015;64(14):148102.
6. Arpita J, Sasankasekhar M. Syntheses, crystal structures, magnetochemistry and electrochemistry of macrocyclic dicopper(II) complexes: monodentate behavior of a potentially chelating ligand. Inorgan Chi Acta. 2013;405:265–73.
7. Wannasina J, Canyooka R, Wisutmethangoon B. Srain refinement behavior of an aluminum alloy by inoculation and dynamic nucleation. Acta Mater. 2013;61:3897–903.
8. Beltyukov AL, Menshikova SG, Ladyanov VI. The viscosity of binary Al–Fe melts in the Al-rich area. J Non-Crystalline Solids. 2015;410(15):1–6.
9. Wang HM, Li GR, Zhao YT, Chen G. In situ fabrication and microstructure of Al_2O_3 particles reinforced aluminum matrix composites. Mater Sci Eng A. 2010;527:2881–5.
10. Heui SR. Thermodynamic chemical energy transfer mechanisms of non-equilibrium, quasi-equilibrium, and equilibrium chemical reactions. Energy. 2015;89:1029–49.
11. Yao ZN, Yang HF, Li JJ. Detection of domain wall distribution and nucleation in ferromagnetic nanocontact structures by magnetic force microscopy. J Magn Magn Mat 2013;342:1–3
12. Rottger A, Lentz J, Theisen W. Boron-alloyed Fe–Cr–C–B tool steels—thermodynamic calculations and experimental validation. Boron-alloyed Fe–Cr–C–B tool steels—thermodynamic calculations and experimental validation. Mater Design 2015;88:420–9.
13. Perez AM, Bruno NM, Pons J. Atomic order and martensitic transformation entropy change in Ni–Co–Mn–In metamagnetic shape memory alloys. Scrip Mat. 2016;110:61–4.
14. Vaucher S, Stir M, Ishizaki K. Reactive synthesis of Ti-Al intermetallics during microwave heating in an E-field maximum. Thermochim Acta. 2011;522:151–4.
15. Jonathan R, Christiane R, Keith A. Oscillating magnetic field effects on chemical reaction yields. Riken Rev. 2002;44:79–81.
16. Biswas K, Hermann R, Das J. Tailored field effects on chemical reaction yields. Riken Rev. 2002;44:79–81.
17. Ruri H, Takafumi T, Hiroshi S. Adhesive behavior of a calcium carbonate particle to solid walls having different hydrophilic characteristics. Int J Heat Mass Transf. 2016;92:603–9.
18. Li GR, Wang HM, Zhao YT. Microstructure of in situ Al3Ti/6351Al composites fabricated with electromagnetic stirring and fluxes. Trans Nonferrous Met Soc China. 2010;20:577–83.

Chapter 4
High-Energy Ultrasonic Synthesis of In-Situ Aluminum Matrix Composites

4.1 Effect of High-Energy Ultrasound on Metal Melt and Reactions

In the early 1927, American scholars Richards and Loomis reported for the first time that ultrasound irradiation promotes chemical reaction. In their experiments in the Chemistry Laboratory of Princeton University, they discovered that the hydrolysis of dimethyl sulfate and the reduction rate of potassium iodide by sulfuric acid under ultrasound irradiation is significantly increased, but this phenomenon did not attract the attention of the scientific community at that time. By the 1940s, reports on ultrasonic energy accelerated chemical polymerization appeared; in the 1960s, reports on the use of ultrasonic energy in biological applications appeared; until the 1980s, under the strong leadership of British scholar Mason and others, sonochemistry was established as an emerging discipline that used ultrasounds to increase the rate of chemical reactions and the rate of compound production and degradation; and it was only in the mid to late 1980s that the sonochemistry gradually attracted the widespread attention of scientific and industrial communities.

4.1.1 *Application of Ultrasonic Chemistry in the Field of Metal Matrix Composites*

Sound energy has the unique advantages of no secondary pollution, simple equipment, and wide applications; as a result, it has received great attention, and sonochemistry has become a flourishing research field. In the field of metallic materials, the introduction of high-energy ultrasound into molten metals has significant effects on improving the reaction rate, refining the solidification microstructure, preventing segregation, and removing inclusions and gases. When ceramic particles/metal melt mixture is ultrasonically processed to prepare metal matrix composites, ultrasonic waves are introduced into the liquid metal through the end of a transducer, and the

Y. Zhao, *In-Situ Synthesis of Aluminum Matrix Composites*,
https://doi.org/10.1007/978-981-16-9120-1_4

wetting and dispersion of reinforcement in the matrix is achieved in a short time. The literature [1, 2] has reported that the energy supplied by ultrasound waves to particle/melt interface greatly promotes the wetting of particles by molten aluminum. High temperature and high pressure caused by the acoustic cavitation effect at microscale level can clean and activate the surface of particles, and reduce the surface energy of melt; and the stirring effect of acoustic current can disperse the particles evenly in the melt.

In recent years, researchers at home and abroad [3, 4] have continuously tried to combine high-energy ultrasound with in-situ reaction to prepare metal matrix composites. Some people have inserted prefabricated blocks of oxide powders (TiO_2, CuO, ZnO, etc.) in the metal melt. Using the instantaneous high temperature and pressure caused by ultrasonic cavitation and at the same time promoting the in-situ reaction of oxide particles with molten metal, a fine distribution of reinforcement phase particles is obtained. The reinforcement phase generated by in-situ reaction has high bonding strength with the matrix, and its size is small which is greatly conducive to improving the performance of metallic materials. In addition, ultrasonic waves have been applied to reaction systems of inorganic salts and molten metals (mainly Al, Mg, and their alloys) to prepare composite materials. For example, the author's group [5–7] has used the in-situ reaction method and ultrasonic waves during the reaction to prepare various particle-reinforced aluminum matrix composites. As an external field with special effects, the application of high-energy ultrasounds in in-situ synthesis of new materials can have an important role in the microstructure and properties improvement of material. However, there are only a few reports related to the coupled effect of high-energy ultrasound and other physical (light, electric, or magnetic) fields on in-situ melt reaction.

4.1.2 Ultrasonic Generator

Figure 4.1 shows a schematic and actual diagram of the device used for synthesizing composite material under an ultrasonic field. The system mainly includes ultrasonic generator, ultrasonic converter, ultrasonic horn, temperature controller, water cooling system, lifting control system, crucible, and resistance heating furnace. The output power of ultrasonic generator is 0–2 kW, the ultrasonic frequency is 20 kHz, and the magnetizing current is 8 A. During the experiment, the power is usually selected by adjusting the power knob. 220 V AC is rectified and filtered and transferred to the power amplifying circuit to amplify the signal, and then output with high frequency and high voltage is applied to both ends of the ultrasonic transducer. In order to radiate required high-frequency oscillations, the ultrasonic transducer converts the high-frequency oscillating voltage into longitudinal high-frequency expansion and contraction along the thickness direction through a piezoelectric wafer and then emits it from the end. The function of ultrasonic horn is to amplify the output amplitude of the transducer and transfer it to the aluminum melt.

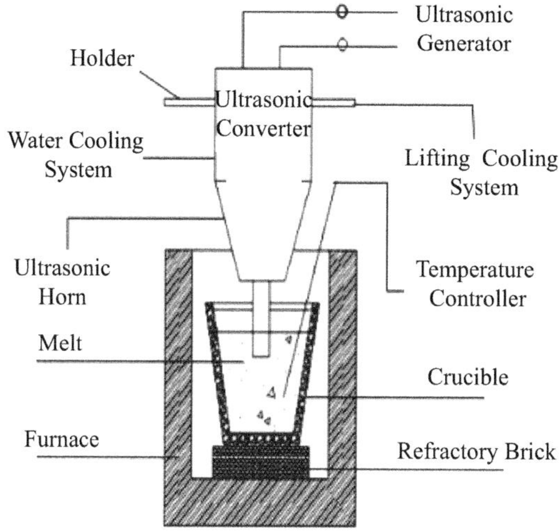

(a) Sketch

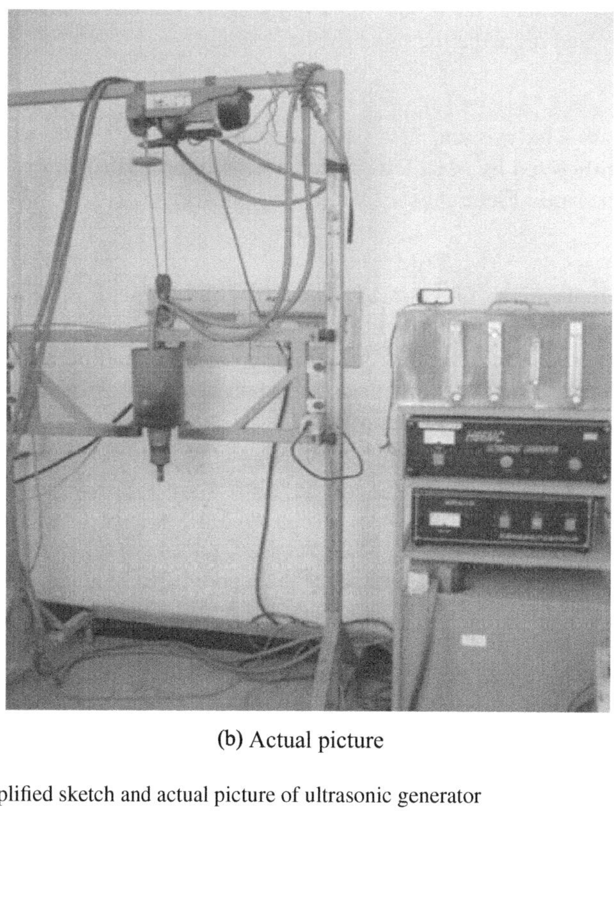

(b) Actual picture

Fig. 4.1 Simplified sketch and actual picture of ultrasonic generator

4.1.3 Effect of High-Energy Ultrasound on the Microstructure of 2024Al Composite

In order to investigate the effect of high-energy ultrasound on the microstructure of composite materials, the ZrB_2(np)/2024Al matrix composites were synthesized by the direct melt reaction using 2024Al-K_2ZrF_6-KBF_4 system. The molar ratio of K_2ZrF_6 and KBF_4 in the reaction was 1:3, the reaction temperature was 870 °C, and the reaction time was 30 min. The amount of reactant addition was 40%, the power and frequency of high-energy ultrasound were 1.2 kW and 20 kHz, respectively, and the application time of high-energy ultrasound was selected as 1, 3, 5, 7, and 9 min. After the application of high-energy ultrasound for a specified period, a quartz tube (7 mm in diameter) was used to extract the melt sample from composite material near the bottom of ultrasonic probe and then water quenched. According to the liquid quenching characteristics of melt, rapid water quenching can well preserve the microstructure of molten composite. Thus, these water-quenched samples were used to evaluate the distribution of nano-ZrB_2 particles in the composite. By observing the microstructure of composite after different application times of ultrasonic vibrations, the effect of high-energy ultrasound on improving the uniform distribution of nano-ZrB_2 particles and reducing the porosity was studied.

4.1.3.1 Phase Analysis and Microstructure of Composite Materials Synthesized by Melt Direct Reaction Under High-Energy Ultrasonic Field

Figure 4.2 shows the SEM, TEM, and HRTEM images of ZrB_2/2024Al composite prepared by application of high-energy ultrasound for 5 min. From Fig. 4.2a, it can be clearly observed that many bright new in-situ generated phases are present in the composite with the size ranging from 200 to 600 nm. In addition, there are many smaller bright new phases around these bright phases with a size of about 100 nm. Figure 4.2b, a TEM image of composite material, shows that there is a dark area of about 500 nm at the center, and some small areas of agglomerated nanoparticles can be seen around. Compared with Fig. 4.2a, the dark area in Fig. 4.2b is the bright white new phase in Fig. 4.2a. The dark area is further enlarged, and it is clearly observed that the dark area is composed of a large number of 30 nm size small particles, as shown in Fig. 4.2c. Diffraction pattern analysis of the marked area *K* in Fig. 4.2c concludes that these particles are nano-ZrB_2 particles. Therefore, the bright white particles in Fig. 4.2a are that of nano-ZrB_2 phase. In summary, nano-ZrB_2 particle-reinforced 2024Al composite was successfully prepared by the direct melt reaction method under high-energy ultrasonic field. Figure 4.2d shows the HRTEM image of nano-ZrB_2/2024Al interface in the composite prepared under ultrasonic field. Compared with the composite prepared without ultrasonic field, the interface experiences no obvious change and it has direct atomic bonding with the matrix; the interface is straight, clear, and clean, and no intermediate phases are produced.

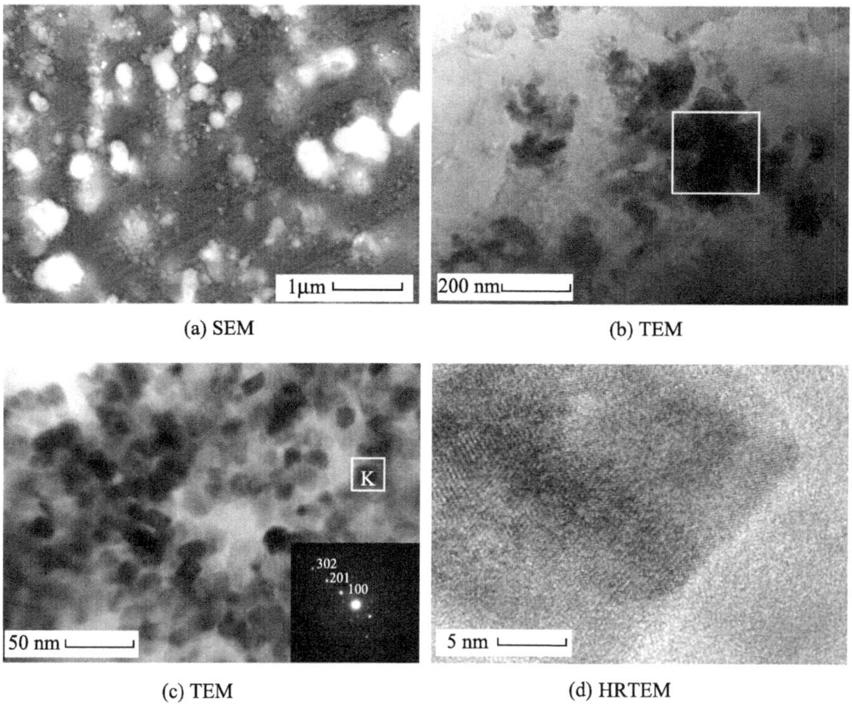

Fig. 4.2 Morphology of ZrB_2 in ZrB_2/2024Al composite prepared by high-energy ultrasound

4.1.3.2 Effect of High-Energy Ultrasounds Time Duration on Particles Distribution and Porosity in Composite Materials

Figure 4.3 is the SEM image of ZrB_2(np)/2024Al composite at different duration times of high-energy ultrasound. The trend of nano-ZrB_2 particle clusters under the action of high-energy ultrasound can be observed. Figure 4.3a shows that the phenomenon of nano-ZrB_2 particles clustering is severe when high-energy ultrasound is not applied to the composite. After 1 $_{\text{min}}$ of high-energy ultrasounds application, the clusters of nano-ZrB_2 particles begin to split, as shown in Fig. 4.3b, which shows that high-energy ultrasound can disperse the clusters of nano-ZrB_2 particles. As the duration time of high-energy ultrasonic continues to increase, the size of nano-ZrB_2 particle clusters is further reduced as shown in Fig. 4.3c, d. Especially when the time is increased to 5 min, the clusters of nano-ZrB_2 particles are disintegrated into small-sized nano-ZrB_2 particles, and these ZrB_2 particles are more evenly distributed in the 2024Al matrix, as shown in Fig. 4.3d. However, when the duration time of high-energy ultrasounds continues to increase, the size of nano-ZrB_2 particles not only ceases to decrease but also becomes coarse, as shown in Fig. 4.3e, f.

In cast aluminum matrix composites, the generation of pores is mainly due to the presence of hydrogen in the melt. In addition, the gas may be introduced into the melt

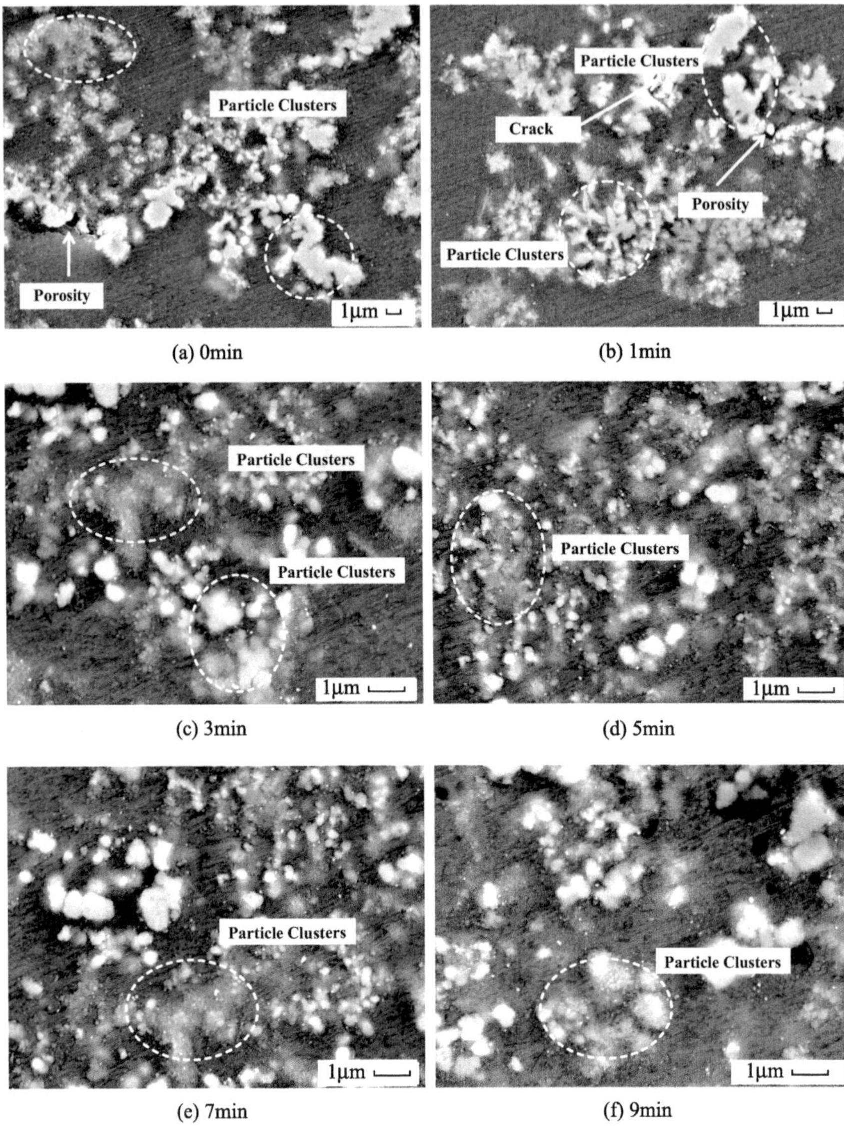

Fig. 4.3 Microstructural changes of ZrB_2/2024Al composite under different application times of ultrasonic field

along with surface slag during the stirring process, resulting in the formation of pores. In the experiment, it was found that some irregular pores were formed in the samples produced without high-energy ultrasound, as shown in Fig. 4.3a. In addition, some nano-ZrB_2 particles were gathered around these pores and the contact with 2024Al matrix was not sufficient. Figure 4.3b–f shows the change of porosity in the composite at different duration times of high-energy ultrasounds. It can be observed that the porosity is significantly reduced with the application of high-energy ultrasound field, and the microstructure of the composite is significantly improved.

4.2 The Principle of High-Energy Ultrasonic Synthesis of Aluminum Matrix Composites

High-energy ultrasourd has been extensively used nowadays. The application and working principle of high-energy ultrasourd in synthesis of in-situ aluminum matrix composites are introduced below.

4.2.1 Effect of High-Energy Ultrasound on A356 Alloy

4.2.1.1 Effect of Ultrasonic Power on A356 Alloy

Figure 4.4 provides the SEM images showing the solidification microstructure of A356 alloy under different ultrasonic powers. Experimental results show that after ultrasonic treatment of A356 alloy melt the morphology and size of α-Al and silicon phase have been greatly improved. Figure 4.4a shows the SEM image of A356 alloy

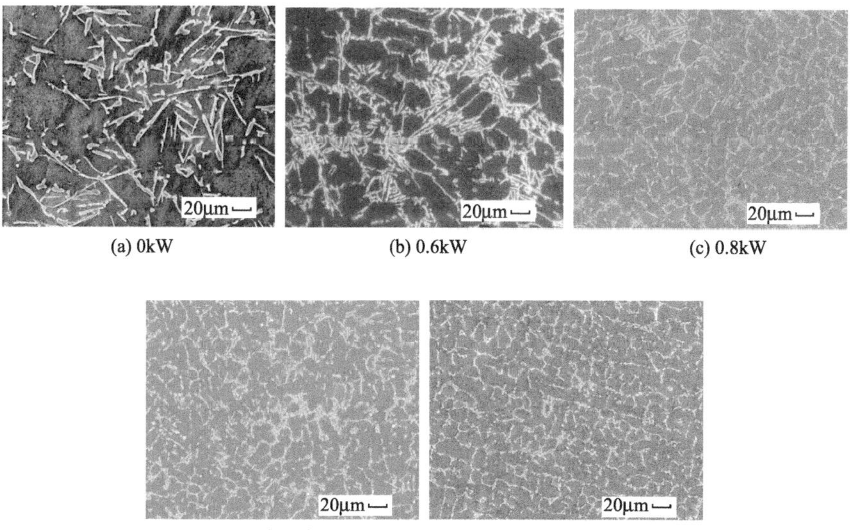

(a) 0kW (b) 0.6kW (c) 0.8kW (d) 1.0kW (e) 1.2kW

Fig. 4.4 SEM images showing solidification microstructure of A356 alloy matrix under different ultrasonic powers without ultrasonic treatment, illustrating that silicon has long dendritic morphology with sharp edges, and α-Al is also comparatively larger. When ultrasonic waves of 0.6 kW power are applied to the melt, the dendritic morphology of silicon is changed into short blocks, and α-Al tends to be round, as shown in Fig. 4.4b. As the ultrasonic power increases, the amount of short silicon fibers gradually increases, their distribution tends to be uniform, and α-Al morphology gradually becomes rounded and finer. Especially when the ultrasonic power is 1.2 kW, silicon has partially granular morphology and partially short fiber morphology, while α-Al becomes finer and more rounded, as shown in Fig. 4.4e. It can be concluded from the above discussion that with the introduction of high-energy ultrasonic waves into the aluminum melt, the microstructure of A356 alloy has changed greatly, silicon size has gradually refined, and α-Al has become finer and more rounded.

The main reason for this is that when the A356 alloy melt is exposed to ultrasonic waves, acoustic cavitation will occur. During the formation and growth of acoustic cavitation bubbles, their sizes rapidly increase, causing the inner liquid to evaporate. The increase of cavitation bubbles and evaporation of inner liquid will absorb heat from the surroundings, which will cause the temperature of molten metal on the surface of cavitation bubble to decrease, resulting in local undercooling. They crystal nuclei are formed near the cavitation bubble, which increases the nucleation rate and makes the microstructure refined. During the collapse of cavitation bubble, the strong shock wave will break the primary crystal and the growing particles; and under the stirring action of acoustic current, they will be dispersed in the melt which is equivalent to the increase of external nuclei, improving the nucleation rate, and inhibiting the growth of crystals at the same time. Therefore, the dendritic silicon in A356 alloy is broken into granules or short fibers after solidification and α-Al grains become fine and nearly round.

4.2.1.2 Effect of Ultrasonic Duration Time on A356 Alloy

Figure 4.5 shows the SEM images of water-quenched sample of A356 alloy under different ultrasonic duration times. It can be seen from the figure that the duration time of ultrasonic treatment has a great impact on the solidification microstructure of A356 alloy. When the duration time is 4 min, most of the silicon in A356 alloy is long needle-shaped with sharp edges and corners, and α-Al is relatively coarse, as shown in Fig. 4.5a. With the increase of ultrasonic duration time, silicon needles are gradually shortened and rounded, and α-Al phase is gradually refined. Especially when the duration time is 10 *min*, silicon is short rod-like and nearly spherical, and α-Al phase is relatively fine and uniform. When the duration time is 12 min, most of the silicon is nearly spherical, between 2 and 5 μm in size, and α-Al phase has a tendency to become elongated and other inclusions also begin to appear, as shown in Fig. 4.5e. These inclusions are identified as Al-Ti phase by EDS analysis, as shown in Fig. 4.5f. The appearance of Al-Ti phase is caused by melting of ultrasonic device's horn into the aluminum melt, which can be confirmed from the observation of appearance of horn after the experiment. So, it does not mean that the longer the

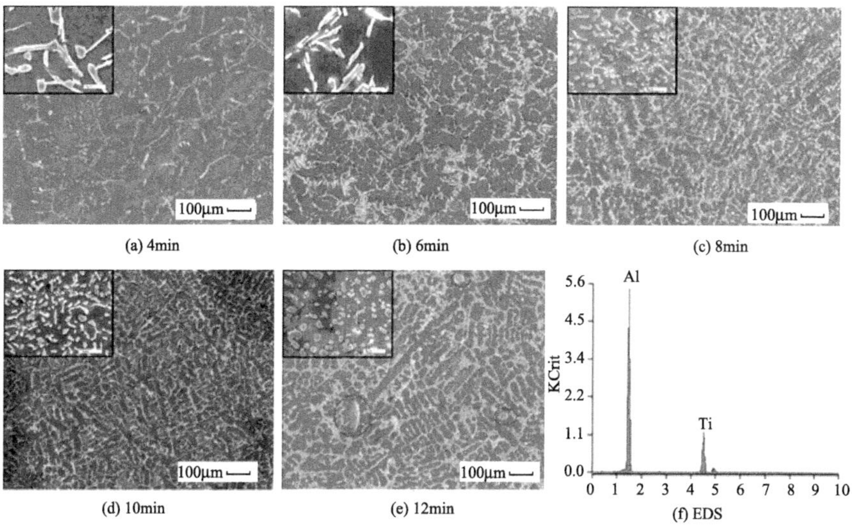

Fig. 4.5 SEM images showing solidification microstructure of A356 alloy matrix under different ultrasonic times

ultrasonic treatment time, the better the effect. If the time is short, the effect is less potent; if the time is longer, other inclusions can produce. Therefore, under these experimental conditions, the best ultrasonic duration time is 10 min.

4.2.2 *Effect of High-Energy Ultrasound on Al-Zr(CO_3)$_2$ Synthetic Composite Material*

When the composite is prepared by in-situ melt reaction under the influence of high-energy ultrasounds, because of the high melt reaction temperature (between 850 and 900 °C), the horn probe is exposed to aluminum melt for a long time, which makes it difficult to investigate the impact of high-energy ultrasounds on in-situ reaction. For this reason, this section describes the intermittent effect of high-energy ultrasound treatment on in-situ reaction. The ultrasonic power is 1.2 kW, the frequency is 20 kHz, and the time of each treatment is 200 s, as shown in Fig. 4.6. From the beginning of melt reaction, the ultrasonic treatment duration is 200 s, after 500 s of melt reaction, the ultrasonic treatment duration is 200 s, and after 1000 s of melt reaction, the ultrasonic treatment duration is 200 s, the total duration of ultrasonic treatment is 600 s.

Figure 4.7 shows the SEM images of the composite prepared from Al-20%Zr(CO_3)$_2$ reaction system under intermittent ultrasonic treatments at an initial temperature of 850 °C, a reaction time of 20 min, and with different ultrasonic powers. It can be seen from Fig. 4.7 that with the increase of high-energy ultrasonic

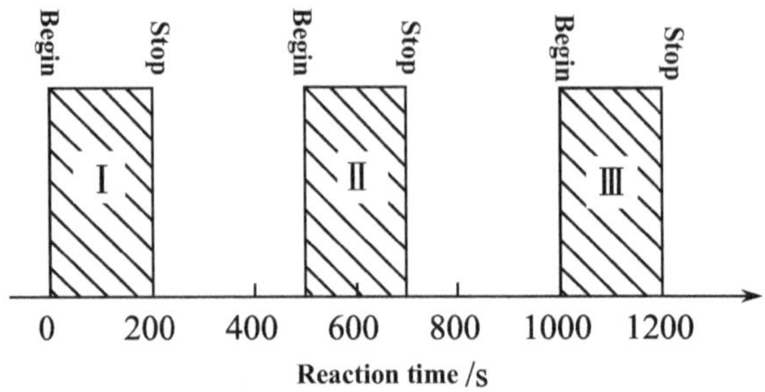

Fig. 4.6 Schematic diagram of intermittent high-energy ultrasonic treatment during in-situ reaction

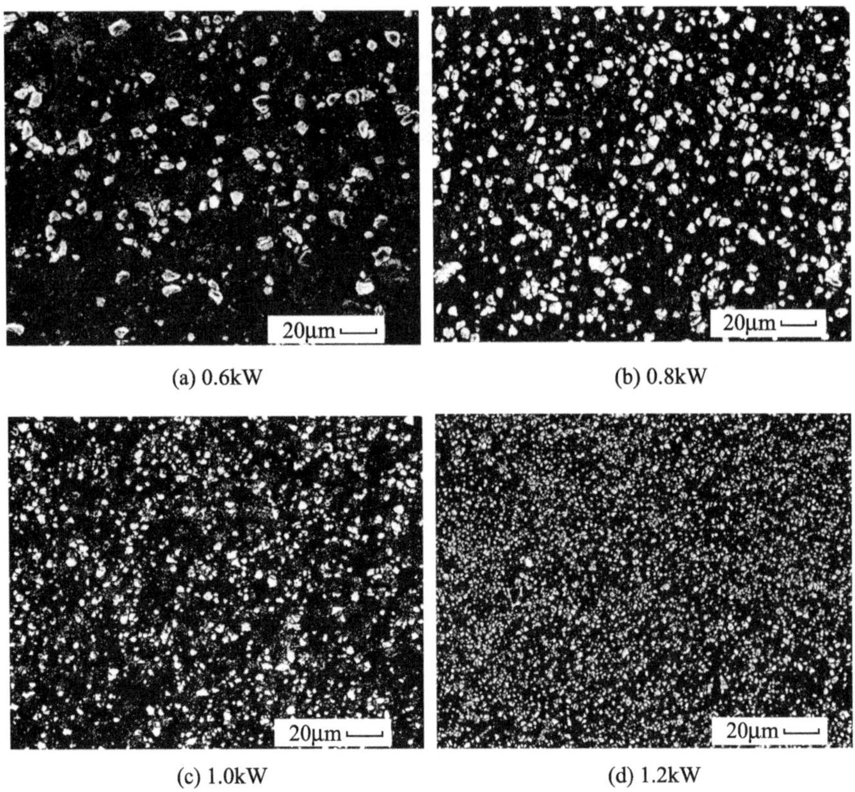

Fig. 4.7 SEM images of composites prepared by in-situ reaction of Al-20%$Zr(CO_3)_2$ system under different ultrasonic powers

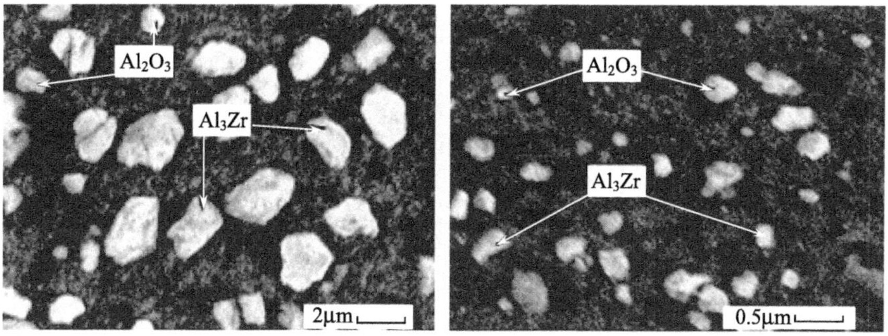

(a) Without high-energy ultrasound (b) With 1.2kW high-energy ultrasound

Fig. 4.8 SEM images showing in-situ particles in the solidification microstructure of composite prepared from Al-20%$Zr(CO_3)_2$ system without high-energy ultrasound and with 1.2 kW high-energy ultrasound

power, the amount of in-situ reinforced Al_2O_3 particles is significantly increased in the composite, and the distribution is more uniform in the matrix. It is especially the case when ultrasonic power is 1.2 kW, the number of in-situ particles in the prepared composite is large, their size is small, and they are evenly distributed in the matrix.

Figure 4.8 shows the SEM images of in-situ particles in the solidification microstructure of composite prepared from Al-20%$Zr(CO_3)_2$ system without and with the application of 1.2 kW high-energy ultrasonic field. It can be seen from Fig. 4.8 that after the application of high-energy ultrasound, the in-situ particles become finer, their shape is rounded, and the particle size is between 0.1 and 0.3 μm. At the same time, the particle/matrix interface is well bonded, the interface is clean, and no reaction products are generated at the interface.

4.2.3 *Effect of High-Energy Ultrasound on Composite Material Synthesized from A356-(K_2ZrF_6 + KBF_4) System*

Figure 4.9 shows the SEM images of composite prepared from A356-20% (K_2ZrF_6 + KBF_4) system at an initial reaction temperature of 850 °C, a reaction time of 20 min, an ultrasonic power of 1.2 kW, and under different modes of ultrasonic treatment. It can be seen from Fig. 4.9 that the microstructure of composite materials prepared by different modes of ultrasonic treatment is different. When the composite material is subjected to ultrasonic treatment continuously for 10 min at the early, middle, and late stages of in-situ melt reaction, microstructure in Fig. 4.9a–c, the number, size, and distribution of reinforcement particles in the solidification microstructure of composite are not as good as those of the composite prepared by intermittent ultrasonic treatments. While using intermittent ultrasonic treatment

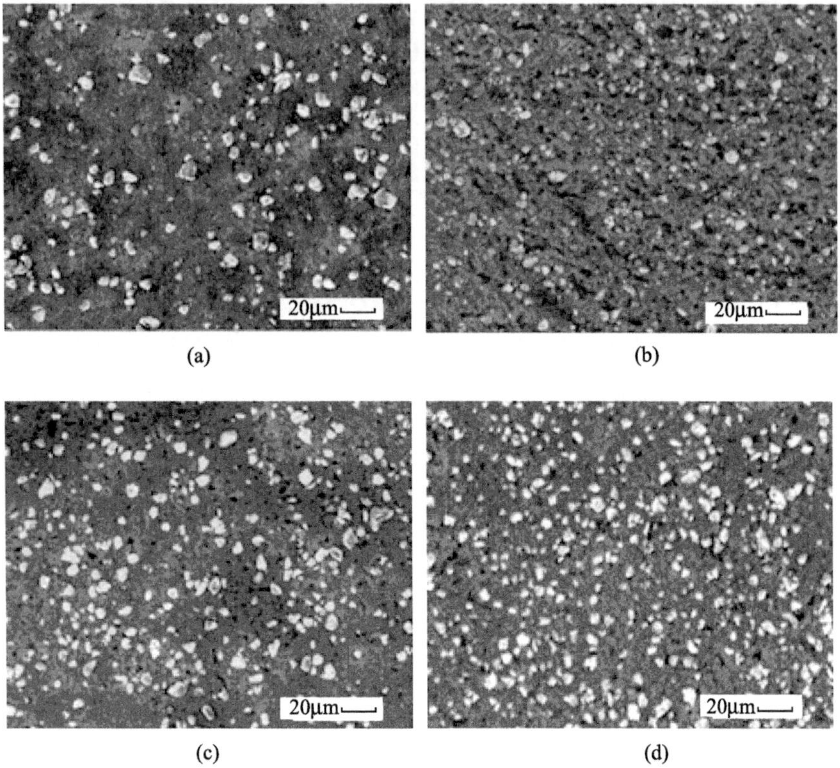

Fig. 4.9 SEM images of composites synthesized by A356-(K_2ZrF_6 + KBF_4) system under different modes of high-energy ultrasound treatment

(shown in Fig. 4.6), large number of reinforcement phase particles are formed in the composite, their size is relatively small, and they are evenly distributed in the matrix, as shown in Fig. 4.9d.

Figure 4.10 shows the SEM image of in-situ particles prepared under intermittent ultrasonic treatment. It can be seen from Fig. 4.10 that under high-energy ultrasonic field, the morphology of in-situ Al_3Zr morphology is square or rectangular, their periphery is relatively round with no sharp edges, the size is between 0.1 and 0.3 μm, the interface is clean, and no inclusions are generated. The morphology of in-situ ZrB_2 particles is square and hexagonal, the interface is clean, the periphery is round with no sharp edges, and the particle size is between 0.1 and 0.2 μm, while some particles are less than 0.1 μm.

The main reason is that when the intermittent treatment is used, the acoustic cavitation effect and the acoustic flow effect of high-energy ultrasound can play a role in the early, middle, and late stages of in-situ melt reaction. In the initial stage of reaction, the reactants K_2ZrF_6 and KBF_4 powders are dispersed in the melt; at the same time, the reaction products quickly diffuse from the reaction layer into the

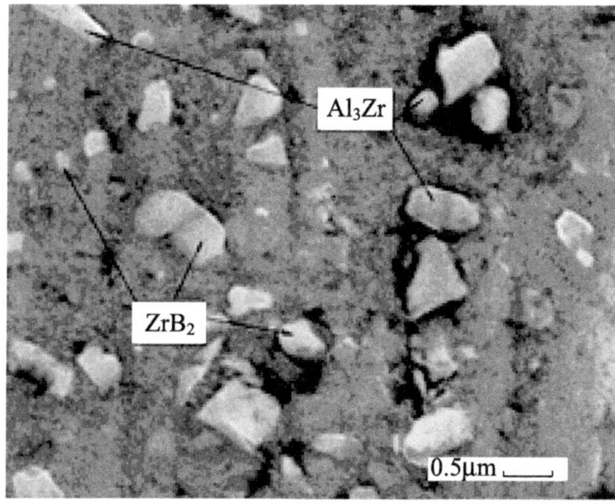

Fig. 4.10 SEM morphology of in-situ Al_3Zr and ZrB_2 particles produced under intermittent ultrasonic treatment

melt, which accelerates the in-situ reaction and increases the number of Al_3Zr and ZrB_2 particles. At the end of reaction, the application of pulsed magnetic field can cause in-situ Al_3Zr and ZrB_2 particles to diffuse uniformly in the aluminum melt.

According to the literature [1, 8, 9], high-energy ultrasonic filed mainly has five basic functions: ① Alternating strings of vibrations are generated; ② the propagation of large-amplitude sound waves in the medium form periodic shock waves of saw-tooth form, resulting in a series of special reactions such as local high temperature and high pressure; ③ nonlinear vibrations cause Bernoulli forces closer to each other; ④ acoustic cavitation in the liquid medium is generated; ⑤ Acoustic flow in the melt is generated. When ultrasonic waves propagate in the liquid or melt, molecules are subjected to a periodic alternating sound field. In the negative pressure phase of sound wave, the liquid is subjected to tensile stress. If the sound pressure is high enough, the liquid is pulled apart to produce cavitation bubbles or cavities. In the subsequent positive pressure phase of sound wave, the cavitation bubbles or cavities close or collapse at a very high speed, thereby generating instantaneous local high temperature, pressure and strong shock waves inside the melt. Although the effect is instantaneous and restrained in a local region, due to high frequency of vibrations, it has inevitably a significant effect on the entire system. At the same time, when ultrasonic waves propagate in the melt, it has a limited amplitude attenuation, which causes a sound pressure gradient to form in the liquid from sound source, resulting in the flow of liquid. In the case of high-energy ultrasounds, when the sound pressure amplitude exceeds a certain value, a jet of fluid can be generated in the liquid. This jet directly leaves the surface of ultrasonic horn and forms circular currents in the entire fluid which is called the acoustic currents. Acoustic current is a combination of circulation and turbulence, the circulation characteristics of acoustic currents cause upside-down movement of the melt so that it is subject to a certain degree of macroscopic stirring effect. Therefore, acoustic currents can significantly improve the uniformities of temperature field and composition of the molten alloy.

When ultrasonic waves are introduced into the aluminum melt, its main functions are the acoustic cavitation and acoustic currents.

The changes caused by acoustic cavitation in liquid A356 alloy can be expressed by Noltingk–Neppiras equation:

$$R\frac{d^2R}{dt^2}+\frac{3}{2}\left(\frac{dR}{dt}\right)^2+\frac{1}{\rho_L}\left[P_0-P_V-P_A\sin(\omega t)+\frac{2\sigma_L}{R}-\left(P_0+\frac{2\sigma_L}{R}\right)\left(\frac{R_0}{R}\right)^{3\gamma}\right]=0 \quad (4.1)$$

where R is the radius of acoustic cavitation bubble; t is the formation time of cavitation bubble; ρ_L is the density of liquid; σ_L is the surface tension of melt; P_v is the vapor pressure inside acoustic cavitation bubble; P_0 is the static pressure; γ is the specific heat; R_0 is the initial radius of acoustic cavitation bubble; and P_A is the sound pressure amplitude in the melt.

Equation (4.1) shows that when the frequency of high-energy ultrasonic waves is 20 kHz and output power is 1.2 kW, the cavitation effect produced by acoustic cavitation bubble's collapse is enough to instantaneously generate a high pressure of about 100 MPa and a temperature of about 10^4 K in the melt at a local level. At the same time, when the acoustic cavitation bubble collapses, a microjet can be generated in the vicinity of bubble. At a certain frequency, the structure of microjet depends on the liquid viscosity and flow velocity. The velocity of microjet is directly proportional to the movement speed of acoustic cavitation bubble wall and inversely proportional to the radius of acoustic cavitation bubble. Because of the collapse of bubble wall, the acoustic cavitation bubbles move very fast, and their radius is very small, so the speed of this microjet is relatively fast. This kind of instantaneous local high-speed acoustic flow promotes a microscopic stirring effect and has a great impact on the evolution and extent of in-situ chemical reaction, refining the grains and improving the microstructure of A356 alloy.

4.2.4 Effect of High-Energy Ultrasound on Composite Material Synthesized from A356-$Ce_2(CO_3)_3$ System

Figure 4.11 shows the SEM images of composites prepared by in-situ reaction of A356-20%$Ce_2(CO_3)_3$ system under intermittent ultrasonic treatment with different ultrasonic powers at an initial reaction temperature of 900 °C. It can be seen from Fig. 4.11 that with the increase of high-energy ultrasonic power, the amount of in-situ Al_2O_3 reinforcement particles is significantly increased in the composite, and their distribution in the matrix is more uniform. Especially, when the ultrasonic power is 1.2 kW, the amount of in-situ particles in the composite is large, their size is small, and the distribution is uniform in the matrix.

Figure 4.12 shows the particle size distribution of in-situ Al_2O_3 reinforcement in the composite prepared from A356-20%$Ce_2(CO_3)_3$ system under 1.2 kW ultrasonic

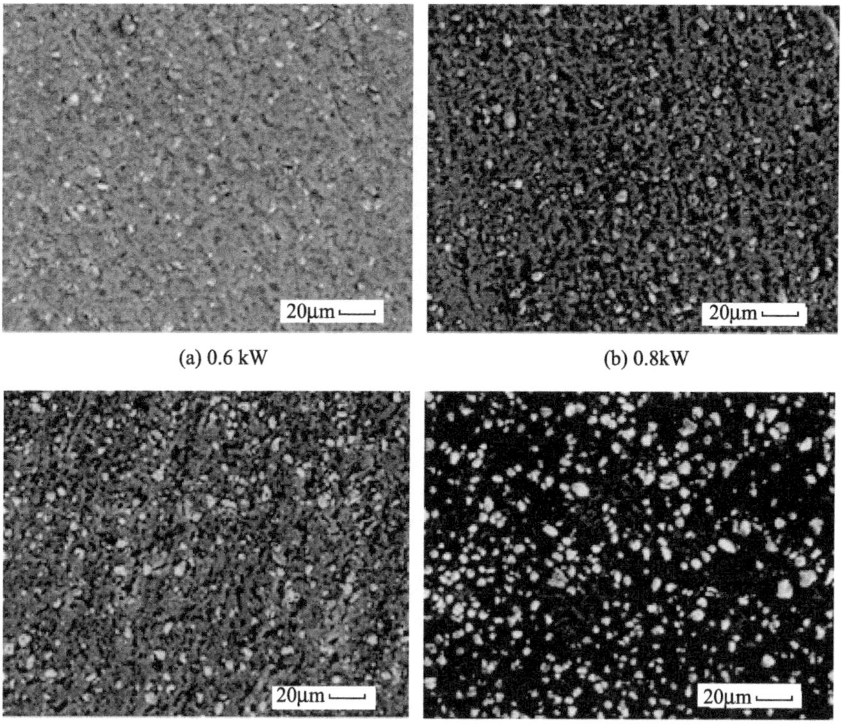

(c) 1.0 kW (d) 1.2kW

Fig. 4.11 SEM images of composites prepared by in-situ reaction of A356-$Ce_2(CO_3)_3$ system under different ultrasonic powers

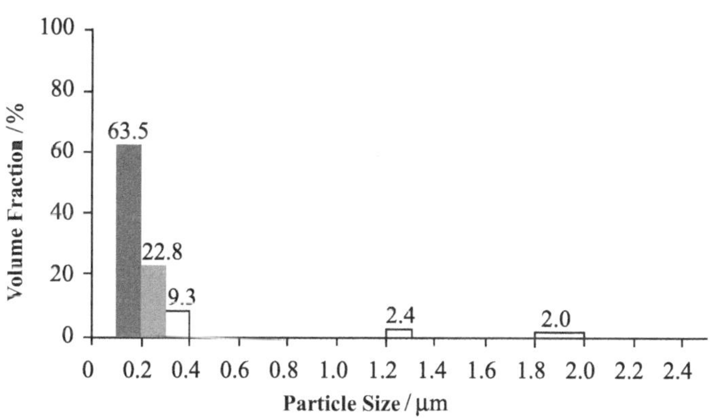

Fig. 4.12 In-situ Al_2O_3 particle size distribution in the composite prepared under 1.2 kW ultrasonic power

power. It can be seen from the figure that about 63.5% of in-situ Al_2O_3 particles have the size in the range of 0.1–0.2 μm, 22.8% of Al_2O_3 particles are in the range of 0.2–0.3 μm, 9.3% of Al_2O_3 particles are in the range of 0.3–0.4 μm, while nearly 2% of particles are relatively large, ranging from 1.8 to 2.0 μm.

The main reason is that after the application of high-energy ultrasound to the melt, the comprehensive combination of high-energy ultrasounds' thermal effect, cavitation effect, and acoustic flow effect accelerates the speed of in-situ chemical reaction in the melt, and simultaneously makes the generated particles distribute uniformly in the melt. However, because of the application of discontinuous high-energy ultrasound, some Al_2O_3 particles may agglomerate during the time with no highenergy ultrasound applied; although the cavitation effect of high-energy ultrasounds can disperse agglomerated particles, due to the radiation effect of ultrasounds, there is still a small number of agglomerated and grown particles of about 2 μm size in the composite.

Figure 4.13 shows the SEM image of composite prepared from in-situ reaction of A356-20%$Ce_2(CO_3)_3$ system at an initial reaction temperature of 900 °C with a reaction time of 20 min, a power of 1.2 kW with intermittent ultrasonic treatment. After etching with HF, it can be observed from the figure that under the influence of high-energy ultrasounds, the morphology of silicon in A356 matrix is round and in-situ Al_2O_3 particles tend to be spherical.

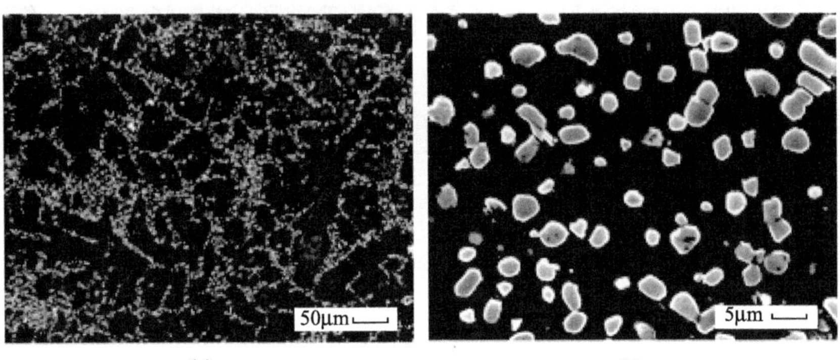

Fig. 4.13 SEM image of composite prepared by in-situ reaction of A356-20% Ce_2 $(CO_3)_3$ system after HF etching

4.2.5 *Effect of High-Energy Ultrasound on Composite Material Synthesized from A356-K_2ZrF_6-KBF_4-$Na_2B_4O_7$ System*

4.2.5.1 Effect of Ultrasonic Powers on the Microstructure of Composites Synthesized from A356-K_2ZrF_6-KBF_4-$Na_2B_4O_7$ System

A 15vol.%(Al_2O_3 + ZrB_2)/A356 composite is prepared from A356-K_2ZrF_6-KBF_4-$Na_2B_4O_7$ reaction system at an initial reaction temperature of 850 °C, under ultrasonic field for 3 min intermittent action. Figure 4.14 shows the effect of different ultrasonic powers (0, 0.5, 1.0, and 1.5 kW) on solidification microstructure of composite prepared under the same ultrasonic time and action mode. It can be observed from the figure that when the ultrasonic frequency is 0.5 kW the volume fraction of particles increases in the solidification microstructure of composite, their volume fraction under three different ultrasonic powers calculated by Image J software is found to be 9.0, 10.2, and 9.1%, and the size of ZrB_2 particle clusters is also reduced. When the ultrasonic frequency is increased to 1.0 kW, the volume fraction of reinforcement phase is the highest, the reaction efficiency is increased by 10% as compared to that without ultrasonic field, and the dispersion of particles is significantly improved.

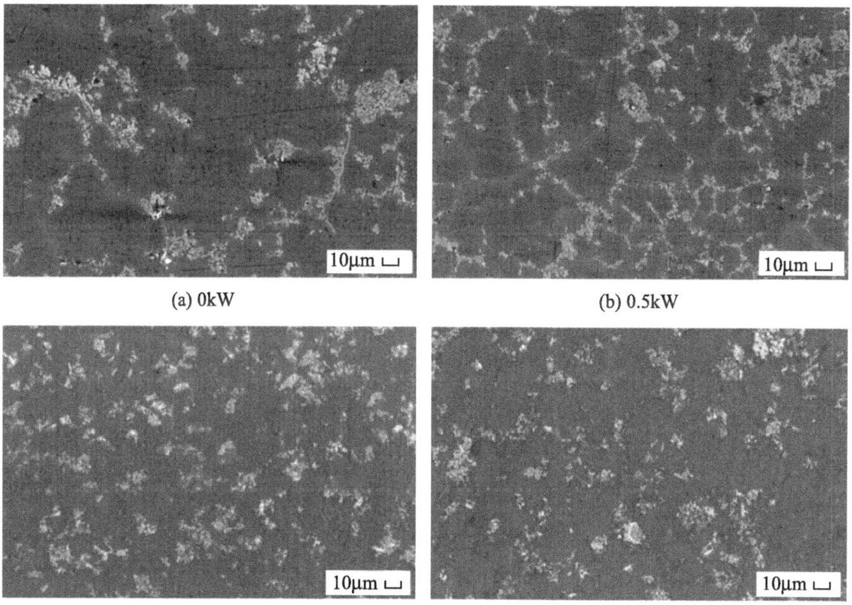

(a) 0kW (b) 0.5kW (c) 1.0kW (d) 1.5kW

Fig. 4.14 SEM images showing solidification microstructure of 15vol.%(Al_2O_3 + ZrB_2)/A356 composite under different ultrasonic powers

However, when the ultrasound frequency is increased to 1.5 kW, the size and distribution of reinforcement phase do not continue to improve; instead, the volume fraction shows a decreasing trend. Therefore, under the same ultrasonic treatment time of 3 min and intermittent action mode, the optimal power for in-situ reaction synthesis of composite from A356-K_2ZrF_6-KBF_4-$Na_2B_4O_7$ system is 1.0 kW.

4.2.5.2 Effect of Ultrasonic Time Durations on the Microstructure of Composites Synthesized from A356-K_2ZrF_6-KBF_4-$Na_2B_4O_7$ System

Figure 4.15 shows the SEM images of sample of water-quenched composite after different times of ultrasonic treatment with 1.0 kW ultrasonic power. It can be observed from the figure that at the beginning of reaction when no ultrasonic filed is applied, a large number of lumpy substances having irregular morphology are generated, which should be the mixture of various intermediate products of Al-Zr-B-O system. When the reaction has taken place for some time and ultrasonic filed is applied for 1 min (Fig. 4.15b), the original lumpy substance has a certain morphology, indicating that a particle-reinforcement phase has been generated at this time, and ZrB_2 reinforcement phase tends to agglomerate. Due to the acoustic

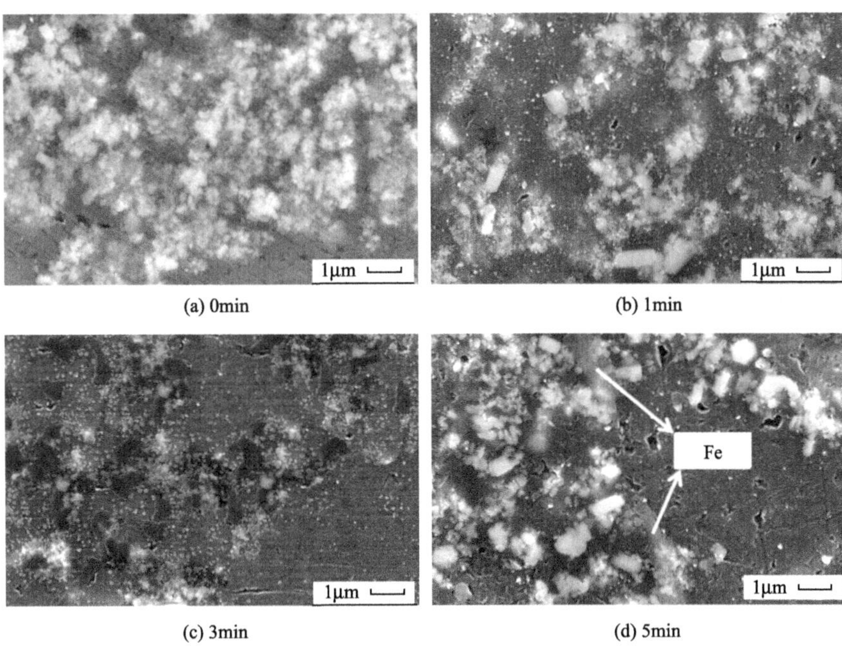

Fig. 4.15 SEM images of water-quenched samples at different reaction times under 1.0 kW ultrasonic power

effect caused by ultrasonic field, the originally agglomerated intermediate mixture is dispersed into smaller areas. When the ultrasonic treatment time increases to 3 $_{min}$, many small ZrB_2 particles peel off from the agglomerates and diffuse into the surrounding matrix. When the ultrasonic treatment time continues to increase, the ultrasonic horn is corroded in the high-temperature melt and forms a needle-like Fe-phase in the composite, as shown in Fig. 4.15d, which is a harmful miscellaneous phase with reduced mechanical properties. Therefore, to control the formation of the harmful miscellaneous phases in the composite melt as much as possible, the best time for A356-K_2ZrF_6-KBF_4-$Na_2B_4O_7$ system is 3 min with intermittent ultrasonic treatment.

When the instant ultrasonic waves propagate in the melt, they cause the alternate compression and expansion of particles in the composite, resulting in pressure gradient, which can affect the mechanical properties. Originally, the instantaneous displacement and velocity of particles are not large, but the particles' acceleration which is proportional to the square root of ultrasonic vibration frequency is very large, exceeding tens of thousands of gravitational acceleration. Such a large acceleration is enough to cause a strong mechanical effect on the particle and can even destroy the particles. The acceleration a of particle's instantaneous motion and the stress caused by this acceleration are as follows:

$$a = \frac{\mathrm{d}v}{\mathrm{d}t} = A_m\omega^2 \sin(\omega t - \phi) \tag{4.2}$$

$$\sigma = \frac{F}{S} = \frac{\rho \Delta V_a}{S} \tag{4.3}$$

where v is the velocity; t is the time; A_m is the maximum displacement amplitude; ω is the frequency; ϕ is the initial phase; and $\Delta m = \rho \Delta V$, which is the mass within the particle's tiny volume element ΔV. In this experiment, the frequency $\omega = 20$ kHz, displacement amplitude $A_m = 20$ μm, volume element $\Delta V = 100\mu m^3$, and $S = 20\mu m^2$. For convenience, assuming that the intermediate state ZrB_2 particles are spherical, and their density $\rho_{ZrB2} = 6.09$ g/cm^3, the stress on the particles is $\sigma = 2.42 \times 10^9$ MPa. Under such a large stress, the intermediate ZrB_2 particles rupture, which promotes the separation of ZrB_2 particles from their parent body and accelerates the in-situ reaction.

Figure 4.16 shows the schematic of the crystal structure of ZrB_2. ZrB_2 is a metalloid compound of space group P6/mmm with hexagonal C32-AlB_2-type structure, and the number of atoms in the Bragg lattice unit cell $m = 1$. This type of crystal structure of ZrB_2 determines that its growth morphology is hexagonal prism-like. As can be seen from the figure, the Zr atoms in the unit cell occupy the corners and bottom center positions of the hexagonal prism, B is in the center of triangular prism composed of Zr atoms, and the projected XY plane is located at the center of gravity of equilateral triangle formed by Zr. The bond distribution between Zr and B atoms shows that a Zr atom and a B atom form 6 equivalent Zr-B bonds on the (0001) plane, and the structure of these Zr-B bonds makes the bonding force between Zr and B

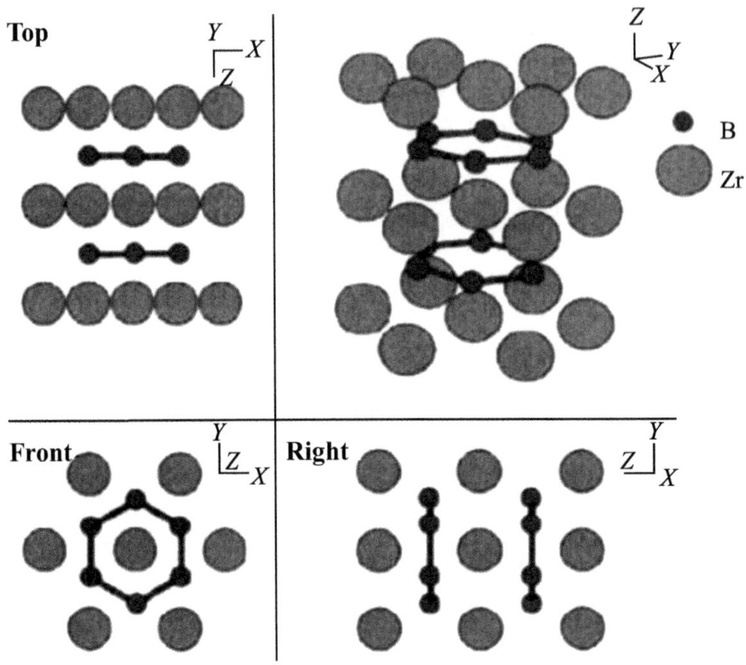

Fig. 4.16 Schematic diagram of the crystal structure of ZrB_2

atoms significantly higher than that between B and B atoms. However, under the combined action of thermal effect, acoustic flow effect, acoustic cavitation effect, and strong mechanical effects caused by ultrasounds in the melt, the weaker bond in ZrB_2 crystal is prone to break, which makes ZrB_2 particles with the small sizes peel off from the agglomerates.

4.2.6 Effect of High-Energy Ultrasound on Composite Synthesized from 6063Al-$Al_2(SO_4)_3$ System

4.2.6.1 Effect of Ultrasonic Power on Reinforcement in $Al_2O_{3(p)}$/6063Al Composites

Figure 4.17 shows the SEM images of solidification microstructure for in-situ 5vol.%Al_2O_{3p} reinforced Al matrix composites synthesized from 6063Al-$Al_2(SO_4)_3$ system at a reaction temperature of 780 °C with a reaction time of 20 min, and with different ultrasonic powers (0.3, 0.5, 0.7, 1.0 kW). The ultrasonic intermittent action time is 2 min. As the ultrasonic power increases, the reinforcement particle size continues to decrease, their shape changes from irregular polygon to small, round, or hexagonal shape. From Fig. 4.17a, it can be observed that when the ultrasonic

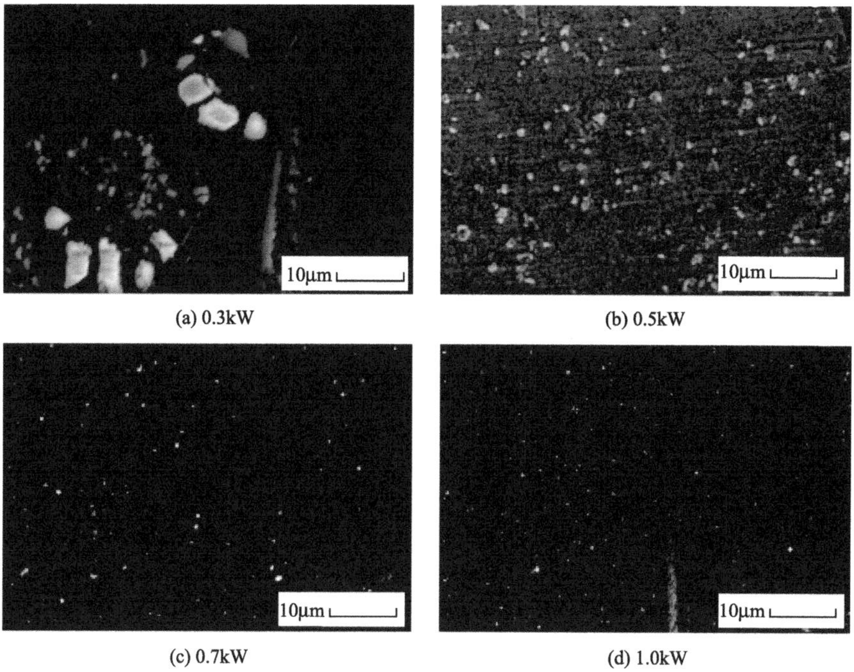

Fig. 4.17 SEM images showing solidification microstructure of Al_2O_3p/6063Al composite prepared under different ultrasonic powers

power is 0.3 kW, the micron-level particles begin to break and refine into submicron particles, the average particle size is around 510 nm, and large-sized particles are still present, the edges of particles are sharp and clear. When the ultrasonic power is 0.5 kW, the large-sized particles are refined to the micron level and some are refined to submicron level, the average particle size is 230 nm, and their shape begin to be round, as shown in Fig. 4.17b. When the ultrasonic power is 0.7 kW, the particles become very fine and reach the nanometer level, the average particle size is about 47 nm, their shape is round or hexagonal, and they are evenly distributed in the matrix, as shown in Fig. 4.17c. When the power is adjusted at 1.0 kW, it can still achieve a very fine and uniform distribution of nanoscale particles with an average size of about 35 nm, as shown in Fig. 4.17d. In addition, the calculation of particles volume fraction with IPP software revealed that the in-situ particles begin to decrease when the ultrasonic power was 1.0 kW, and the impurity phases begin to appear, so the optimal ultrasonic power for synthesis of composite from 6063Al-$Al_2(SO_4)_3$ system is 0.7 kW.

4.2.6.2 Effect of Ultrasonic Time Durations on Reinforcement in $Al_2O_{3(p)}$/6063Al Composite

Figure 4.18 shows the microstructure of in-situ 5vol.%Al_2O_{3p}/6063Al composite synthesized from 6063Al-$Al_2(SO_4)_3$ system at a reaction temperature of 780 °C, with ultrasonic power of 0.7 kW and different ultrasonic treatment times (i.e., 1, 2, 3, 4 min). After 1 min of ultrasonic treatment, the particles begin to refine, the average size is 1 μm and their shape is still irregular polygons, as shown in Fig. 4.18a. After 2 min of ultrasonic treatment, particles are further refined and evenly distributed in the matrix, the average size is 54 nm and their shape becomes nearly spherical, as shown in Fig. 4.18b. After 3 min of ultrasonic treatment, fine particles begin to agglomerate, the particle size becomes coarse from nanoscale to submicron level (nearly 1 μm) and their number density decreases sharply, as shown in Fig. 4.18c. After 4 min of ultrasonic treatment, the particles volume fraction is further reduced, their size distribution becomes very uneven and ranges from 1 to 6 μm, which deteriorates the microstructure of composite, so the optimal ultrasonic treatment time is 2 min, and nanoscale reinforcement particles can be obtained at this time.

Under different ultrasonic treatment times, the size of reinforcement phase particles first decreases and then increases, the shape of particles changes frequently and becomes round. The main reason is that when the particle size is refined to the

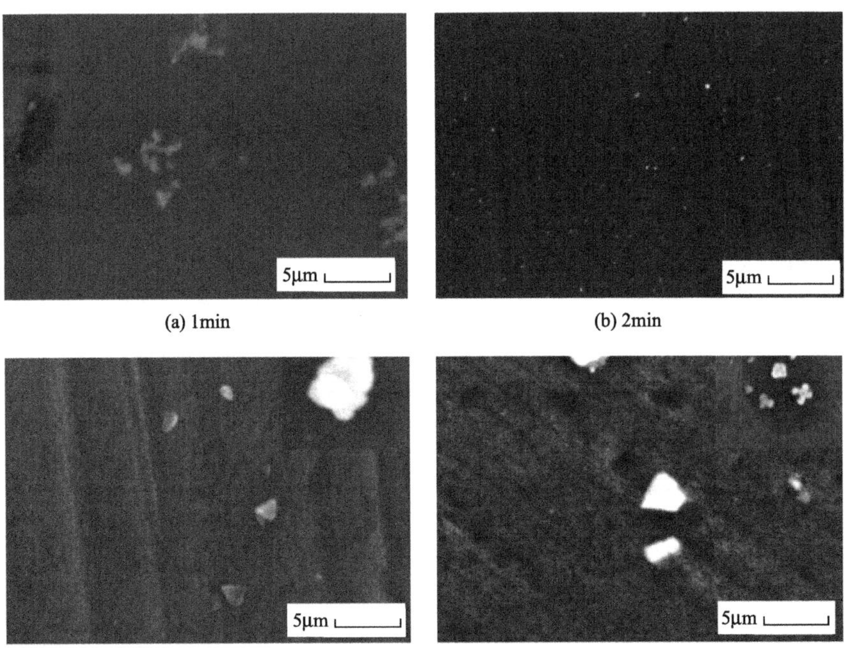

Fig. 4.18 SEM images showing microstructure of water-quenched composite samples with different ultrasonic times

nanometer level, the total surface area is greatly increased and the surface energy is also increased, the small particles tend to come closer and form agglomerates. The granular aggregates of grown/coarsened particles in a short time easily settle down, resulting in a decrease of particles' yield in the final composite.

4.2.6.3 Kinetics of Melt Reaction Under Ultrasound Field

Figure 4.19 shows the microstructure of water-quenched Al_2O_{3p}/6063Al composite samples extracted at different stages of the melt reaction under 0.7 kW ultrasonic power. At the beginning of reaction without ultrasonic treatment, the $Al_2(SO_4)_3$ anhydrous powder rapidly reacts with the melt to produce nearly spherical particles, but their size is relatively large, having a diameter of about 5 μm, as shown in Fig. 4.19a. After the application of ultrasonic treatment for 45 s, the coarse reinforcement phase particles begin to break, as shown in Fig. 4.19b. With continuous application of ultrasonic treatment, the broken particles gradually disperse in the surroundings, as shown in Fig. 4.19c. After 2 min of ultrasonic treatment, the particles are distributed in the matrix with an average size of 51 nm, and the microstructure is ideal, as shown in Fig. 4.19d.

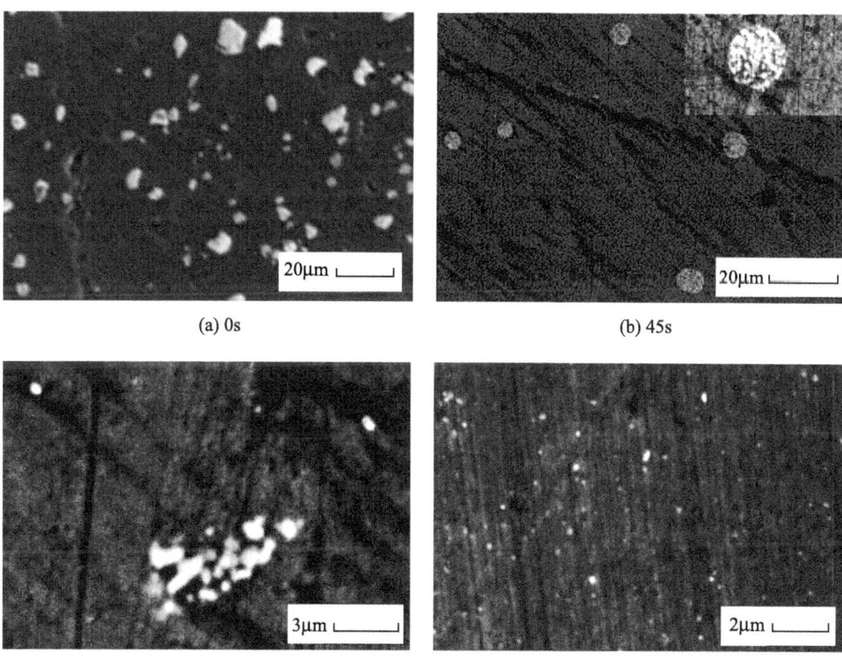

(a) 0s (b) 45s

(c) 75s (d) 120s

Fig. 4.19 SEM images showing microstructure of water-quenched samples of Al_2O_3p/6063Al composite with different reaction times during ultrasonic action

Ultrasonic treatment has a refining effect on the particles because it generates periodic vibrations in molten aluminum. Tiny bubbles in the melt are activated by disturbance, they expand and contract with vibrations separating it from the surrounding environment to form a cavity, and continually combine and grow into relatively large bubbles. The vibrational frequency of grown bubbles is different from that of small bubbles, when the vibrational frequency is synchronized with ultrasonic vibrations, resonance occurs, the amplitude suddenly increases, and the tension on bubbles suddenly becomes stronger. It applies pressure on surrounding melt, the bubbles themselves collapse and produce new strong pressure waves under the action of tension accompanied by local high temperature and high pressure, which have a strong impact on surroundings and break up the larger particles in the melt into smaller ones. After the large particles are broken, they are dispersed and distributed as fine particles. In addition, when ultrasounds propagate in the melt, due to interaction between acoustic waves and molten aluminum, a certain sound pressure gradient is formed in the melt, which accelerates the flow of liquid aluminum. The broken particles are dispersed by the acoustic flow and spread throughout the molten matrix. At the same time, local high temperature generated by acoustic cavitation effect is homogenized by acoustic flow such that the temperature of the entire melt rises, thereby reducing the viscosity of local melt, which further improves the wettability of particles and facilitates uniform dispersion of reinforcement particles in the matrix.

4.2.6.4 Effect of Ultrasonic Power on the Grain Size of Al_2O_3p/6063Al Composite Matrix

Figure 4.20 shows the optical micrographs of matrix grain size in Al_2O_{3p}/6063Al composite under different ultrasonic powers after 2 min of intermittent action, and *r* represents the calculated average grain radius. It can be seen from the figure that as the ultrasonic power increases the grain size starts decreasing and the effect of grain refinement becomes more and more significant. Thus, ultrasonic action refines the grains by making use of ultrasonic cavitation and acoustic flow effects.

According to the relationship between ultrasonic power and average grain size of matrix in Al_2O_{3p}/6063Al composite (Fig. 4.20), a best fitting curve is constructed in Fig. 4.21. From Fig. 4.21, it can be seen that as the ultrasonic power increases from 0, the grain size decreases significantly. However, when ultrasonic power reaches about 1 kW, the decrease in matrix grain size slows down.

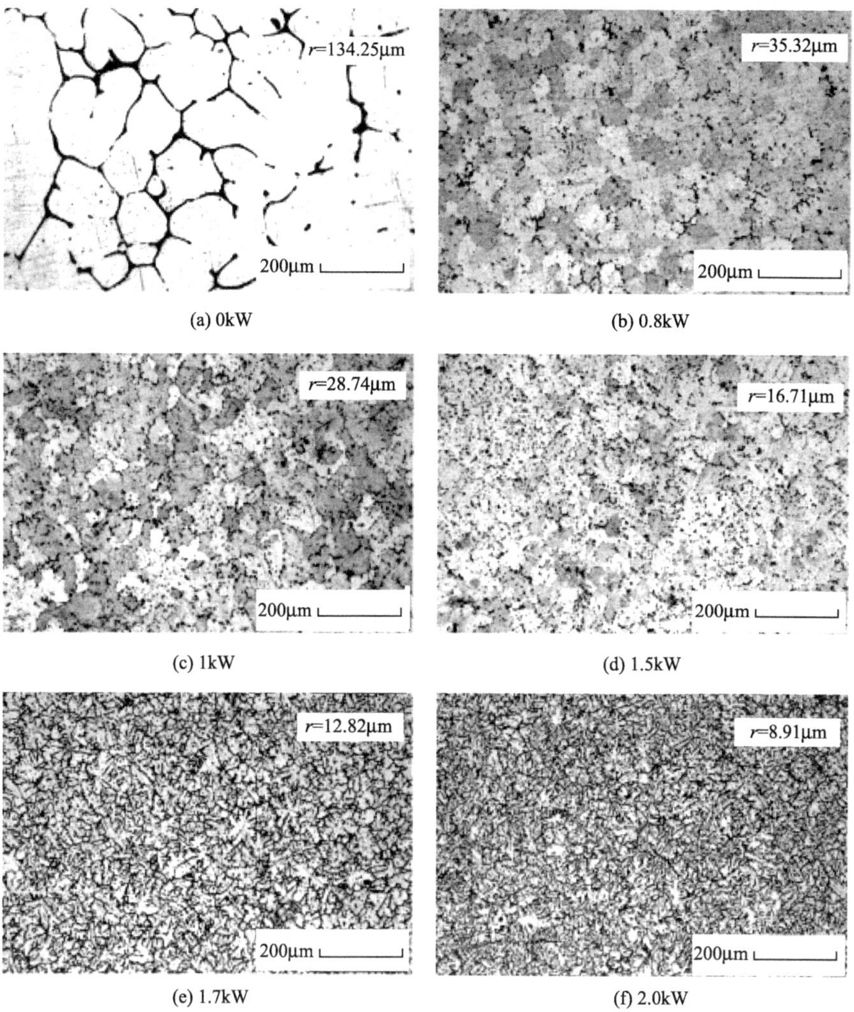

Fig. 4.20 Matrix grain size images of Al_2O_{3p}/6063Al composite under different ultrasonic powers

4.2.7 *Effect of High-Energy Ultrasonic on Composite Synthesized from 7075Al-(Al-3B) Alloy-Ti System*

4.2.7.1 Effect of Ultrasonic Powers on TiB_2/7055Al Composite

Figure 4.22 shows the SEM image of in-situ 5vol.%TiB_2/7055Al composite synthesized from 7055Al-(Al-3B) alloy-Ti system at a reaction temperature of 850 °C with 30 min reaction time under different ultrasonic powers (0, 0.5, 1.0, and 1.5 kW). The ultrasonic action time is 3 min, and the intermittent action method is adopted. As can

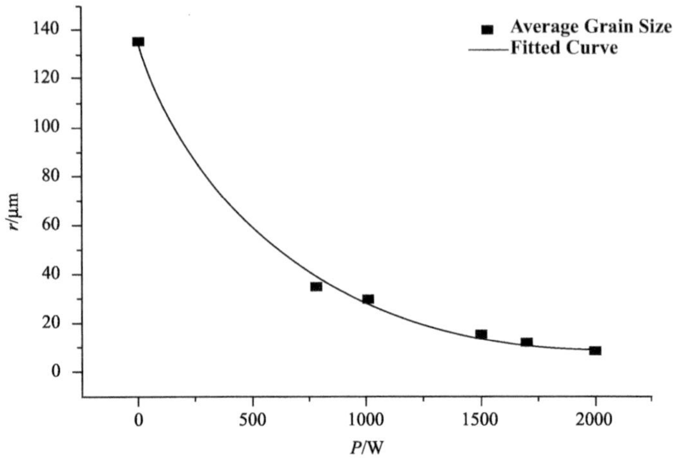

Fig. 4.21 Relationship between average grain size of matrix and ultrasonic power in Al_2O_{3p}/6063Al composite

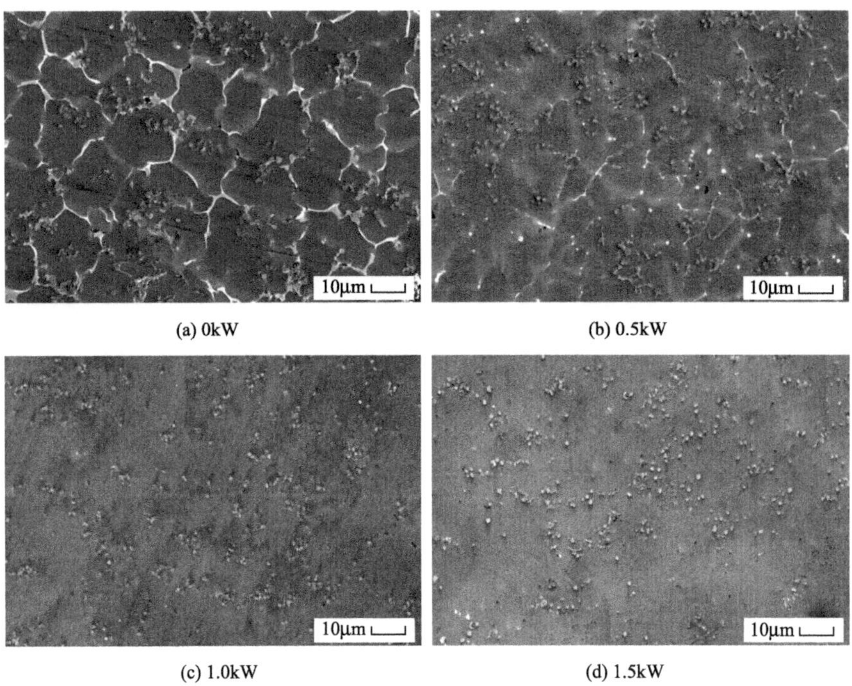

Fig. 4.22 SEM images showing solidification microstructure of 5vol.%TiB_2/7055Al composite under different ultrasonic powers

be seen from Fig. 4.22a, when no magnetic field is applied, the generated particles are polygonal in shape, large in size, and uneven in distribution. When the applied ultrasonic power is 0.5 kW, Fig. 4.22b shows that the TiB_2 particles are polygonal or nearly spherical in shape, their size is decreased, distribution is relatively uniform, and particles volume fraction is increased; in addition, the second phase changes from a continuous grid-like distribution to a discontinuous distribution; when ultrasonic power is 1.0 kW, the volume fraction of reinforcement phase in the composite reaches the highest; at this time, the TiB_2 particles are small in size, mostly nearly spherical in shape, and evenly distributed in the matrix as shown in Fig. 4.22c. However, when the ultrasonic power reaches 1.5 kW, it is found that the size and distribution of reinforcement particles do not continue to improve, instead, their volume fraction tends to decrease, as shown in Fig. 4.22d. Therefore, under the same ultrasonic action time and action mode, the best ultrasonic power for synthesis of TiB_2/7055Al composites is 1.0 kW. After the analysis of the overall scenario, it is concluded that the application of ultrasonic field accelerates in-situ reaction process, reduces the size of generated particles, and increases their volume fraction, thus improving the wettability and distribution of particles throughout the matrix.

4.2.7.2 Kinetics of Melt Reaction in 7055Al-(Al-3B) Alloy-Ti System Under Ultrasound Power

Figure 4.23 shows the microstructure of composite sample water quenched at different stages of melt reaction under 1.0 kW ultrasonic power. It can be seen from Fig. 4.23a that when no ultrasonic filed is applied at the beginning of reaction, the added Ti agent reacts quickly with liquid aluminum to form massive or short rod-shaped Al_3Ti particles. When reaction has taken place for quite some time and the ultrasonic field is applied for 1 min, Fig. 4.23b shows that a large number of Al_3Ti particles have started to fracture, break, and reaction has started at the edges of Al_3Ti particles. At this time, significant TiB_2 particles are generated near the reaction particles, and they continuously peel off from the particle agglomerates. The ultrasonic effect improves the heat and mass transfer rate of the melt such that the products at the reaction interface gradually diffuse into the surrounding melt, and evenly distribute in the matrix, which accelerates the in-situ reaction rate. As the reaction continues and ultrasonic time is increased to 3 min, no obvious Al_3Ti particles are found in the melt, and the generated TiB_2 particles increase significantly, their size is finer, and they are more uniformly dispersed as shown in Fig. 4.23c. When the ultrasonic action time is further increased, the particle size and distribution are not significantly improved, and the corrosion phenomenon of ultrasonic horn in high-temperature melt is increased forming a long strip of Fe-rich phase (shown in Fig. 4.23d) in the composite, which is a harmful impurity phase that can reduce the mechanical properties of composite. Therefore, to control the formation of impurity phase in the composite as much as possible, the optimal ultrasonic action time for synthesis of composite from 7055Al-(Al-3B) alloy-Ti system is 3 min with intermittent action.

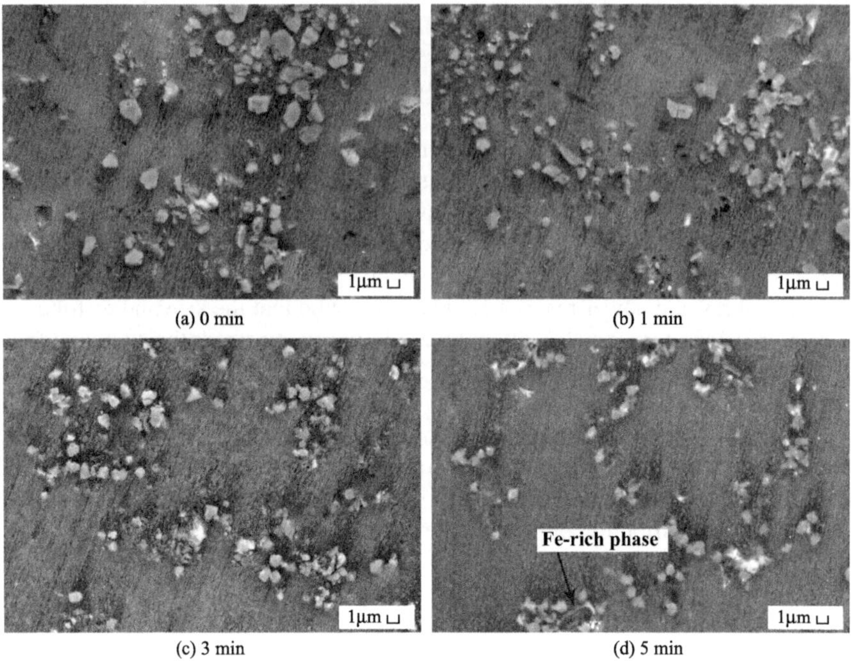

Fig. 4.23 SEM images of TiB_2/7055Al composite samples water-quenched after different times of synthesis process under 1.0 kW ultrasonic power

4.3 Mechanism of In-Situ Aluminum Matrix Composites Synthesis Under High-Energy Ultrasound

4.3.1 The Characteristics and Principle of Ultrasound

Ultrasounds refer to mechanical waves with a sound frequency of 2×10^4–2×10^9 Hz. When the intensity of ultrasounds is low, it can be used as a carrier medium for detection and loading of information. When its intensity exceeds a certain value, its interaction with sound media can affect or even destroy the state, nature, and structure of the latter; at this time, ultrasonic waves appear as a form of energy, called power ultrasound. It is generally believed that the ultrasonic effect accompanies the appearance of the following phenomena: Firstly, when sound pressure changes, the solvent is compressed and diluted, causing the fluid to move frequently; the second is that when cavitation and temperature changes are small, introducing a large amount of vibrational energy into small volume produces the micro-jets effect. When ultrasonic waves propagate in a metal melt, they produce acoustic cavitation and acoustic flow effects, and at the same time, some of these mechanisms cause changes in flow field, pressure field, and temperature field in the melt, which produces some special effects [10, 11].

(1) Acoustic cavitation effect

Ultrasonic cavitation means that tiny bubble nuclei are activated in the liquid under the action of ultrasonic waves, which is the manifestation of a series of dynamic processes such as oscillation, growth, contraction, and collapse of bubble nucleus. In the negative pressure phase of acoustic wave, the amount of force on coal is P_h and P_a, where P_h is the hydrostatic pressure and P_a is the sound pressure. If the sound intensity is large and the corresponding negative pressure on the liquid is also strong enough, the average distance between nano-molecules exceeds a limit distance, which destroys the integrity of liquid structure resulting in creation of cavities and holes. Once the cavity is formed, it grows until the negative sound pressure reaches its maximum value (P_a). These voids are compressed again in the positive pressure phase of sound wave, causing some cavitation bubbles to continuously oscillate while the others completely collapse. This high-speed collapsed liquid produces an instantaneous local high temperature and high pressure. Ma and Chen [12], through theoretical calculations and actual measurements, found that this instantaneous local high temperature and high pressure can be as high as 10^4 K and 10^3 MPa, respectively.

(2) Acoustic flow effect

Due to the interaction of ultrasound and viscous force in the liquid, a limited amplitude attenuation can form a certain pressure gradient in the liquid from the source, thereby forming a jet of fluid. The jet leaves the end face of horn to form a circulation in the entire fluid (Fig. 4.24). The speed of sound flow caused by ultrasound in the melt is fast, which can reach 10–10^3 times the fluid heat convection speed.

The combined effect of acoustic cavitation and acoustic flow makes ultrasonic treatment one of the effective ways to improve the solidification microstructure and mechanical properties of metals and alloys. In addition, high pressure generated by short-term cavitation can break agglomerated particles and disperse them (Fig. 4.25), and high temperature generated at the same time can improve wettability between liquid metal and the reinforcement particles. The acoustic flow destroys the boundary

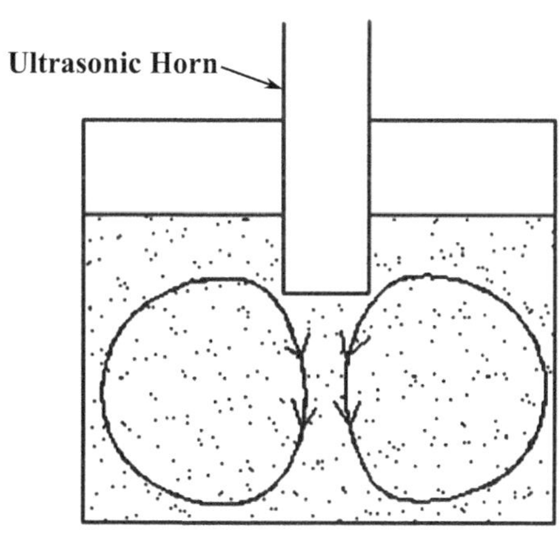

Fig. 4.24 Schematic diagram of acoustic flow

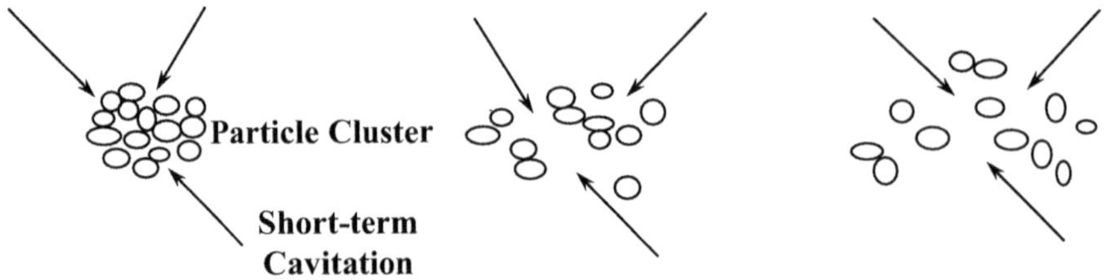

Fig. 4.25 Dispersion of agglomerated particles by cavitation of high pressure shock wave

layer, accelerates mass and heat transfer, and plays a key role in promoting the dispersion of reinforcement phase and cleaning the surface impurities.

4.3.2 Action Mechanism of High-Energy Ultrasound During In-Situ Melt Reaction

4.3.2.1 Acoustic Flow Effect

When ultrasonic waves propagate in the melt, due to the interaction of sound waves and viscous force of the liquid, the limited amplitude attenuation causes a certain sound pressure gradient to form in the liquid from the source, causing the liquid to flow and form a circulation in the entire liquid. This flow is the acoustic flow.

Since it is impossible to observe acoustic flow in a high-temperature melt, a transparent liquid is used instead of high-temperature melt for observation to verify the existence of acoustic flow. Pan et al. [1] used pure glycerol (at 20 °C temperature, the viscosity coefficient $\mu = 1.94$ Pa s) to observe the acoustic flow. Figure 4.26 shows the schematic diagram of acoustic flow pattern in glycerol when ultrasound is applied for 15 s. It is observed that the scope of acoustic flow is basically throughout the liquid and its total flow is circular. The flow under the horn is vertically downward and fast, after reflection at the bottom of the container, it moves upward where the speed is slowed, and a small vortex is formed near the end of the horn. Results show that due to the circular characteristics of acoustic flow, the particles can flip up and

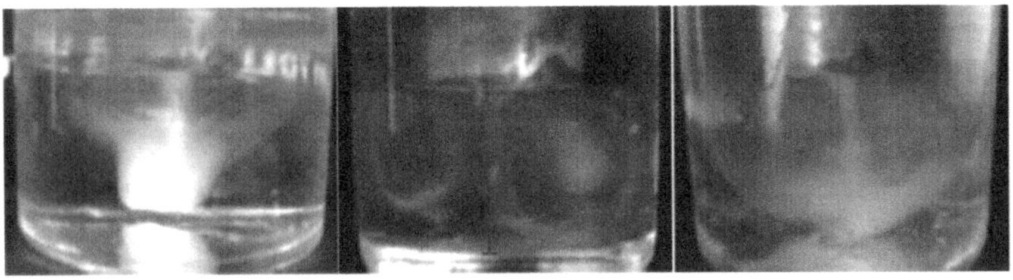

Fig. 4.26 Schematic diagram of acoustic flow in glycerol

down in the molten pool, and the finite amplitude acoustic flow is turbulent in the viscous melt, so the particles are subjected to a certain amount of agitation, causing the original clusters of particles to collapse. Therefore, the acoustic flow has a stirring effect on the particles and hinders the agglomeration of particles.

It is assumed that the ultrasonic waves have no reflection at the melt-crucible bottom interface, the melt-air interface, and the melt-crucible wall interface. According to Huygens principle, the end face of the horn can be divided into countless facets, each facet element can be regarded as an equivalent sound source. Then, the sound pressure at any point in the sound field is the linear superposition of sound pressure at all point sources.

On the basis of the model established on the end face of horn shown in Fig. 4.27, *OT* is the tangential direction of end surface, *ON* is the normal direction of end surface, the sound pressure generated at point *A* in the melt field by a point source (point B in Fig. 4.27) can be expressed as

$$P_A = \frac{\mathrm{j}\omega\rho V_0}{2\pi}\int_0^{2\pi}\int_0^{a}\frac{x\mathrm{e}^{-jkr}}{r}\mathrm{d}x\mathrm{d}r \tag{4.4}$$

$$r = \sqrt{R^2 + x^2 - 2Rx\sin\,\alpha\,\sin\,\beta} \tag{4.5}$$

where k is the wavenumber, $k = 2\pi f/C$; f is the ultrasonic vibration frequency, and its value is 20 kHz; C is the propagation speed of ultrasonic waves in the melt, and $C = 3300$ m/s in A356 alloy melt; ω is the angular frequency, $\omega = 2\pi f$; V_0 is the effective vibration speed of end face of the horn, $V_0 = 2\pi fA$; A is the amplitude of end face of the horn, taken as 32.3 μm; and ρ is the density of A356 melt, which is

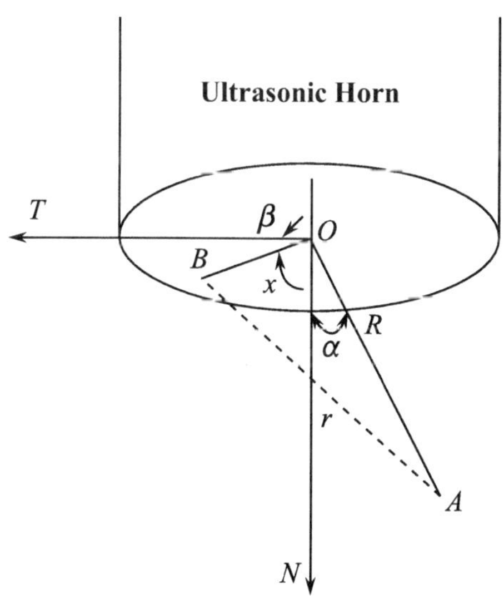

Fig. 4.27 Schematic diagram for calculation of sound pressure distribution in the melt

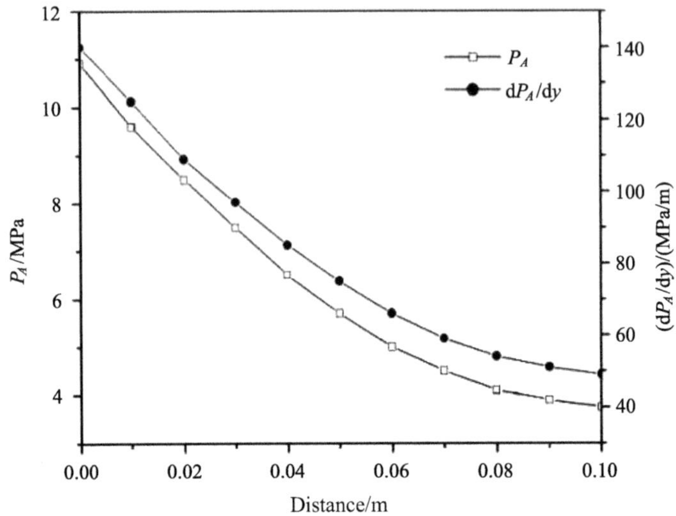

Fig. 4.28 Distribution diagram of sound pressure and sound pressure gradient along the normal direction of end face of the horn

2.665 g/cm^3. $R = OA$; $x = OB$; $r = AB$. While a is the radius of the contact end of the horn and the melt. α is the angle between the OA line and normal to the end face ON. β is the angle between the OB end face and the end face tangent OT.

Calculated results of the sound pressure, P_A, the distribution along the normal direction (distance range y = 0–0.1 m), and the distribution of sound pressure gradient, dP_A/dy, are shown in Fig. 4.28. It can be seen from the figure that the sound pressure amplitude is maximum at the centered face of the horn, which is 11 MPa. The sound pressure gradient at each point in the normal direction ON of the end surface also decreases with distance from the end of horn, which shows that the acoustic flow effect caused by acoustic pressure gradient gradually decreases.

4.3.2.2 Acoustic Cavitation Effect

When ultrasonic waves propagate in the melt, the melt molecules are subjected to the periodic alternating acoustic field and are torn apart to produce cavitation bubbles or holes. These cavitation bubbles or holes collapse at a very high speed, which then generates instantaneous high temperature and high pressure in the melt.

The maximum bubble wall velocity U_{m} during the adiabatic collapse of acoustic cavitation bubble and the radius R_{m} of corresponding cavitation bubble satisfy the following equation.

$$U_{\mathrm{m}} = \left.\frac{\mathrm{d}R}{\mathrm{d}t}\right|_{R=R_{\mathrm{m}}} = \frac{2\left[P_0\left(\frac{R_0}{R_{\mathrm{m}}}\right)^{3\gamma} - (P_A + P_0)\right]}{R_{\mathrm{m}}\rho} \tag{4.6}$$

where P_0 is the static pressure of the melt, $P_0 = 1.013 \times 10^5$ Pa; P_{A} is the sound pressure amplitude, taking the value of 10^5–6×10^6 Pa; R_0 is an initial radius of cavitation bubble, for convenience in calculation, the range of R_0 is set as 5×10^{-8}–10^{-6} m; and γ is the gas constant inside the bubble, taking $\gamma = 4/3$.

According to Rayleigh hypothesis, the kinetic energy of each fluid particle becomes deformation elastic energy of the particle calculated according to the fluid's bulk elastic modulus. When the cavitation bubble collapses, we get the shock wave pressure amplitude as:

$$P_{\mathrm{m}} = \rho C U_{\mathrm{m}} \tag{4.7}$$

Calculation shows that with the increase of sound pressure amplitude, the shock wave pressure amplitude generated on the collapse of acoustic cavitation bubble gradually increases, and it can reach to a maximum value of 10^5 MPa.

According to the formula given in the literature [1, 13], the maximum temperature when acoustic cavitation bubble collapses is expressed as:

$$T_{\max} = T_0 \frac{(P_0 + P_A)(\gamma - 1)}{P_0} \tag{4.8}$$

where T_0 is the temperature of the melt. Equation (4.8) shows that there is a linear relationship between the maximum temperature of acoustic cavitation bubble when it collapses and the sound pressure amplitude. Under experimental conditions, it is calculated that the instantaneous high temperature generated in the melt on collapse of cavitation bubble can be as high as 2.255×10^4 K.

The above calculations and analysis show that at the beginning of the reaction the local high temperature generated in the aluminum melt by ultrasounds can reduce the surface tension of melt, improve the wettability of reactant powder by melt, promote the in-situ reaction, and improve the nucleation rate of reinforcement phase. At the same time, the stirring effect of ultrasounds brings more reactants into the melt, increasing the reaction contact surface and the extent of reaction. In the middle and late stages of the reaction, many tiny nuclei of bubbles and cavities produced by ultrasounds play a major role, these bubble nuclei stretch and grow under negative circulating pressure, and are then compressed under the action of positive pressure. The analysis shows that the local instantaneous high temperature and high pressure generated by acoustic cavitation bubble can reach 10^4 K and about 10 GPa, respectively. And because the ultrasonic frequency is very high and acoustic cavitation bubbles are widely distributed in the melt, almost the entire melt is covered by the instantaneous high temperature and high pressure. When the particle size is small, the specific surface area of particles and their interfacial energy are relatively large, at this time, the particles tend to agglomerate to reduce their interfacial energy. These kinds of particles that rely on van der Waals forces to reunite last for a period of time. The strong pressure generated by acoustic cavitation bubbles can break up agglomerated particles, hindering the aggregation and growth of particles to some extent. In addition, ultrasounds can remove impurities in the melt and purify the melt.

In order to investigate the effect of high-energy ultrasound on the microstructure of composite, a ZrB_2(np)/2024Al composite is synthesized from 2024Al-K_2ZrF_6-KBF_4 system using direct melt reaction method under high-energy ultrasound field. The molar ratio of K_2ZrF_6 and KBF_4 in the reaction is 1:3, the reaction temperature is 870 °C, and the reaction time is 30 min. The amount of reactant salts is 40% of the melt, the power and frequency of high-energy ultrasound are 1.2 kW and 20 kHz, respectively, and the time of high-energy ultrasound treatment is selected as 1, 3, 5, 7, and 9 min. According to Young's equation:

$$\cos\theta = \frac{\sigma_{SV} - \sigma_{SL}}{\sigma_{LV}} \tag{4.9}$$

where θ is the contact angle; σ_{SV} is the solid/air interfacial energy; σ_{SL} is the solid/liquid interfacial energy; and σ_{LV} is the gas/liquid interfacial energy. During high-energy ultrasonic treatment, the stirring effect caused by acoustic flow and the instantaneous high temperature and high pressure caused by acoustic cavitation effect can significantly reduce the surface tension and friction of 2024Al melt and reduce the surface energy of the melt. The contact angle θ between the particle and the melt is reduced, and the wettability of ZrB_2 particles in 2024Al melt is enhanced. In addition, acoustic flow and acoustic cavitation effects of high-energy ultrasounds produce vibrations inside the melt, generating alternating shear stresses which effectively crush the nano-ZrB_2 particle clusters.

A schematic diagram in Fig. 4.29 shows the principle of high-energy ultrasonic dispersion of nano-ZrB_2 particle clusters. As shown in Fig. 4.29a, the introduction of high-energy ultrasound causes nonlinear effects in molten metal such as acoustic cavitation and acoustic flow. As the action time of high-energy ultrasound increases, some nano-ZrB_2 particle clusters are broken into small nano-ZrB_2 particle clusters by the shock wave generated due to the collapse of cavitation bubbles, as shown in Fig. 4.29c. With further increase of action time, more and more nano-ZrB_2 particle clusters enter the ultrasonic action zone and are broken into fine nano-ZrB_2 particles. When the high-energy ultrasonic action time reaches 5 $_{\text{min}}$, fine nano-ZrB_2 particles are evenly dispersed in the matrix. However, the cavitation effect increases the surface activity of particles and provides a driving force for the reunion of the particles. Therefore, as the action time of high-energy ultrasound further increases, the nano-ZrB_2 particles reunite and settle again, as shown in Fig. 4.29d. Thus, the action time duration of high-energy ultrasound should not be too long. In this experiment, the optimal application time duration of high-energy ultrasound is 5 min.

The acoustic cavitation effect produced by high-energy ultrasounds in the melt refines the melt by degassing. In general, high-energy ultrasound degassing can be

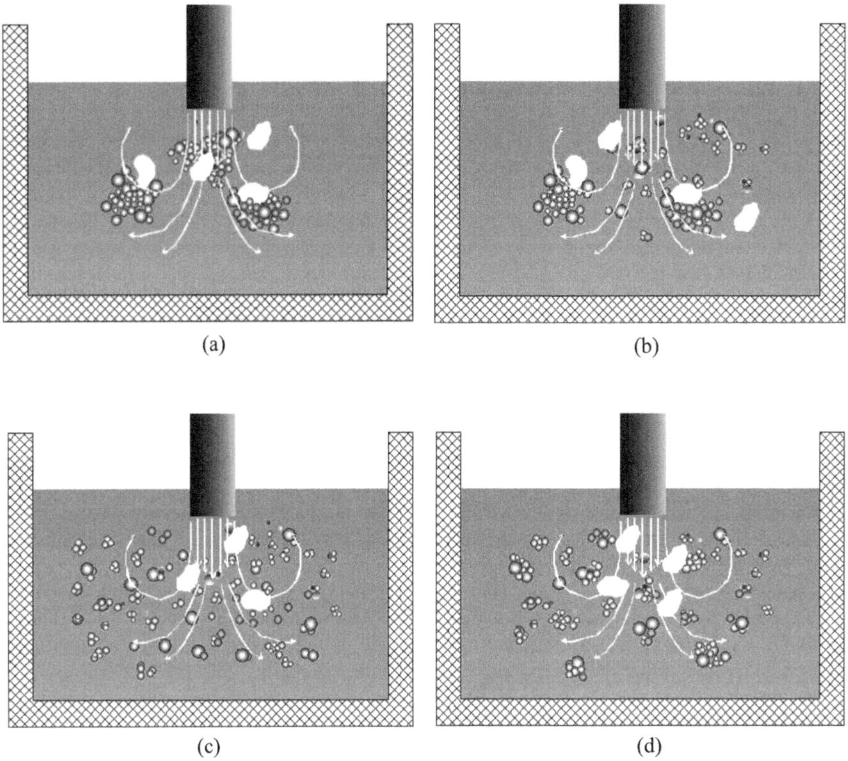

Fig. 4.29 Schematic diagram showing the principle of high-energy ultrasound dispersion of nano-ZrB_2 particle clusters

divided into three stages: ① cavitation bubbles; ② cavitation bubbles adsorb small bubbles in the melt to form large bubbles; and ③ large bubbles float out of the surface of molten metal.

This study shows that when high-energy ultrasounds are introduced into the aluminum melt, cavitation bubbles are generated near the end of ultrasonic horn. In this process, cavitation bubbles can directly capture dissolved hydrogen in the aluminum melt. In addition, when nano-ZrB_2 particle clusters are broken, the bubbles in the clusters can also be captured by cavitation bubbles. Large bubbles float on the surface of molten metal to escape, resulting in degassing of the composite. When the action time of high-energy ultrasonic treatment is increased, the porosity in the composite can be significantly reduced.

References

1. Lei P, Jie T, Zhaofeng C, et al. Acoustic effect of high-energy ultrasound in particle/metal melt system. Mater Eng. 2006;34:35–7, 42 (in Chinese).
2. Songli Z, Yutao Z, Gang C, et al. Mechanical behavior of $(Al_3Zr+ZrB_2)$/A356 composites synthesized in-situ by sonochemical. Chinese J Nonferrous Metals. 2008;18(12):2140–4 (in Chinese).
3. Yu P, Mei Z, Tjong SC. Structure, thermal and mechanical properties of in situ Al-based metal matrix composite reinforced with Al_2O_3, and TiC submicron particles. Mater Chem Phys. 2005;93(1):109–16.
4. Saravanan C, Subramanian K, Sivakumar DB, et al. Fabrication of aluminum metal matrix composite—A review. In: Proceedings of national conference on recent trends and developments in sustainable green technologies, 2015.
5. Jiashan D, Yutao Z, Songli Z, et al. Sonochemical in-situ synthesis $(Al_2O_3 + Al_3Zr)$ p/A356 composite material. Foundry. 2008;57(4):354–8 (in Chinese).
6. Lei L, Yutao Z, Zhihong J, et al. Microstructure and mechanical properties of ZrB_2/A356 composites synthesized in-situ by sonochemical. Thermal Processing Technology. 2010;39(10):98–101 (in Chinese).
7. Dengbin C, Yutao Z, Guirong L, et al. The effect and mechanism of high-energy ultrasound on the solidification structure of in-situ synthesized Al_3Ti/6070 composites. Chinese J Nonferrous Metals. 2009;19(11):1956–61 (in Chinese).
8. Zhang S, Zhao Y, Cheng X, et al. High-energy ultrasonic field effects on the microstructure and mechanical behaviors of A356 alloy. J Alloy Compd. 2009;470(1–2):168–72.
9. Qiu B, Ke L, Ting G, et al. The effect of ultrasonic vibration power on the solidification structure of AZ31B magnesium alloy ingot. Special Cast Nonferrous Alloys. 2009;29(6):576–79, 489–90 (in Chinese).
10. Jian XG. The effect of ultrasonic vibration on the solidification of light alloys. Knoxiville: The University of Tennessee; 2005.
11. Xu H, Jian X, Meek TT, et al. Degassing of molten aluminum A356 alloy using ultrasonic vibration. Mater Lett. 2004;58(29):3669–73.
12. Liqun M, Feng C. Preparation of fine particle reinforced metal matrix composites by high-energy ultrasound. J Mater Res. 1995;9(4):372–5 (in Chinese).
13. Lei J, Yutao Z, Tianping W. In-situ sonochemical preparation of (ZrB_2+Al_3Zr) p/2124 composite and its wear properties. Func Mater. 2013;44(7):988–92 (in Chinese).

Chapter 5
Synthesis of In-Situ Aluminum Matrix Composites by Acoustomagnetic Coupling Field

From Chaps. 3 and 4 on the application of electromagnetic method and high-energy ultrasonic method for synthesis of in-situ aluminum matrix composites, it can be seen that external fields (electromagnetic field, ultrasonic field) significantly affect the reaction process (thermodynamics, kinetics) of metal melt with reactants, and affect the size, morphology, and distribution of reaction products (i.e., reinforcement). Therefore, the external field is an important controlling parameter to prepare in-situ aluminum matrix composites with fine reinforcements, uniform distribution, and good performance. This chapter focuses on the application of acoustomagnetic coupling method (ultrasonic field and magnetic field coupling) in the synthesis of in-situ aluminum matrix composites combined with typical reaction systems (Al-Ti system, Al-Ti-B system, Al-Zr-B-O system), in order to study the effect of acoustomagnetic coupling on the microstructure and properties of in-situ aluminum matrix composites, as well as their synthesis mechanism.

5.1 Application of Acoustomagnetic Coupling Method on Metal Melt and Reaction

5.1.1 Influence of Acoustic-Magnetic Field on Metal Melt and Reactions

Dynamic or changing magnetic field can generate electric currents in the conductive molten metal and cause the melt to agitate, which affects the transport and dispersion of reactants and products (reinforcement) in the melt, thereby causing a concentration gradient of elements near the reactive mass in the melt, which in turn affects the progress of melt reaction [1]. Therefore, the synthesis and forming process of metal matrix composites is controlled by the non-contactable force of electromagnetic field. This technology has been highly valued by researchers in recent years for its

Y. Zhao, *In-Situ Synthesis of Aluminum Matrix Composites*,
https://doi.org/10.1007/978-981-16-9120-1_5

effectiveness in the improvement of microstructure and performance of composites (see Sect. 3.1 for details).

At the same time, the introduction of high-energy ultrasonic field into the liquid metal generates periodic stresses and sound pressure and thus produces many nonlinear effects such as ultrasonic cavitation and acoustic streaming [2]. These effects can produce high temperature and high pressure in the melt within a very short time (tens of seconds), which have obvious effects on removing the inclusions and gases in the melt, refining the alloy solidification microstructure, reinforcement particles and liquid alloy wetting, promoting the uniform dispersion of reinforcement particles and the rate of reaction (see Sect. 4.1 for details) [3].

Acoustomagnetic coupling field has the advantages of ultrasonic field and magnetic field; at the same time, it compensates for their deficiencies. This field makes full use of ultrasonic cavitation, acoustic flow, and magnetic field stirring to change the dynamics of reaction and crystallization, affecting the nucleation and growth of in-situ reinforcement particles in order to optimize their morphology, size, and distribution [4].

5.1.2 Application of Acoustic-Magnetic Coupling Field in Preparation of Alloys and Composite Materials

Zhang et al. [5] of Dalian University of Technology first studied the effect of acoustomagnetic coupling field (ultrasonic field and electromagnetic field) on the microstructure of aluminum alloy, namely A356. Results showed that under the combined action of ultrasonic cavitation, acoustic flow, and magnetic field stirring, α-Al and eutectic Si have been significantly refined, especially the eutectic Si has changed its morphology from original lath shape to fine rod shape. Later, Shao et al. [6] of Northeastern University studied the effect of acoustomagnetic coupling field (ultrasonic field and low-frequency electromagnetic field) on as-cast microstructure and properties of semi-continuously cast AZ80 magnesium alloy. Results showed that under the action of acoustomagnetic coupling field, the as-cast microstructure of AZ80 magnesium alloy has been significantly refined from both the microscopic and macroscopic aspects, and the alloy properties have been significantly improved. By the comparison of microstructure and performance, it is believed that acoustic and electromagnetic coupling field is more effective than a single external field.

The application of acoustomagnetic coupling field in synthesis of aluminum matrix composites was earlier reported by Tsunekawa et al. [7], who combined the ultrasonic field and electromagnetic stirring effect to synthesize Al–Mg alloy matrix composite using SiO_2 powder as the reactant. Results showed that the two physical fields together greatly improve the wettability of SiO_2 powder by alloy melt and promote the in-situ reaction, thus preparing $MgAl_2O_4$ and Al_2O_3 particle-reinforced aluminum matrix composites. There are relatively few reports on the application of acoustomagnetic coupling field in the preparation of aluminum matrix composites, but from the above literature, it is observed that the acoustomagnetic coupling field has the advantages of both the ultrasonic field and the electromagnetic field, showing a better effect than a single field. Therefore, the study of process parameters optimization and the mechanisms of acoustomagnetic coupling field is of great importance for the preparation of in-situ aluminum matrix composites with excellent performances.

5.2 The Principle of Synthesis of In-Situ Aluminum Matrix Composites by Acoustomagnetic Coupling Field

Figure 5.1 shows the acoustomagnetic coupling field device and its schematic diagram. Under acoustomagnetic coupling field, the reaction between reinforcement-forming compound and the melt is not only affected by acoustic current and cavitation generated by high-energy ultrasonic field but also affected by the high-speed melt stirring under the action of electromagnetic field; thus, the reaction process and final microstructure of composite are different from those of a single external field. The process parameters and microstructure characterization of composites prepared under a single field are different from those under the acoustomagnetic coupling field used in this chapter. Aiming at typical reaction systems (Al-Ti, Al-Ti-B, and other systems), the effect of acoustic and electromagnetic coupling fields on the microstructure and performance of in-situ aluminum matrix composites is studied.

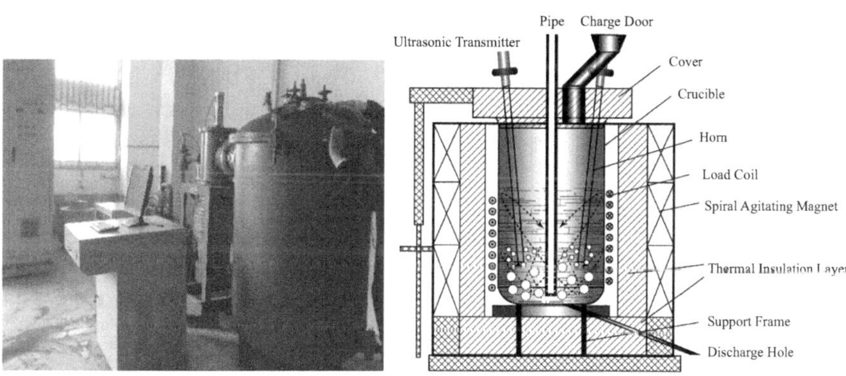

(a) Acoustic-magnetic Field Coupling Device

(b) Schematic Diagram

Fig. 5.1 Acousomagnetic coupling field device

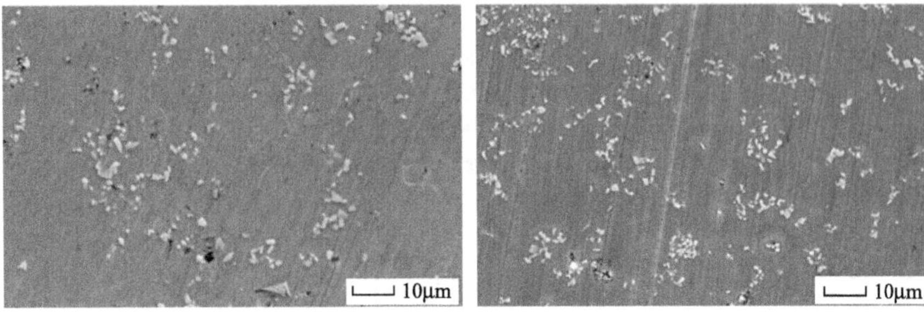

(a) 1.2kW (b) 1.6kW

Fig. 5.2 SEM images of Al_3Ti/6070Al composites produced under different ultrasonic powers in acoustomagnetic coupling field

5.2.1 *Reactive Synthesis of Al_3Ti/6070Al Composites Under Acoustic-Magnetic Field Coupling*

5.2.1.1 Microstructure of Al_3Ti/6070Al Composites Synthesized Under Acoustomagnetic Coupling Field

Based on 6070Al-K_2TiF_6 system, in-situ Al_3Ti/6070Al composites are prepared under the acoustomagnetic coupling field [8]. The ultrasonic parameters are: ultrasonic power 1.2 and 1.6 kW, action time 3 min; magnetic field parameters are: excitation current 200 A, frequency 5 Hz action time 3 min. Figure 5.2 shows the SEM images of Al_3Ti/6070Al composite synthesized under acoustomagnetic coupling field using K_2TiF_6 powder as reactant equal to 10% of 6070Al melt. It can be seen from the figure that the Al_3Ti particles obtained under the coupling field are extremely small. When the ultrasonic power is 1.2 kW, about 62.5% of Al_3Ti reinforcement particles have a size between 0.2 and 0.5 μm, only about 3.7% of Al_3Ti-reinforced particles have a size between 0.8 and 1.2 μm. When the ultrasonic power is increased to 1.6 kW, about 72.4% of Al_3Ti reinforcement particles are between 0.2 and 0.5 μm, and the particle size between 0.8 and 1.2 μm is reduced to about 1.6%. Furthermore, the distribution of Al_3Ti reinforcement particles in the composite obtained under the acoustomagnetic coupling field is relatively uniform.

5.2.1.2 Influence of Powder Addition on the Microstructure of Al_3Ti/6070Al Composite Prepared Under Acoustomagnetic Coupling Field

For the same reaction system, the ultrasonic parameters are: ultrasonic power 1.2 kW, action time 3 min; magnetic field parameters are: excitation current 200A, frequency 5 Hz, action time 3 min. Figure 5.3 shows the SEM images of Al_3Ti/6070Al composite synthesized under the acoustomagnetic coupling field with different

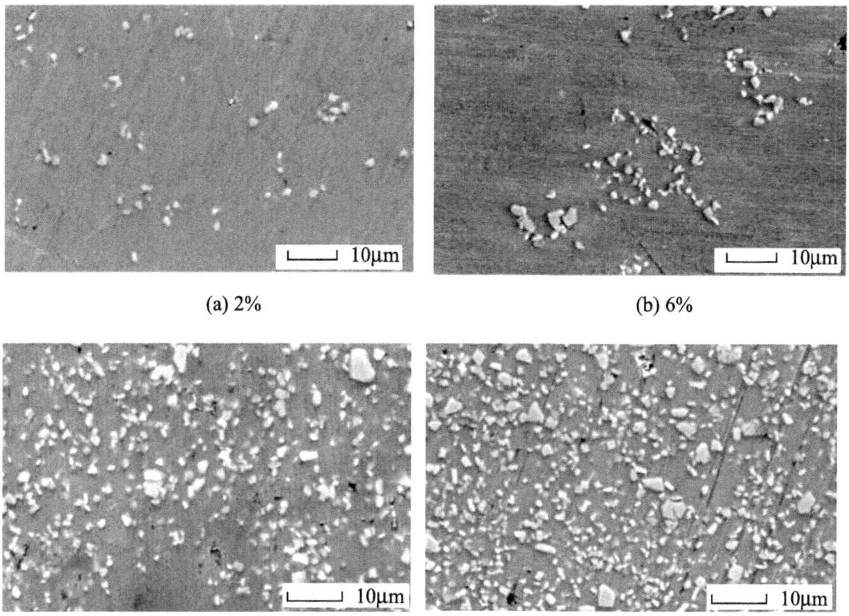

(a) 2% (b) 6%

(c) 18% (d) 54%

Fig. 5.3 SEM images of Al_3Ti/6070Al composites prepared with different amounts of K_2TiF_6 powder additions in acoustomagnetic coupling field

amounts of K_2TiF_6 powder (2%, 6%, 18%, 54%, respectively, of melt mass). It can be seen from the figure that the particle size increases to a great extent with the amount of K_2TiF_6 powder, and the morphology of particles gradually changes from granules and small blocks to lumps, lamellae, or short rods. When the addition level is less than or equal to 18%, the number of particles is linearly related to the amount of powder, and when the amount of powder is further increased, this linear relationship weakens sharply. In addition, it can be observed from Fig. 5.3 that the porosity defects in the solidified microstructure of composite gradually increase with increasing the amount of powder; thus, the volume fraction of reinforcement in the composite should not be very high.

The reason for the above-mentioned changes is when the added amount of reactant is too large, the reactant powder and the generated particles increase the viscosity of liquid aluminum, reduce the fluidity of melt, and make the diffusion process difficult. And a large amount of powder that has not participated in the reaction is transformed into lumps so it cannot have good contact with molten aluminum, preventing the reaction from proceeding further, and finally, it is removed along with molten slag. At the same time, despite the combined action of ultrasonic and electromagnetic fields, generated particles in the reaction layer cannot diffuse out quickly, causing themselves to agglomerate and grow in size, resulting in the phenomenon of particle growth. In addition, from the viscosity formula [9], it can be seen that the viscosity of liquid aluminum rapidly increases as the volume fraction of Al_3Ti particles increases. When the volume fraction exceeds to a great extent, the fluidity of composite melt becomes poor, and it cannot feed the melt during the solidification process, resulting in more porosity defects in samples.

5.2.1.3 Interface Analysis of Al_3Ti/6070Al Composites Synthesized by Reaction Under Acoustomagnetic Coupling Field

Figure 5.4 shows the TEM image of interface structure of Al_3Ti reinforcement particles and Al-matrix in Al_3Ti/6070Al composite synthesized under acoustomagnetic coupling field. Selected area electron diffraction (SAED) shows that the particles in Fig. 5.4a are Al_3Ti. Particles are smaller with near-spherical shape and round corners, their size is 0.2–0.5 μm, and the interface between reinforcement and the substrate is

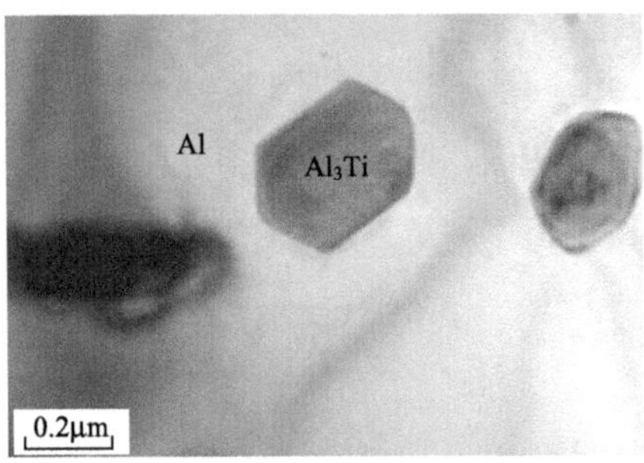

(a) TEM Morpholygy of The Al_3Ti Particles

(b) Interfacial Electron Diffraction Patterns

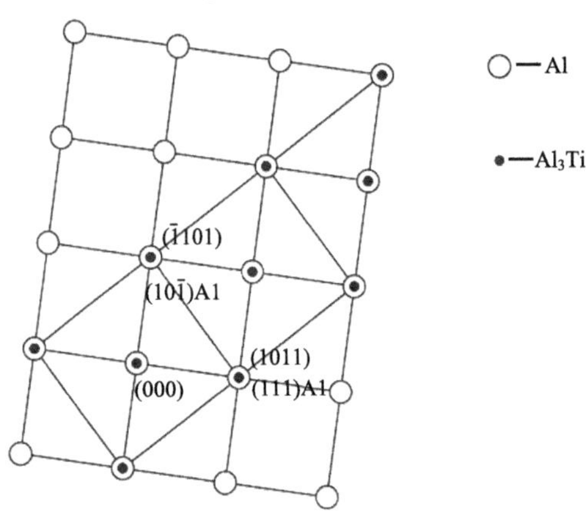

(c) Interfacial Electron Diffraction Calibration

Fig. 5.4 TEM image of Al_3Ti/6070Al composite synthesized under acoustomagnetic coupling field

Fig. 5.5 Morphology and distribution of dislocations in Al_3Ti/6070Al composite synthesized under acoustomagnetic coupling field

smooth and intact. In addition, the electron diffraction pattern at the Al_3Ti/Al interface, given in Fig. 5.4b, shows that it consists of diffraction spots of only Al_3Ti and Al phases, indicating that there is no reaction at the interface. The indexed spots of diffraction pattern are shown in Fig. 5.4c. It is observed that Al_3Ti and Al have the following orientation relationship: $Al_3Ti(1\ 0\ 11)//Al(1\ 1\ 1)$, $Al_3Ti[0\ 1\ 0]//Al[1\ \bar{2}\ 1]$. The interplanar spacing of $Al_3Ti(1\ 0\ 11)$ is 0.2409 nm, and the interplanar spacing of Al(1 1 1) is 0.2338 nm. According to the interface mismatch formula [10], the mismatch between Al_3Ti and Al is calculated to be $\delta = 2.99\%$. The degree of interface mismatch between the two phases is very small; therefore, it can be regarded as a coherent interface and is well bonded between Al_3Ti reinforcement particles and α-Al phase.

Figure 5.5 shows the distribution and morphology of dislocations in Al_3Ti/6070Al composite prepared under the acoustomagnetic coupling field. It can be seen that the dislocation density is very high at the particle/matrix interface in the composite, and dislocations interaction appears at the particles. This is mainly due to the difference in thermal expansion coefficient of Al_3Ti particles and Al-matrix as the solidification shrinkage in Al-matrix is greater than that of Al_3Ti particles.

5.2.2 *Reaction Synthesis of TiB_2/7055Al Composites Under Acoustomagnetic Coupling Field*

5.2.2.1 Microstructure of TiB_2/7055Al Composites Prepared Under Acoustomagnetic Coupling Field

In-situ TiB_2/7055Al composite is synthesized using 7055Al-K_2TiF_6-KBF_4 as the reaction system. The ultrasonic parameters are: ultrasonic power of 0.6 and 0.8 kW, action time 5 min; electromagnetic field parameters are: excitation current 200A, frequency 5 Hz, action time 5 min, and acoustomagnetic coupling field is used in continuous action mode. Figure 5.6 shows the SEM image of TiB_2/7055Al composite synthesized under acoustomagnetic coupling field using K_2TiF_6 + KBF_4 powder as 20% of the 7055Al melt. It can be observed from the figure that under the acoustomagnetic coupling field with ultrasonic power of 0.8 kW, a large number of nanoscale TiB_2 particles with 80–100 nm size are formed. These nano-TiB_2 particles primarily have hexagonal morphology with obvious "round" corners, some particles have "spherical" appearance, as shown in Fig. 5.6c.

The reason is: surface curvature of reinforcement phase particles and corresponding solute concentration around them are both different. At the place where the curvature is large, the equilibrium solute concentration in the liquid phase is high; and at the place where curvature is small, the equilibrium solute concentration in the liquid phase is low. According to the crystal growth principle, solute will diffuse from the place with larger curvature to the place with smaller curvature, so that the sharp corners of particles are blunted, which is conducive to the spheroidization of reinforcement phase particles. TiB_2 particles generated by the reaction are abraded and collide under the action of ultrasonic flow and magnetic field stirring; the abrasion and collision cause the particles, originally agglomerated or adhered, to be separated and even break at defective places, resulting in "refinement" effect. At the same time,

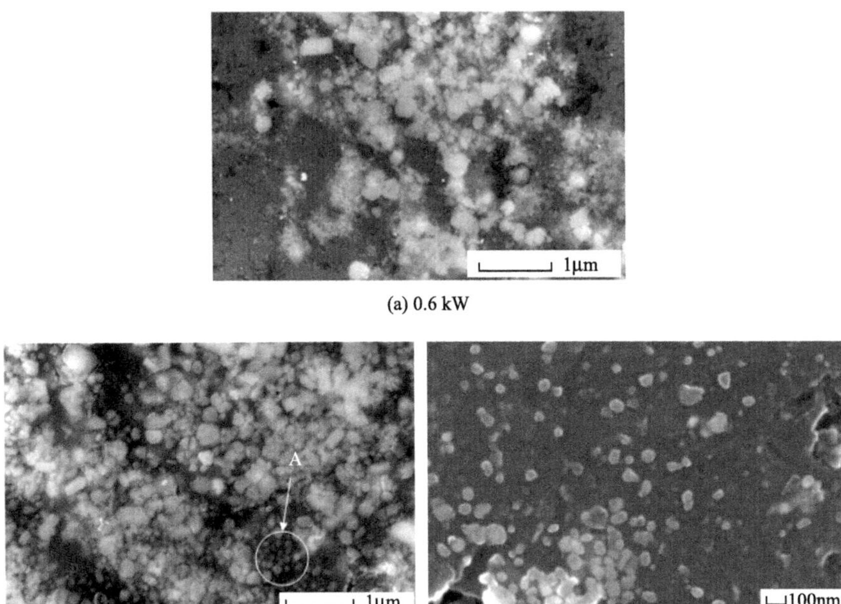

(a) 0.6 kW

(b) 0.8kW

(c) Magnification of A area

Fig. 5.6 SEM images of TiB_2/7055Al composites prepared with different ultrasonic powers under acoustomagnetic coupling field abrasion and collision also cause the particles to collide and grind against each other so that the sharp corners of particles are "blunted" and "spheroidization" effect is produced; therefore, it promotes "ripeness" of reinforcement phase particles.

The ultrasonic power of 0.8 kW under the acoustomagnetic coupling field can obtain the refinement effect of 1.6 kW unltrasonic power under a single ultrasonic field, Therefore, the critical ultrasonic power consumed in the formation of nano-TiB_2 particles is greatly reduced, thereby reducing corrosion damage to the ultrasonic horn during usage. At the same time, the percentage of total in-situ nano-TiB_2 particles in the composite obtained under acoustomagnetic coupling field is greatly increased, which further refines the in-situ composite.

Figure 5.6 shows that although TiB_2 reinforcement particles are obtained under continuous action of acoustomagnetic coupling field, nano-TiB_2 particles still exist in the form of submicron-size TiB_2 particles clusters. In order to further effectively disperse the TiB_2 particles under acoustomagnetic coupling field, a reasonable match between the size and the distribution of TiB_2 particles is sought, and the action mode of acoustomagnetic coupling field is changed from continuous to intermittent. Specifically, once in every minute, each time lasts for 30 s with a total of six times. The ultrasonic parameters: ultrasonic power 0.8 kW, and magnetic field parameters: excitation current 200 A, frequency 5 Hz. Figure 5.7 shows the SEM image of TiB_2/7055Al composite synthesized using K_2TiF_6 + KBF_4 powders as 20% of the melt and with intermittent acoustic and magnetic coupling fields. It can be seen from the figure that under the intermittent action of acoustomagnetic coupling field, the size of TiB_2 particles is the same, all of which are submicron size, and their distribution is more uniform.

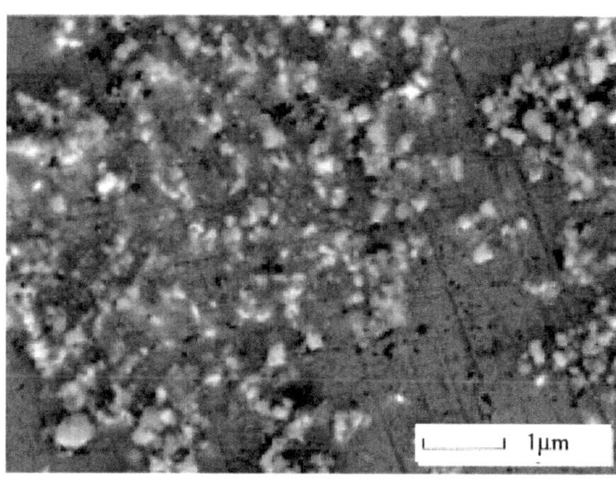

Fig. 5.7 SEM image of TiB_2/7055Al composite obtained under the intermittent action of acoustomagnetic coupling field

5.2.2.2 Mechanical Properties of TiB_2/7055Al Composite Prepared Under Acoustomagnetic Coupling Field with Different Volume Fractions of Reinforcements

Using the process parameters of TiB_2/7055Al composite synthesized under acoustomagnetic coupling field described in the previous section, submicron-nanosize hybrid TiB_2 particles and submicron TiB_2 particles with different volume fractions are obtained. Figure 5.8 shows the mechanical properties of two types of composites in T7 state with different volume fractions of reinforcements. It can be seen from the figure that the tensile strength of TiB_2/7055Al composite gradually increases with volume fraction under the intermittent effect of acoustomagnetic coupling field, and elongation decreases to a certain extent. When the volume fraction is 4%, tensile strength of the composite is 690 MPa with 20.4% increase and the elongation is 7.5% with 21.9% decrease. Under the continuous action of acoustomagnetic coupling field, the tensile strength of TiB_2/7055Al composite first increases with the increase of volume fraction and then decreases, while elongation decreases rapidly. It is especially the case when the volume fraction is 1%, tensile strength of the composite is 607 MPa, 5.9% higher than 7055Al matrix, and elongation is 6.3%, 34.4% lower than 7055Al matrix.

The main reason is when the applied acoustomagnetic coupling field is kept continuously, a large number of nanoparticles are produced in the composite and they are distributed in agglomerated form. When the volume fraction is small, the nanoparticles clusters are small and have a strengthening effect on 7055Al matrix. When

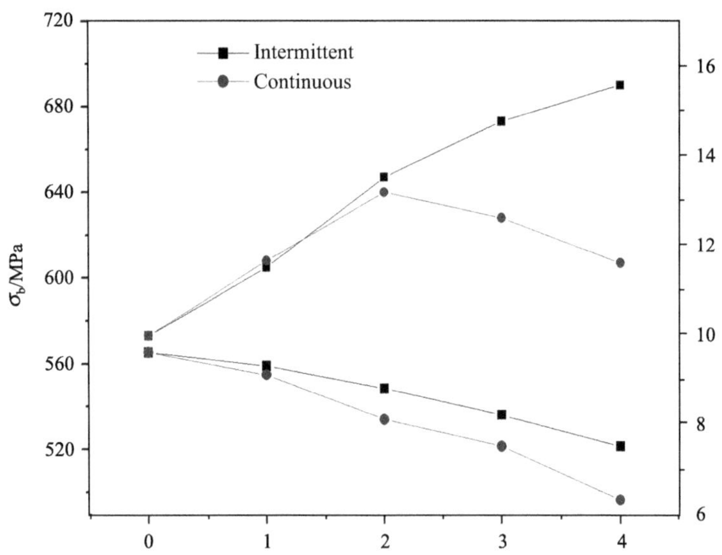

Fig. 5.8 Influence of reinforcement volume fraction on mechanical properties of TiB_2/7055Al composite under coupling fields with different action modes

the volume fraction gradually increases, the nanoparticles clusters also gradually increase which fracture and destroy the matrix, and the strength of the composite decreases, the elongation decreases rapidly, as shown in Fig. 5.8. This result shows that when nanoparticles are not effectively dispersed, their strengthening effect is not good as compared to submicron particles. In addition, existing research shows that the smaller the particle size, the lower the reduction in elongation of the composite, and the larger the size, the better the wear performance of the composite. The compromised particle size can ensure not only the better elongation of composite but also the wear performance of the composite, so the composite has superior comprehensive performance. The submicron size is just an ideal particle size. Therefore, in the situation when nano-TiB_2 particles cannot be effectively dispersed, it is of significant practical importance to control the size of TiB_2 particles at a submicron level.

5.2.3 (Al_2O_3 + ZrB_2)/A356 Composite Prepared by Acoustomagnetic Coupling Field

When only ultrasonic field acts on in-situ melt reaction of A356-K_2ZrB_6-KBF_4-$Na_2B_4O_7$ system, the optimum parameters are [11] the ultrasonic power 1.0 kW, the action time 3 min with intermittent action; and the parameters for the best action of single magnetic field are magnetic field strength 0.1 T with intermittent action. Thus, the optimum parameters of both the individual physical fields are taken, magnetic field is applied to the entire system in an intermittent mode, and the ultrasonic horn is continuously moved in the melt during this process, placing the entire reaction system under the joint action of acoustic-magnetic coupling field to produce (Al_2O_3 + ZrB_2)/A356 composite.

5.2.3.1 Microstructure of (Al_2O_3 + ZrB_2)/A356 Composite Synthesized by acoustomagnetic Coupling Field

Figure 5.9 shows low-magnification SEM images of (Al_2O_3 + ZrB_2)/A356 matrix composites synthesized under individual ultrasonic field and magnetic field, and under the combined action of acoustic-magnetic fields, respectively. It can be seen from Fig. 5.9a that when no physical field is applied, ZrB_2 particles in the prepared composite have a severe agglomeration, and clusters are connected, forming a network structure. In the composite, synthesized under ultrasonic field, the particles' cluster size becomes smaller and the degree of dispersion is significantly improved. When both the physical fields are applied to the reaction system (Fig. 5.9d), the dispersion of particles is more uniform than that under the ultrasonic field alone. In addition, the volume fraction of particles (calculated by ImageJ software) in composite prepared under acoustomagnetic coupling field, reached 12.1% and reaction yield reached 80.7%. Compared with those under ultrasonic field alone and

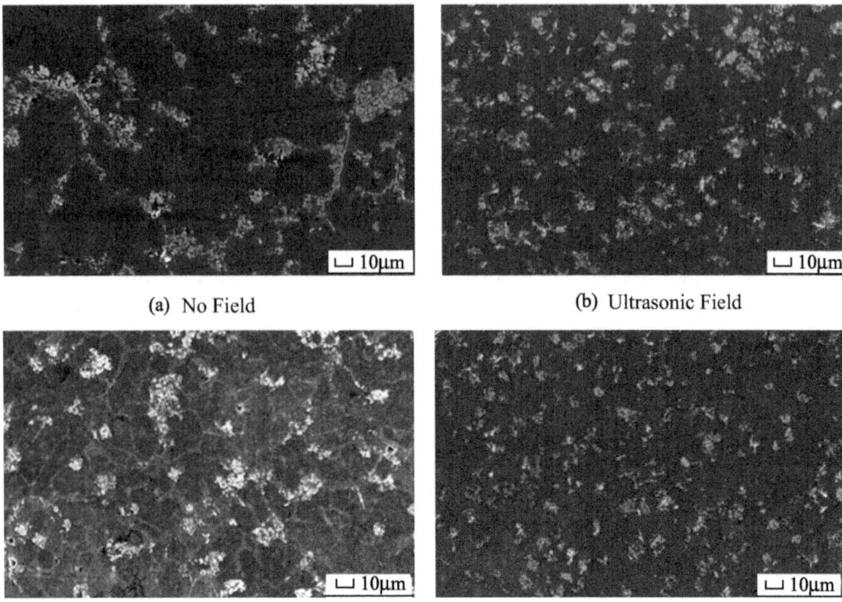

(a) No Field (b) Ultrasonic Field

(c) Magnetic Field (d) Coupled Field

Fig. 5.9 SEM images of $(Al_2O_3 + ZrB_2)$/A356 composites under ultrasonic, magnetic, and acoustomagnetic coupling fields

magnetic field alone, it increased 18.7% and 11.0%, respectively, and increased by 39.1% compared to the situation when no physical field is applied. Therefore, the effect of acoustomagnetic coupling field on the efficiency of in-situ melt reaction is significant.

Figure 5.10 shows high-magnification SEM images of ZrB_2 particles in the composites synthesized under those three physical fields. It can be seen from the figure that under the action of ultrasonic field alone, fine ZrB_2 phase in the composite peels off from its cluster and diffuses in the surrounding matrix. Under high-frequency magnetic field alone, the size of ZrB_2 particles in the composite is reduced, but the degree of dispersion is not significantly improved. However, under the coupling action of ultrasonic and magnetic field (Fig. 5.10d), ZrB_2 particles are more uniformly dispersed in the matrix, and their size is smaller (about tens under nanometers) than that of ultrasonic field alone or magnetic field alone.

It is because in the melt reaction, when ultrasonic and high-frequency magnetic fields are coupled, the coupling method possibly combines two ultrasonic waves: one is the ultrasonic wave generated by powerful ultrasonic field that acts on the melt reaction through the transducer and the horn; the other one is a longitudinal wave formed by periodic vibrations directly generated in conductive metal by local action of high-frequency electromagnetic force, also called the electromagnetic ultrasonic wave. The intensity of electromagnetic ultrasonic waves generated by high-frequency

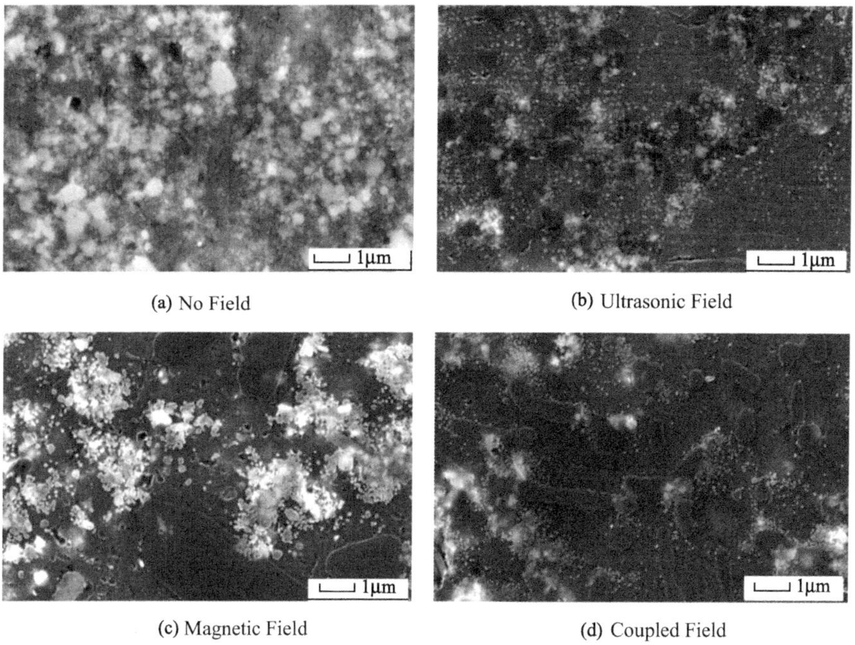

Fig. 5.10 SEM images of ZrB_2 particles in composites under ultrasonic, magnetic, and acoustomagnetic coupling fields

alternating magnetic field alone is low, and the dispersion of agglomerated particles is limited. When ultrasonic and high-frequency magnetic fields are applied together on the melt reaction, the ultrasonic waves and electromagnetic ultrasonic waves increase the strength of single ultrasonic field, so the dispersion of particles in coupled fields is higher than that under one ultrasonic field alone.

5.2.3.2 Microstructure of A356 Matrix of Composite Synthesized Under Acoustomagnetic Coupling Field

Figure 5.11 shows the microstructure of matrix after corrosion of $(Al_2O_3 + ZrB_2)$/A356 matrix composite synthesized under different external fields. Figure 5.11a shows the microstructure of A356 matrix of composite prepared without physical field, the morphology of primary α-Al is rose like, the morphology is coarse and uneven. When the ultrasonic field is applied during synthesis (Fig. 5.11b), α-Al grains in the matrix become fine and rounded, which may be due to remelting of rose-like crystals or fracture of the petals caused by thermal and circulation effects of ultrasounds. Under the action of magnetic field, α-Al grains in the composite are smaller than those in Fig. 5.11b; this is because the electromagnetic stirring effect of magnetic field is conducive to the diffusion of

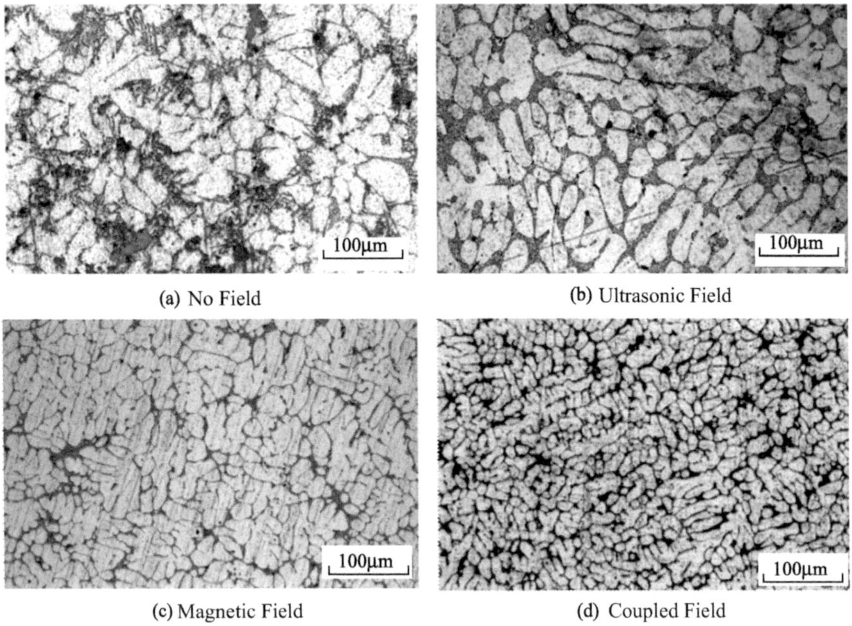

Fig. 5.11 Morphology of primary α-Al in A356 matrix composites under ultrasonic, magnetic, and acoustomagnetic coupling fields

heat in the system which increases the solidification rate, and this effect is greater than the thermal effect caused by ultrasounds, so α-A1 grains under the magnetic field are smaller than those under the ultrasonic field, as shown in Fig. 5.11c. The microstructure of composite synthesized under the combined action of the above two physical fields is shown in Fig. 5.11d, indicating that the primary α-A1 grains are more refined. The morphology and size of the eutectic silicon can no longer be explained by this figure, so SEM images with a larger magnification are used to illustrate the microstructure.

Figure 5.12 shows high-magnification SEM images of Si phase in $(Al_2O_3 + ZrB_2)$/A356 composites prepared under different external fields. In the absence of physical field, the eutectic silicon phase in the composite is needle like with a size of about tens of microns (Fig. 5.12a). Under the action of ultrasonic field alone, the eutectic silicon becomes finer, the size is less than 10 μm, and it has a tendency to continue breaking into short rods (Fig. 5.12b). Under the action of magnetic field alone, the morphology of eutectic silicon phase is similar to that under ultrasounds, and the effect of magnetic field on silicon phase is slightly inferior to the effect of ultrasonic field. In hypoeutectic Al-Si alloy, the eutectic nucleates on primary α-A1 matrix, and therefore local liquid between α-A1 dendrites is Si-rich and nucleates many small Si particles, and β (Al, Si, Fe) or Al phases in the melt can act as nucleation substrate for eutectic silicon. Ultrasonic cavitation can induce pressure pulses in the liquid metal, increasing the wettability of heterogeneous particles

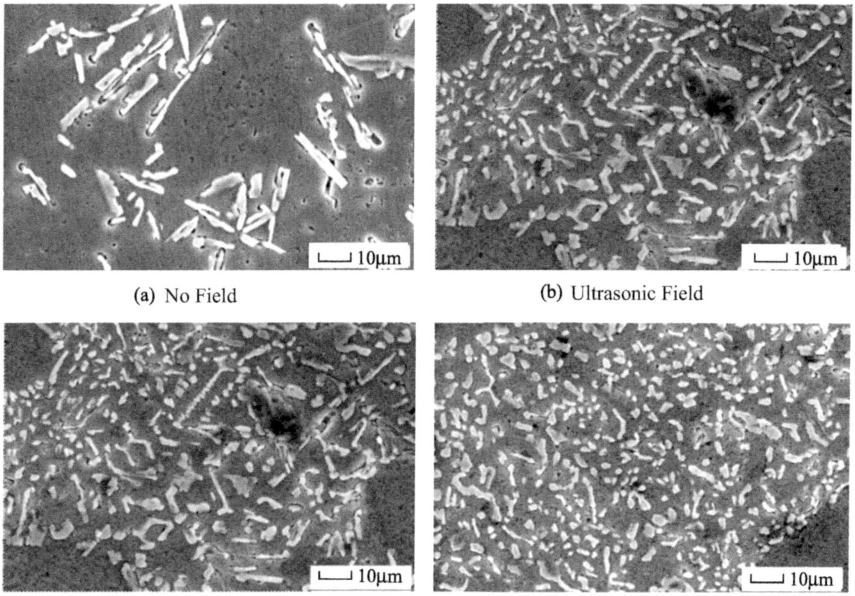

(a) No Field (b) Ultrasonic Field

(c) Magnetic Field (d) Coupled Field

Fig. 5.12 Morphology of eutectic silicon in A356 matrix composites under ultrasonic, magnetic, and acoustomagnetic coupling fields

with matrix, which are all conducive to the nucleation of eutectic silicon. Compared with the magnetic field, the ultrasonic field can increase the probability of eutectic nucleation between dendrite, so under the ultrasonic field, the size of eutectic silicon is correspondingly smaller. The ultrasonic field can cause melt pressure pulsation and increase the melting point of melt, and under the joint action of magnetic field, a very fine eutectic silicon structure is obtained (Fig. 5.12d), showing granular morphology of Si phase with a size of about 1 μm.

5.2.3.3 Mechanical Properties of ($Al_2O_3 + ZrB_2$)/A356 Composite Prepared Under Acoustomagnetic Coupling Field

Figure 5.13 shows tensile properties of 15vol.%($Al_2O_3 + ZrB_2$)/A356 matrix composite at room temperature prepared from A356-K_2ZrF_6-KBF_4-$Na_2B_4O_7$ system at an initial reaction temperature of 850°C, under individual ultrasonic filed (1.0 kW, 3 min, intermittent action mode), individual high-frequency magnetic field (0.1 T, 8 min, intermittent action mode) and under the combined action of these two fields in 20 min, respectively. It can be seen from Fig. 5.13 that when ultrasonic and magnetic fields are applied together to the melt reaction system, the prepared as-cast ($Al_2O_3 + ZrB_2$)/A356 composite has a tensile strength (σ_b) as high as 302 MPa, and the elongation (δ) is as high as 9.1%. Compared with the optimal performance

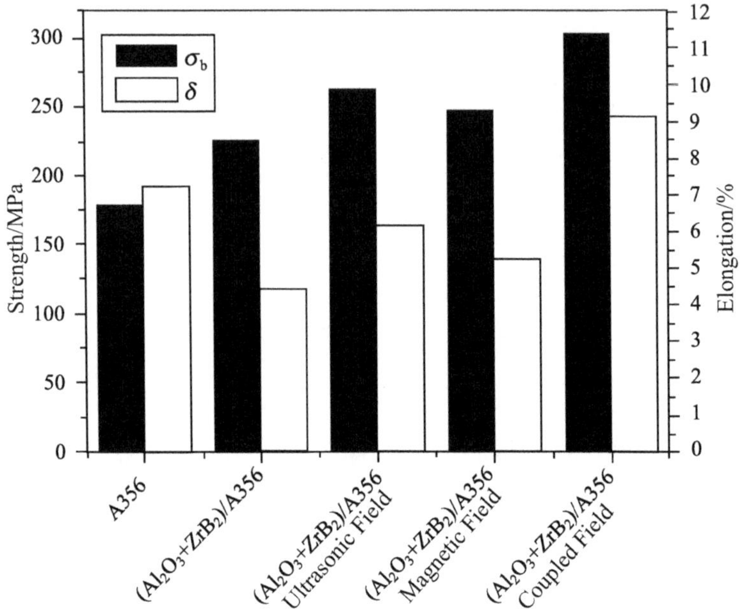

Fig. 5.13 Comparison of tensile properties of composites synthesized under different conditions

of composite under ultrasonic, it increases by 14.8% and 48.5%, respectively; and compared with the properties of composite when no physical field is applied, the tensile strength increases by 34.22%, and elongation is doubled. In addition, due to the agglomeration of ZrB_2 particles, the fracture elongation of synthesized composite under individual fields is lower than that of as-cast A356 alloy. However, under the action of acoustomagnetic coupling field, the nanometer size ZrB_2 particles peel off from the agglomerates and are well dispersed in the matrix, the primary α-Al grains in the matrix are refined, and the eutectic silicon changes from needle like to granular. All these factors cause the fracture elongation of composite to be higher than that of A356 alloy.

5.2.3.4 Tensile Fracture Morphology of $(Al_2O_3 + ZrB_2)$/A356 Composite

The fracture process of materials includes two stages of crack initiation and crack propagation. The fracture characteristics of particle-reinforced aluminum matrix composites are usually related to the microstructure of matrix alloy, the morphology, size, distribution of particles, and the bonding strength of matrix alloy. Figure 5.14 shows the tensile fracture morphology of 15vol.%$(Al_2O_3 + ZrB_2)$/A356 matrix composite synthesized from A356-K_2ZrF_6-KBF_4-$Na_2B_4O_7$ system at an initial

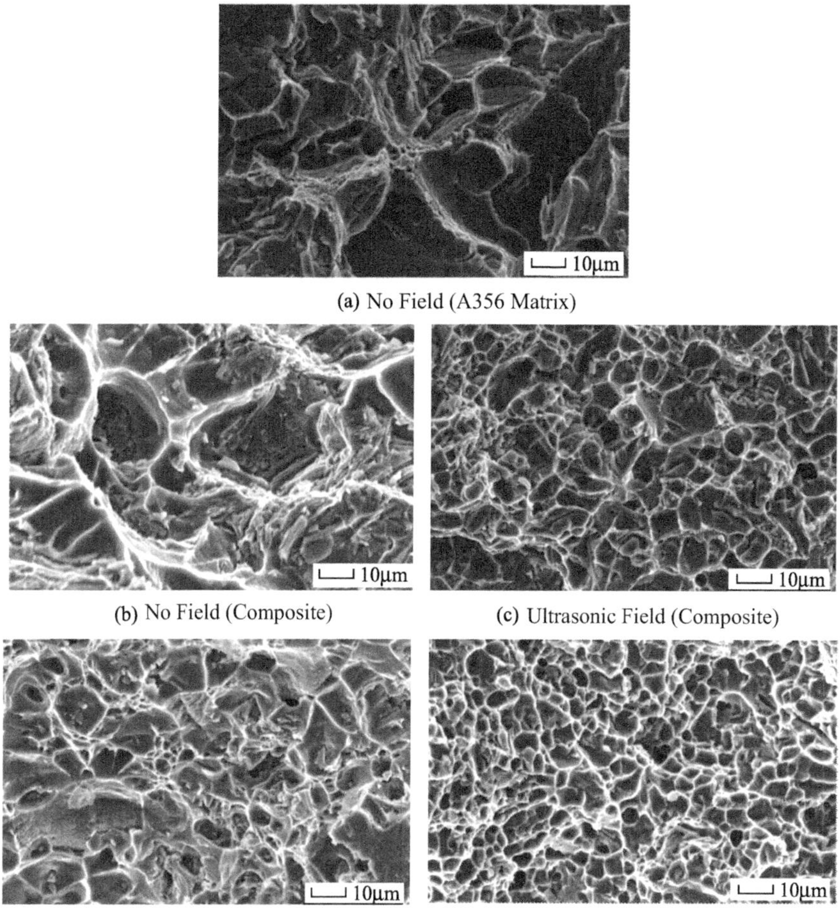

(a) No Field (A356 Matrix)

(b) No Field (Composite)

(c) Ultrasonic Field (Composite)

(d) Magnetic Field (Composite)

(e) Coupled Field (Composite)

Fig. 5.14 Tensile sample fracture morphology of A356 alloy and $(Al_2O_3 + ZrB_2)$/A356 matrix composites synthesized under different external fields

reaction temperature of 850°C, under individual ultrasonic field, individual high-frequency magnetic field, and under the combined action of these two fields for 20 min, respectively. It can be seen from Fig. 5.14a that there is a large area of brittle fracture on the surface with dimples and tear ridges locally, the fracture mechanism is brittle-plastic fracture, and the elongation is 7.2%. Figure 5.14b corresponds to the fracture morphology of composite with 15vol.% reinforcement. The number of dimples on the fracture surface is reduced, there are more particles and brittle flat areas with some tear ridges on the surface, some of the dimples are deep and large, some fine particles are internally distributed, the phenomenon of particle agglomeration is serious. So the elongation is significantly reduced from 7.2 to 4.4%. With a decrease of 38.9%, and the fracture mechanism is brittle-ductile mixed mode. When the ultrasonic power of 1.0 kW is applied, the brittle flat areas of composite decrease, the number of dimples increases, the morphology of dimples becomes smaller, and the fracture characteristics are weakened, showing more obvious dimple-type fracture characteristics (Fig. 5.14c). Figure 5.14d shows the tensile fracture morphology of composite synthesized under high-frequency magnetic field. The brittle flat areas of composite are decreased while the dimples are increased, their size becomes smaller and shallower, showing more obvious dimple-type fracture characteristics. Although the application of high-frequency magnetic field improves the performance of composite to a certain extent demonstrated by an increase in the number of dimples and a decrease in the size of the dimples, the phenomenon of particles agglomeration in dimples is still serious. When both the physical fields jointly act on the reaction system in their best mode, the brittle flat zone on fracture is disappeared, the number of dimples increases to a large extent, the size of dimples is small with uniform morphology, showing the characteristics of dimple-type fracture, and the elongation is 9.1% (Fig. 5.14e).

5.3 Mechanism of Acoustomagnetic Coupled Synthesis of Aluminum Matrix Composite

5.3.1 Flow of Molten Aluminum in Ultrasonic Field

When ultrasonic waves propagate in the melt, they produce finite amplitude attenuation such that a certain sound pressure gradient is formed in the liquid from the sound source, resulting in the flow of melt. When the sound pressure amplitude exceeds a certain value, a jet of fluid is generated in the liquid, which is the sound flow. Therefore, while investigating the motion of melt under the ultrasonic field, the fluid motion caused by the acoustic current is mainly considered [12].

In the aluminum melt, according to the general Navier–Stokes (N-S) equations:

$$\frac{\partial \rho}{\partial t} + \nabla \cdot (\rho u) = 0 \tag{5.1}$$

$$\rho \frac{\partial u}{\partial t} + \rho(u \cdot \nabla u) = -\nabla P + \left(\zeta + \frac{n}{3}\right)\nabla\nabla \cdot u + \eta \nabla^2 u \tag{5.2}$$

Assuming that the fluid is under a positive pressure, that is, $P = P(\rho)$, then

$$\nabla P = C^2 \nabla \rho \tag{5.3}$$

where $C = C(\rho)$ is the local speed of sound. By expanding the fluid field (omit the perturbation parameters), we have

$$\rho=\rho_0+\rho_1+\rho_2+\cdots \tag{5.4}$$

$$u=u_1+u_2+\cdots \tag{5.5}$$

where ρ_0 is the density of melt at rest, which is a constant. When there is no ultrasonic effect, the melt is in a static state, so $u_0=0$, u_1 is the first-order quantity of sound field, ρ_2 and u_2 are the second-order quantities of sound field. Substituting these quantities into the equations and making the quantities of each order equal, we have

$$\frac{\partial\rho_1}{\partial t}+\rho_0\nabla\cdot u_1=0 \tag{5.6}$$

$$\rho_0\frac{\partial u_1}{\partial t}=-C_0^2\nabla\rho_1+\left(\frac{4}{3}\eta+\zeta\right)\nabla\nabla\cdot \mathrm{u}_1-\eta\nabla\times\nabla\times \mathrm{u}_1 \tag{5.7}$$

$$\frac{\partial\rho_2}{\partial t}+\rho_0\nabla\cdot u_2+\nabla\cdot(\rho_1u_1)=0 \tag{5.8}$$

$$\begin{aligned}&\rho_0\frac{\partial u_2}{\partial t}+\frac{\partial}{\partial t}(\rho_1u_1)+\rho_0[(u_1\cdot\nabla)u_1+u_1\nabla\cdot u_1]\\&=-C_0^2\nabla\rho_2-C_0\left(\frac{\mathrm{d}C}{\mathrm{d}\rho}\right)_0\nabla\rho_1^2+\left(\frac{4}{3}\eta+\zeta\right)\nabla\nabla\cdot\mu_2-\eta\nabla\times\nabla\times\mu_2\end{aligned} \tag{5.9}$$

where $\left(\frac{\mathrm{d}C}{\mathrm{d}\rho}\right)_0$ is the equilibrium value of local sound velocity gradient with density. Suppose further that the first-order field is not circular, then the second-order field satisfies the following equations.

$$\frac{\partial\rho_2'}{\partial t}+\rho_0\nabla\cdot\mu_2=\frac{1}{\rho_0C_0^2}+\left(\frac{4}{3}\eta+\zeta\right)\mu_1\cdot\nabla\frac{\partial\rho_1}{\partial t} \tag{5.10}$$

$$\begin{aligned}\rho_0\frac{\partial\mu_2}{\partial t}=&\ C_0^2\nabla\rho_2'-\nabla(\rho_0\mu_1^2)+\left(\frac{4}{3}\eta+\zeta\right)\nabla\nabla\cdot\mu_2\\&-C_0\left(\frac{\mathrm{d}C}{\mathrm{d}\rho}\right)_0\nabla\rho_1^2-\eta\nabla\times\nabla\times u_2-\left(\frac{4}{3}\eta+\zeta\right)\frac{\rho_1}{\rho_0}\nabla\nabla\cdot u_1\end{aligned} \tag{5.11}$$

After eliminating ρ_2', we have

$$\begin{aligned}&\rho_0\frac{\partial u_2}{\partial t^2}-\rho_0C_0^2\nabla\times\nabla u_2-\rho_0C_0^2\nabla\times\nabla\times\nabla u_2\\&-\left(\frac{4}{3}\eta+\zeta\right)\frac{\partial}{\partial t}(\nabla\nabla\cdot u_2)+\eta\frac{\partial}{\partial t}(\nabla\times\nabla\times u_2)\\&=-\frac{\partial}{\partial t}\left\{\nabla\left[\rho_0\mu_1^2+C_0\left(\frac{\mathrm{d}C}{\mathrm{d}\rho}\right)_0\rho_1^2\right]\right\}\end{aligned}$$

$$-\frac{1}{\rho_0}\left(\frac{4}{3}\eta+\zeta\right)\left\{\frac{\partial}{\partial t}(\rho_1\nabla\nabla\cdot u_1)+\nabla\left[u_1\cdot\nabla\left(\frac{\partial\rho_1}{\partial t}\right)\right]\right\} \tag{5.12}$$

If defining

$$R_i=\nabla\times u_i,\ i=1,2 \tag{5.13}$$

and performing curl operation on Eq. (5.12), the equation of R_2 can be obtained.

$$\frac{\partial R_2}{\partial t}-\frac{\eta}{\rho_0}\nabla^2 R_2=\frac{1}{\rho_0^3}\left(\frac{4}{3}\eta+\frac{\zeta}{\eta}\right)\nabla\rho_1\times\nabla\left(\frac{\partial\rho_1}{\partial t}\right) \tag{5.14}$$

Steady-state vortex satisfies the equation below.

$$\nabla^2 R_2=\frac{1}{\rho_0^2}\left(\frac{4}{3}+\frac{\zeta}{\eta}\right)\nabla\rho_1\times\nabla\left(\frac{\partial\rho_1}{\partial t}\right) \tag{5.15}$$

Assuming that the radius of ultrasonic horn is r_1, the radius of crucible is r_0, the wall of crucible is rigid, and ignoring the reflection of sound waves at the crucible bottom-melt interface, the pressure change of first-order sound field can be expressed as a traveling wave propagation along the axis of crucible [13]:

$$P_1=C_0^2\rho_1=P(r)\sin(kz-\omega t) \tag{5.16}$$

Substituting Eq. (5.16) into Eq. (5.15), we can get

$$\nabla^2 R_2=K\frac{\mathrm{d}P^2(r)}{\mathrm{d}r}(-i\sin\varphi+j\cos\varphi) \tag{5.17}$$

where

$$K=\frac{k^2}{2\rho_0^2C_0^3}\left(\frac{4}{3}+\frac{\zeta}{\eta}\right) \tag{5.18}$$

where φ is the polar angle on the crucible section; and i and j are the unit vectors along x and y axis on the section, respectively. For Eq. (5.17), the specific solution can be expressed as

$$R_2=f(r)(-i\sin\varphi+j\cos\varphi) \tag{5.19}$$

where $f(r)$ satisfies the equation:

$$\left[\frac{1}{r}\cdot\frac{\mathrm{d}}{\mathrm{dr}}\left(\mathrm{r}\frac{\mathrm{d}}{\mathrm{dr}}-\frac{1}{\mathrm{r}^2}\right)\right]\mathrm{f(r)}=\mathrm{K}\frac{\mathrm{dP}^2}{\mathrm{dr}} \tag{5.20}$$

Integral transformation of Eq. (5.20) is

$$f(r) = \frac{K}{r}\int_0^r rP^2\mathrm{d}r + 2Nr + \frac{M}{r} \tag{5.21}$$

where N and M are the integral constants. In order to make $f(r)$ continuous at $r = 0$, set $M = 0$. Since the flow of molten aluminum in the crucible only has an axial velocity, that is, $u_{2r} = u_{2\varphi} = 0$, $u_{2z} = g(r)$, we have

$$R_2 = \nabla \times u_2 = \frac{\mathrm{d}g(r)}{\mathrm{d}r}(i \sin\varphi - j \cos\varphi) \tag{5.22}$$

It can be concluded from Eq. (5.19) that,

$$\frac{\mathrm{d}g(r)}{\mathrm{d}r} = -f(r) \tag{5.23}$$

When $r = r_0$, that is, at the crucible wall $u_{2z} = 0$, we have

$$g(r_0) = 0 \tag{5.24}$$

Then the integral of Eq. (5.23) is

$$\begin{aligned} g(r) = \int_r^{r_0} f(t)\mathrm{d}t &= K\int_r^{r_0}\int_0^t \frac{x}{t}P^2(x)\mathrm{d}x\mathrm{d}t + \beta\left(r_0^2 - r^2\right) \\ &= KW(r) + \beta(r)(r_0^2 - r^2) \end{aligned} \tag{5.25}$$

where $W(r)$ is the following double integral.

$$W(r) = \int_r^{r_0}\int_0^t \frac{x}{t}P^2(x)\mathrm{d}x\mathrm{d}t \tag{5.26}$$

Exchanging the order of points, we have

$$\begin{aligned} W(r) = &\int_0^r xP^2(x)\int_r^{r_0}\frac{1}{t}\mathrm{d}t\mathrm{d}x + \int_r^{r_0} xP^2(x)\int_x^{r_0}\mathrm{d}t\mathrm{d}x \\ &\times \int_0^r xP^2(x)\ln\frac{r_0}{r}\mathrm{d}x + \int_r^{r_0} xP^2(x)\ln\frac{r_0}{r}\mathrm{d}x \end{aligned} \tag{5.27}$$

If defined

$$\Gamma(x, r) = \begin{cases} x\ln\dfrac{r_0}{r}, 0 < x < r \\ x\ln\dfrac{r_0}{r}, r < x < r_0 \end{cases} \tag{5.28}$$

Then $W(r)$ can be expressed as

$$W(r) = \int_0^{r_0} \Gamma(x, r)P^2(x)\mathrm{d}x \tag{5.29}$$

Then

$$g(r)=\mathrm{K}\int_0^{\mathrm{r}_0} \Gamma^2(\mathrm{x}, \mathrm{r})\mathrm{P}^2(\mathrm{x})\mathrm{dx}+\beta(\mathrm{r})\left(\mathrm{r}_0^2-\mathrm{r}^2\right) \tag{5.30}$$

Total amount of molten aluminum in the crucible is constant, so the net flow through any section of the crucible must be zero, that is

$$\int_0^{r_0} rg(r)\mathrm{d}r = 0 \tag{5.31}$$

Substituting Eq. (5.30) into Eq. (5.31), and exchanging the order of integration, we can finally have

$$\beta(r)=\frac{\mathrm{K}}{\mathrm{r}_0^4}\int_0^{r_0}\mathrm{P}^2(\mathrm{x})\left(\mathrm{x}^3-\mathrm{x}\mathrm{r}_0^2\right)\mathrm{d}_\mathrm{x} \tag{5.32}$$

Assuming that the first-order sound field is a collimated beam,

$$P=\begin{cases} P_0,\ r\leq \mathrm{r}_1 \\ 0,\ r>\mathrm{r}_1 \end{cases} \tag{5.33}$$

and substituting it into $W(r)$, $\beta(r)$ can be obtained.

$$W(r) = P_0^2\left(\int_0^r \ln\frac{r_0}{r}x\mathrm{d}x + \int_r^{r_1} x\ln\frac{r_0}{x}\mathrm{d}x\right) = \frac{P_0^2 r_1^2}{2}\left[\frac{1}{2}\left(1-\frac{r}{r_1}\right)+\ln\frac{r_0}{r_1}\right] \tag{5.34}$$

$$\beta(r) = \frac{KP_0^2}{r_0^4}\left(x^3 - xr_0^2\right)\mathrm{d}x = \frac{KP_0^2}{2}\left(\frac{r_1}{r_0}\right)^2\left[\frac{1}{2}\left(\frac{r_1}{r_0}\right)^2 - 1\right], r\leq r_1 \tag{5.35}$$

If $r > r_1$, then

$$W(r) = \int_0^{r_1} P_0^2\ln\frac{r_0}{r}x\mathrm{d}x = \frac{P_0^2 r_0^2}{2}\ln\frac{r_0}{r} \tag{5.36}$$

Let $x = \frac{r}{r_0}$, $y = \frac{r_1}{r_0}$, we have

$$\begin{cases} \frac{1}{2}\left(1-\frac{x^2}{y^2}\right)-(1-x^2)\left(1-\frac{1}{2}y^2\right)-\ln y, \ 0<x\leq y \\ g(r)=G(r_1)-\left(1-\frac{1}{2}y^2\right)\left(1-x^2\right)-\ln x, \ y<x<1 \end{cases} \tag{5.37}$$

where

$$G(r_1) = \frac{1}{2}K P_0^2 r_1^2 \tag{5.38}$$

The theoretical diagram of acoustic flow calculated according to Eq. (5.38) is shown in Fig. 5.15. It can be seen from the figure that the acoustic flow is downward along the crucible axis under the horn, and after being reflected by the surface of crucible, it moves upward, thereby forming a circulation in the entire melt. The circular characteristics of acoustic current accelerate solute exchange within the entire melt and inhibit the growth of particles after nucleation. In addition, acoustic flow within the molten aluminum is turbulent, which has a certain stirring effect on the particles and can disperse the originally loose particle clusters. Acoustic flow has both macroscopic and microscopic stirring effects on the entire melt and has a certain impeding effect on the agglomeration of reinforcement particles [14].

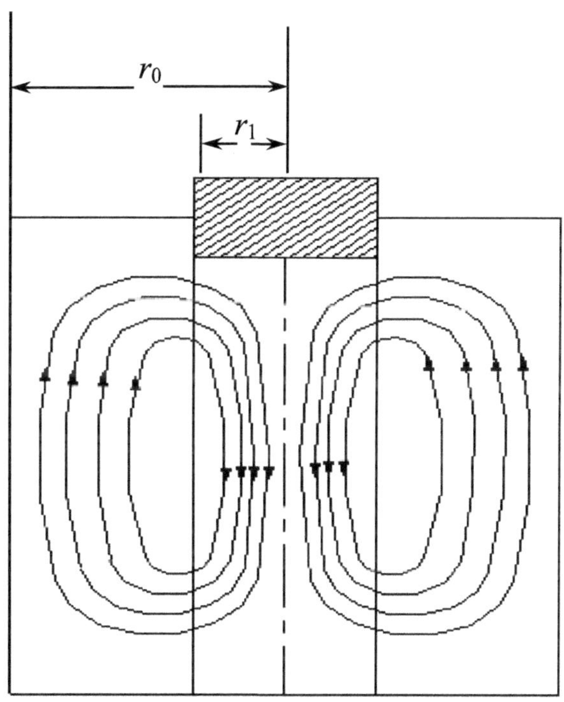

Fig. 5.15 Theoretical diagram of acoustic flow

5.3.2 *Flow Field Analysis in Electromagnetic Stirring Process*

Symbols of physical quantities (SI units) used in this section and their meanings are listed below:

H is the magnetic field strength (vector quantity), A/m;

B is the magnetic flux density, or magnetic induction intensity (vector quantity), T;

E is the electric field strength, or induced electromotive force (vector quantity), V;

D is the electric displacement (vector quantity);

J is the induced current, conductive current density (vector quantity), A;

V is the fluid velocity, m/s;

t is the time, s;

μ is the permeability, and the value for molten aluminum is approximately $4\pi \times 10^{-7}$ H/m;

σ is the conductivity, S/m;

η is the kinematic viscosity of fluid, Pa·s;

ρ is the fluid density, kg/m^3;

ω is the angular frequency of alternating current, and its relationship with current frequency f (unit: Hz) is $\omega = \frac{2\pi}{f}$.

The controlling equations of electromagnetic stirring effect include Maxwell's equations, Ohm's law of electric current, the equation of continuity, and the equation of motion. In order to simplify the analysis, the following assumptions are made for the electromagnetic stirring process in molten aluminum: ① Liquid aluminum is an incompressible conductive fluid; ② the density ρ, kinematic viscosity v, conductivity σ and other physical parameters of molten aluminum are all constants; ③ the permeability of molten aluminum is taken as the permeability of vacuum ($4\pi \times 10^{-7}$H/m); ④ the flow rate of molten aluminum is very small, so Coulomb force of electric field on the charge is ignored; ⑤ since the frequency of electromagnetic stirring is very low, the displacement current is ignored.

(1) Electromagnetic field governing equation

Maxwell's equations comprehensively summarize the basic laws of electromagnetic fields, which are the basic equations applicable to macroscopic electromagnetic fields, and describe the relationship between vector electromagnetic field, the charge, and the current. The differential form of this set of equations is as follows [15, 16, 17]:

$$J = \nabla \times H \text{ (Ampere's circuiter l theorem)} \tag{5.39}$$

$$\nabla \times E = -\frac{\partial B}{\partial t} \text{ (Faraday's law of electromagnetic induction)} \tag{5.40}$$

$$\nabla \times B = 0 \text{ (Flux continuity equation)} \tag{5.41}$$

$$B = \mu \cdot H \text{ (Constitutive equation)} \tag{5.42}$$

Ohm's law for the moving fluid is

$$J = \sigma(E + \nabla \times B) \text{ (Ampere's circuiter l theorem)} \tag{5.43}$$

The conductive fluid is driven by electromagnetic field, and velocity field, electric field, and magnetic field emerge at the same time and affect each other. It is generally believed that the electromagnetic field affects the velocity field; in fact, the velocity field has a certain effect on electromagnetic field. However, considering the magnetic field Reynolds number $R_m < 1$, while examining the melt velocity field and electromagnetic field distribution characteristics in the in-situ synthesis process, the effect of the velocity field on electromagnetic field is ignored:

$$\nabla \times B = 0 \tag{5.44}$$

Combining Eqs. (5.39)–(5.44), we have

$$\nabla^2 B = -\mu\sigma \frac{\partial B}{\partial t} \tag{5.45}$$

By solving the equation, the current density J and the magnetic flux density B are obtained, so that the electromagnetic volume force can be obtained according to the following equation:

$$F = J \times B \tag{5.46}$$

The core stirring force is the volume force that drives the flow, that is, the Lorentz force. By adding solved magnetic force F as a source term to the fluid motion equation, the velocity of melt flow can be obtained, and then the flow field of melt can be obtained.

(2) Fluid motion governing equation

The continuity equation (i.e., mass conservation equation) of incompressible fluid is

$$\nabla \times V = 0 \tag{5.47}$$

For Newtonian fluids, the momentum conservation equation is

$$\rho \frac{\mathrm{d}V}{\mathrm{d}t} = \rho F + \nabla \cdot P \tag{5.48}$$

The velocity derivative of a particle in the fluid and the local derivative of velocity have the following relationship:

$$\frac{\mathrm{d}V}{\mathrm{d}t} = \frac{\partial V}{\partial t} + (V \cdot \nabla)V \tag{5.49}$$

Combining Eqs. (5.46)–(5.49), we get the equation of motion of Newtonian fluid under the action of magnetic field as:

$$\rho\frac{\partial V}{\partial t} + \rho(V \cdot \nabla)V = -\nabla P + \eta\nabla^2 V + J \times B + \rho g \tag{5.50}$$

Ren of Northeastern University [18] used the above-mentioned electromagnetic field equations and fluid motion equations to establish related model and calculated the distribution of molten steel flow field in continuous casting mold with rotating electromagnetic field through ANSYS software, as shown in Fig. 5.16. He found that the flow field can be divided into two types and three areas: ① main flow area: the circumferential rotating flow generated by the action of tangential magnetic field force F_θ; ② secondary flow zone: divided into upper circulation zone and lower circulation zone, the combined action of radial electromagnetic force F_r and the gradient of electromagnetic force from top to bottom produces a circulation from the center to the edge of the melt, and the circulation direction of lower zone is opposite to that of upper zone.

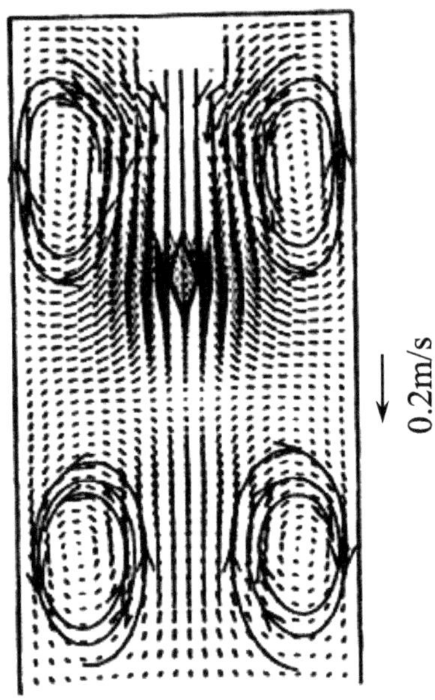

Fig. 5.16 Flow field distribution in the longitudinal section of molten steel in a continuous casting mold with rotating electromagnetic field [18]

5.3.3 Analysis of the Coupling Effect of Ultrasonic Field and Magnetic Field

5.3.3.1 The Coupling Equation of Ultrasonic Field and Magnetic Field

When ultrasounds propagate in molten aluminum, their pressure P can be expressed as

$$\nabla\left(\frac{1}{\rho}\right)\nabla P - \frac{1}{\rho c^2}\cdot\frac{\partial^2 p}{\partial t^2} = 0 \tag{5.51}$$

where c is the propagation speed of ultrasounds in molten aluminum; and ρ is the density of molten aluminum.

Taking the second derivative of Eq. (5.16), we have

$$\frac{\partial^2 P}{\partial t^2} = -\omega^2 P(r)\sin(kz - \omega t) = -\omega^2 P \tag{5.52}$$

Assuming that molten aluminum is an incompressible ideal fluid and substituting Eq. (5.52) into Eq. (5.51), we have

$$\nabla^2 P = -\frac{1}{c^2}\omega^2 P \tag{5.53}$$

Taking the divergence of both sides of Eq. (5.50), omitting the higher-order terms and the influence of fluid gravity field, we have

$$\rho\nabla\cdot\frac{\partial V}{\partial t} + \rho\nabla(V\cdot\nabla)V = -\nabla^2 P + \nabla\cdot(J\times B) \tag{5.54}$$

Substituting Eq. (5.53) into Eq. (5.54), we have

$$\rho\nabla\cdot\frac{\partial V}{\partial t} + \rho\nabla(V\cdot\nabla)V = \frac{1}{c^2}\omega^2 P + \nabla\cdot(J\times B) \tag{5.55}$$

Equation (5.55) is the coupled equation for ultrasonic field and magnetic field, the first term on the right side shows the effect of ultrasonic field on the fluid, and the second term on the right side is the effect of magnetic field shows the fluid.

It is generally believed that the ultrasonic transducer can convert 20% of ultrasonic power W into ultrasonic oscillations, so the radiated sound pressure of ultrasonic horn can be expressed as

$$P_A = \sqrt{\frac{0.2W}{\pi r_A^2}(2\rho_A c_A)} \tag{5.56}$$

where r_A, ρ_A, c_A are the radius, the density of the ultrasonic horn, and the propagation speed of ultrasonic waves in the horn, respectively. Therefore, according to Snell's law, the sound pressure P_0 at the ultrasonic horn and the melt interface can be expressed as

$$P_0 = \frac{2\rho c}{\rho c + \rho_A c_A} P_A \tag{5.57}$$

When solving Eq. (5.55), the boundary conditions for ultrasonic field term of the equation are: ① at liquid–gas boundary, $P = 0$; ② at the ultrasonic horn-melt interface, $P = P_0$.

5.3.3.2 Analysis of Melt Movement in Acoustomagnetic Coupling Field

Figure 5.17 shows a schematic diagram of forced convective movement of aluminum melt under the acoustomagnetic coupling field. It can be seen from the figure that the acoustic current generated by ultrasonic field mainly moves along the axis of the horn, while the melt agitation generated by low-frequency rotating magnetic field is mainly along the tangential direction of the horn. The two fields reinforce each other, so that the entire melt is vigorously agitated. This strong stirring effect has two functions: on the one hand, it accelerates mass and heat transfer processes, reduces the solute concentration near the bell jar, and inhibits the growth of reinforcement particles; on the other hand, it effectively refines the particles by expanding the scope of ultrasounds.

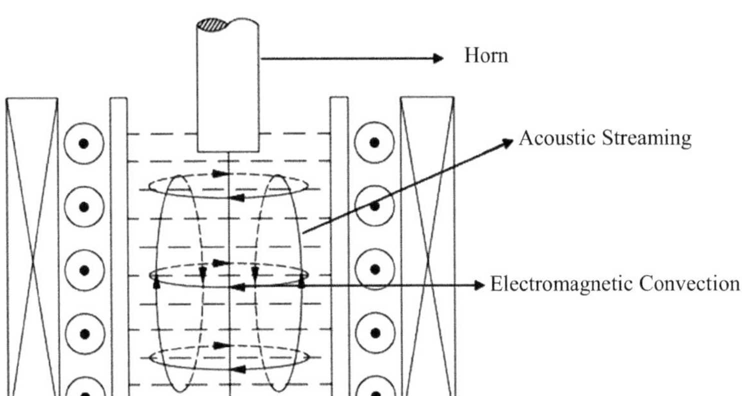

Fig. 5.17 Schematic diagram of forced convection of melt caused by acoustomagnetic coupling field

5.3.3.3 Synergy Mechanism During Coupling

In the conventional synthesis process, the reactant particles are pressed into the molten aluminum by bell jar, and their conditions for entering into the molten aluminum are poor. The electromagnetic force generated during low-frequency electromagnetic stirring is mainly demonstrated as the shear force, which pushes the melt to make circular motion in the crucible with constant angular velocity. Reactant particles are forced into the liquid aluminum and evenly distributed, which increases the extent of in-situ reaction, reduces the number of losses, and increases the number of reinforcement particles in the melt. The vortex phenomenon is observed on the surface, the core of vortex is formed in the center, the melt outside the vortex core performs free vortex motion, and rest of the melt is the combination of forced vortex and free vortex region. The presence of forced vortex current makes the reactant materials enter the melt and the pressure difference is always greater than zero, which is the driving force for particles to enter the melt. When the electromagnetic force increases, the tangential velocity of eddy current boundary increases, the pressure difference also increases, and the particles enter the melt more easily. From this perspective, the greater the electromagnetic force, the faster the tangential speed, and more favorable the conditions for reactant particles to enter the molten aluminum [19].

When the same amount of reactant is added, the volume fraction of corresponding reinforcement phase particles will increase. When reactant particles enter the melt, there will be two processes, i.e., macro-mixing and micromixing. Macro-mixing is the crushing process which represents the shearing force and makes the reactant powder to be separated in granular form, preventing local sintering under high-temperature conditions; micromixing is the diffusion process, which can eliminate the concentration gradient between adjacent regions of the mixed melt, due to the internal convection of melt generated by stirring force, the reactant particles are evenly distributed inside the melt, increasing the contact area with molten aluminum and promoting the melt reaction. The effect of alternating Lorentz force on in-situ chemical reaction is mainly reflected in two aspects, and the second aspect is to improve the emulsification conditions for reactant and liquid aluminum so that more and more reactants enter the melt to participate in the reaction, as shown in Fig. 5.18.

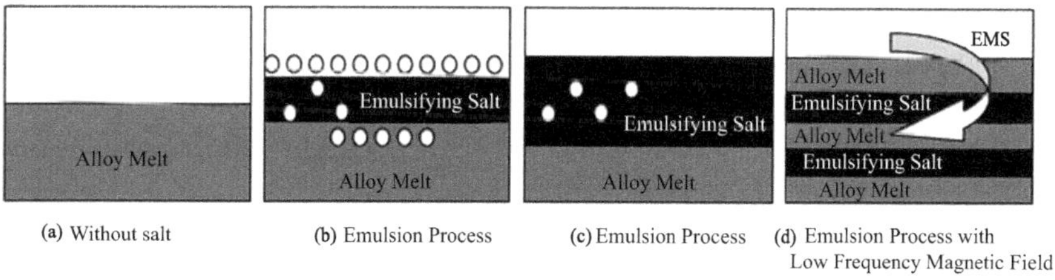

Fig. 5.18 Emulsification process of inorganic reactant salt in the melt before and after applying a magnetic field

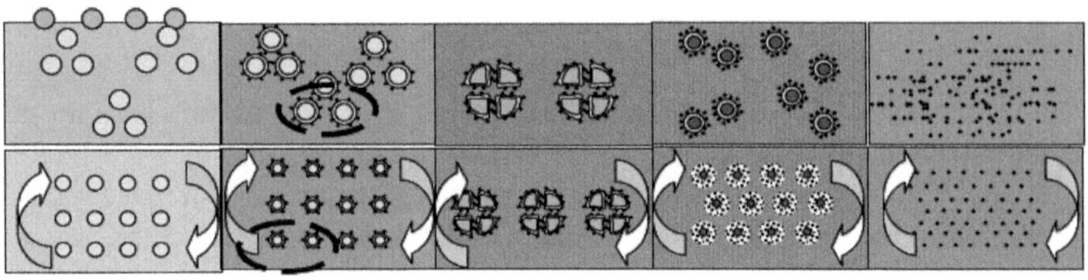

Fig. 5.19 Schematic diagram showing evolution of in-situ chemical reaction and synthesis mechanism under the action of low-frequency magnetic field

The selected reactant salt has a low melting point, and emulsification reaction occurs after its contact with molten aluminum, that is, the powder melting and chemical reaction proceeds simultaneously, and the macroscopic appearance expresses emulsification characteristics. When no magnetic field is applied, the emulsification process depends on the concentration gradient; after applying a magnetic field, the emulsification process is improved by the electromagnetic force, and emulsified layer comes in full contact with liquid aluminum, which is conducive to the full entry of reactants into the melt and increases the reinforcement particles' volume fraction. Figure 5.19 shows the schematic diagram of the evolution of in-situ chemical reaction under Lorentz force in a low-frequency magnetic field and magnetic field action mechanism.

When the ultrasonic field and magnetic field are coupled, their synergistic mechanism is shown in Fig. 5.20. Under the action of ultrasonic field alone, a large number of cavitation bubbles are generated beneath the horn, at this time, the cavitation bubbles are mainly concentrated in the vicinity of the horn and cannot migrate to the depth of the melt. Therefore, the sound pressure amplitude reaches its maximum value near the surface of the horn and decreases with the increase of the distance from the face of the horn. Professor Pan Lei of Nanjing University of Aeronautics and Astronautics [20] established the correlation model which is confirmed by theoretical calculations. Under acoustomagnetic coupling field, cavitation bubbles generated by ultrasounds migrate to the entire melt under the action of magnetic field, thereby reducing the speed at which the sound pressure amplitude decays with the distance from the surface of the horn, ensuring the effect outside of the crucible and the bottom of the melt.

A comprehensive analysis of results and mechanism of acoustomagnetic coupling field suggests that the microstructure of particle-reinforced aluminum matrix composites prepared under ultrasonic filed alone is characterized by small particle size but contains clusters of small particles; the microstructure of particle-reinforced aluminum matrix composites prepared under electromagnetic field alone contains particles of larger size than that under ultrasonic field, the particle dispersion is better than that under ultrasonic field, and the clustering phenomenon is reduced. When the acoustomagnetic coupling field is applied, benefits of both the fields are utilized to overcome shortcoming and optimize the as-cast microstructure of composite. The

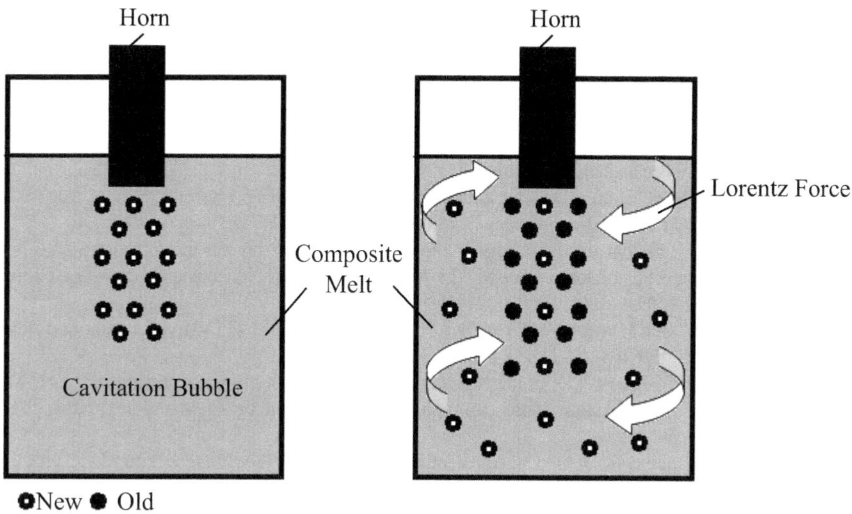

Fig. 5.20 Schematic diagram showing the movement of acoustic cavitation bubbles in the ultrasonic field driven by electromagnetic Lorentz force

size of particles in the microstructure of composite prepared under acoustomagnetic coupling field is smaller and the dispersion is significantly improved with a high volume fraction. The acoustomagnetic coupling field not only combines the advantages of separate fields and overcomes the weaknesses of individual ultrasonic field and magnetic field, but also promotes particles dispersion.

From the perspective of the mechanism of acoustomagnetic coupling field, the sonochemical effect originates from the acoustic cavitation, the formation, growth, and rapid collapse of cavitation bubbles in the liquid. This sudden collapse of bubbles caused by adiabatic compression or shock waves in the gas phase of collapsed bubbles generates local hot spots. These hot spots formed in solution meet the requirement that the thermodynamic crystallization system must meet a certain amount of energy, it increases the rate of crystalline nucleation and thus facilitates the formation of tiny particles. But, the scope of ultrasonic field is limited.

The main effects under electromagnetic field are the force effect and the thermal effect. The force effect is mainly the Lorentz force; it forms eddy currents on the surface, enabling more reactant particles to enter the melt to participate in the reaction, increasing the yield of reactants and the actual volume fraction of particles. At the same time, it accelerates the emulsification of potassium fluorotitanate in molten aluminum and accelerates the in-situ reaction. Simultaneously, the electromagnetic force intensifies the concentration and temperature fluctuations in the melt, thereby promoting the nucleation.

References

1. Chufang X, Kejin R. Electromagnetic fields and electromagnetic waves. Beijing: People's Education Press; 1979. (in Chinese).
2. Chunxiang C. Material synthesis and preparation. Shanghai: East China University of Science and Technology Press; 2010. (in Chinese).
3. Liu R, Li X, Hu S. Simulation of ultrasonic cavitation effect and bubble motion process in aluminium alloy melt. China Foundry Equipment Technol. 2007;6:16–9. (in Chinese)
4. Yao L, Hao H, Gu SW, et al. Effects of electromagnetic stirring on the microstructure and mechanical properties of super light Mg-Li-Al-Zn alloy. Trans Nonferrous Metals Soc China. 2010;20(2):S388–92.
5. Zhang ZT, Li J, Yue HY, et al. Microstructure evolution of A356 alloy under compound field. J Alloy Compd. 2009;484(1–2):458–62.
6. Shao ZW, Le QC, Zhang ZQ, et al. A new method of semi-continuous casting of AZ80 Mg alloy billets by a combination of electromagnetic and ultrasonic fields. Mater Des. 2011;32(8–9):4216–24.
7. Tsunekawa Y, Suzuki H, Genma Y. Application of ultrasonic vibration to in-situ MMC process by electromagnetic melt stirring. Mater Des. 2001;22(6):467–72.
8. Dengbin C. Study on the microstructure and properties of aluminium-matrix in-situ composites synthesized under ultrasound/magnetic field. Zhenjiang: Jiangsu University; 2012. (in Chinese).
9. Zhicheng K, Dayuan M, Carling D. Research on the measurement of aluminium melt viscosity. Hunan Nonferrous Metals. 2001;17(2):18–20 (in Chinese).
10. Xia Q, Li Z, Qiu X, et al. Study on the dispersion mechanism of sodium hexametaphosphate on serpentine. Min. Metall. Eng. 2002;22(2):51–4. (in Chinese)
11. Huajing Y. Research on melt reaction synthesis of (Al_2O_3+ZrB_2)/A356 nanocomposite under acousto-magnetic field. Zhenjiang: Jiangsu University; 2012. (in Chinese).
12. Jun W, Feng C, Baode S. The role of high-energy ultrasound in the preparation of particle-reinforced metal matrix composites. J Shanghai Jiaotong Univ (Chin Ed). 1999;33(7):813–6 (in Chinese).
13. Ruo F. Sonochemistry and its applications. Hefei: Anhui Science and Technology Press; 1990. (in Chinese).
14. Jiang R, Li X, Liu R, et al. Study on fine-grain mechanism and action area of pure aluminium by power ultrasound. Special Casting Nonferrous Alloys. 2008;28(7):560–3. (in Chinese)
15. Moreau R. Magneto hydrodynamics. Dordrecht: Kjuwer Academic Publisher; 1990.
16. Hu W. Magnetohydrodynamics of the universe. Beijing: Science Press; 1987. (in Chinese)
17. Wu Q, Li H. Magnetohydrodynamics. Changsha: National University of Defense Technology Press; 2007. (in Chinese)
18. Bingzhi R. Numerical simulation of the coupling process of electromagnetic field, flow field and temperature field in electromagnetically stirred continuous casting mold. Shenyang: Northeastern University; 2008. (in Chinese).
19. Yamaguchi M, Yamamoto I, Ishikawa F, et al. Thermodynamic theory of magnetic field effects on chemical equilibria and applications to metal-hydrogen systems. J Alloy Compd. 1999;253–254(6):191–4.
20. Pan L, Tao J, Wu S, et al. Preparation and research of two zinc-based composite materials under high-energy ultrasound. Foundry. 2005;54(12):1219–22. (in Chinese)

Chapter 6
Interface Structure of Matrix/in-Situ Reinforcement

6.1 Morphology and Growth Mechanism of In-Situ Al_3zr

6.1.1 TEM Morphology and Crystal Structure of In-Situ Al_3Zr

The TEM morphology and electron diffraction pattern of in-situ Al_3Zr crystal in composites fabricated from Al-$Zr(CO_3)_2$ system with the pulsed magnetic field strength of 0.07 T under different reaction temperatures for 20 min are shown in Figs. 6.1 and 6.2, respectively. It can be seen from Fig. 6.1 that the morphologies of Al_3Zr are mainly granular, bulk-shaped, lath-shaped, and spiculate. It shows that the fabrication process, especially the melt temperature, is closely related to the morphology of Al_3Zr crystal. As shown in Fig. 6.1a, the main morphology of Al_3Zr, fabricated at 850 °C, is granular with a round and smooth surface without sharp edges and corners. There are several facets on the surface of Al_3Zr showing the substrate with crystalline growth. The size of Al_3Zr particle is 0.2–0.4 μm with a clear interface and no impurity between particle and matrix exists. As shown in Fig. 6.1b, the Al_3Zr crystal produced at 900 °C is mainly bulk shaped with the size of 0.2 ~ 0.3 μm and the growth steps are clearly observed. Figure 6.1c shows that at the reaction temperature of 950 °C, the main morphology of Al_3Zr is lath-shaped with a width of 0.1–0.2 μm and an aspect ratio of 4.0. When the reaction temperature of melt reaches 1,000 °C, the Al_3Zr crystal is mainly spiculate or fiber shaped, shown in Fig. 6.1d, with a diameter of about 200 nm [1].

Figure 6.2 shows the electron diffraction pattern of Al_3Zr crystal. Observation of Fig. 6.2 demonstrates that the Al-Zr compounds generated by the reaction of Al-$Zr(CO_3)_2$ system under pulsed magnetic field are Al_3Zr crystals rather than Al_2Zr, Al_3Zr_2, $AlZr_3$, or other structures and the crystalline structure of them belongs to tetragonal (quadrate) system.

Y. Zhao, *In-Situ Synthesis of Aluminum Matrix Composites*,
https://doi.org/10.1007/978-981-16-9120-1_6

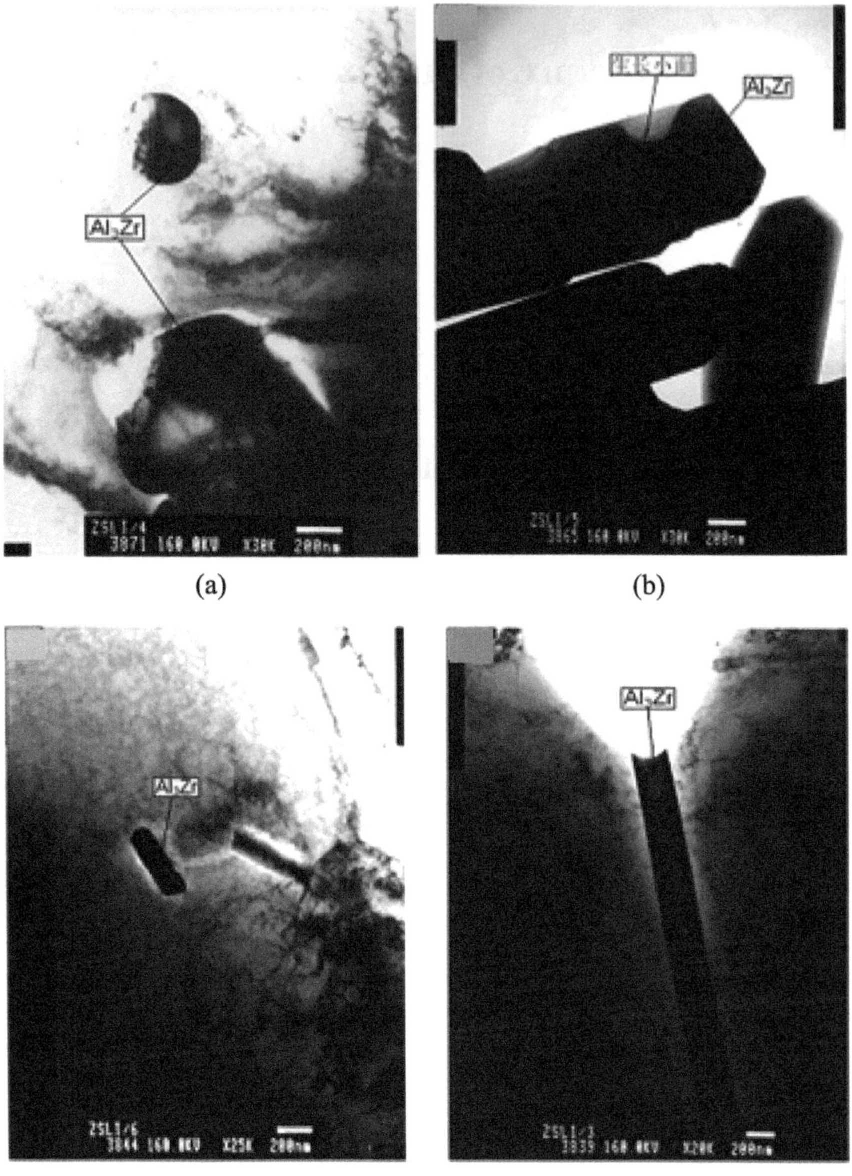

Fig. 6.1 TEM morphology of in-situ Al_3Zr particles at different temperatures

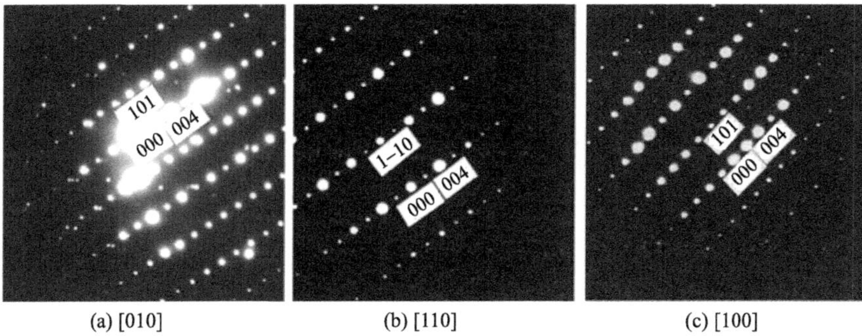

Fig. 6.2 Electron diffraction pattern of Al_3Zr crystal

According to the crystallographic data [2], Al_3Zr crystal is DO_{23} structure with the lattice constants of $a = 0.4013$~0.4015 nm, $c = 1.732$~1.735 nm. The coordinates of the included 16 atoms in the single unit cell are:

The location of the origin: (0, 0, 0); (1/2, 1/2, 1/2).

+4Zr(C_{4v}): (0, 0, $\pm z$); $z = 0.122 \approx 1/8$.

+4Al(D_{2h}): (0, 1/2, 0); (1/2, 0, 0).

+4Al(D_{2h}): (0, 1/2, 1/4); (1/2, 0, 1/4).

+4Al(C_{4v}): (0, 0, $\pm$ z); $z = 0.361 \approx 3/8$.

It can be seen from the structural model shown in Fig. 6.3 that the stacking sequence of Al_3Zr crystal is ABCABCABC….

Fig. 6.3 Structural model of Al_3Zr crystal [3]

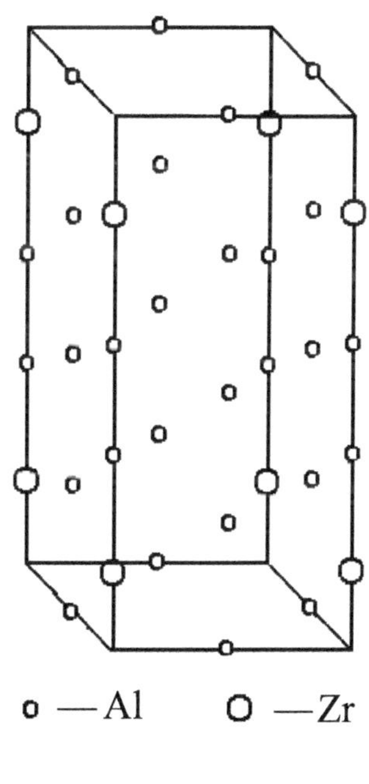

6.1.2 Formation and Growth Mechanism of In-Situ Al_3Zr Phase

The in-situ chemical reaction, $Al + Zr \rightarrow Al_3Zr$, generates Al_3Zr crystals at a specific temperature under physical field with Zr atoms coming from inorganic salt (such as Zr $(CO_3)_2$, K_2ZrF_6) and Al from the melt. According to the binary phase diagram of Al-Zr system, Zr is saturated and the surplus Zr reacts with Al to generate Al_3Zr when the temperature is over 800 °C. The Al-Zr phase diagram and the relation [4] between formation trend of Al_3Zr and temperature are shown in Figs. 6.4 and 6.5, respectively.

In general, when the crystal grows in a free growth system, there is a difference between the growth rate of each crystalline plane, which means the growth rate of crystal is anisotropic. The commonly used crystalline growth rate is referred to as the distance which crystalline plane (hkl) moves outward parallel to the normal direction in unit time, which is then called the linear growth rate.

The motivation of crystalline growth stems from supersaturation degree ΔC or supercooling degree ΔT of the growing phase in an environment. The variation in crystalline growth morphology results from the change in the relative growth rate of each surface. The two-dimensional model of crystalline growth is now taken as an example to illustrate the relationship between the variation in the relative growth

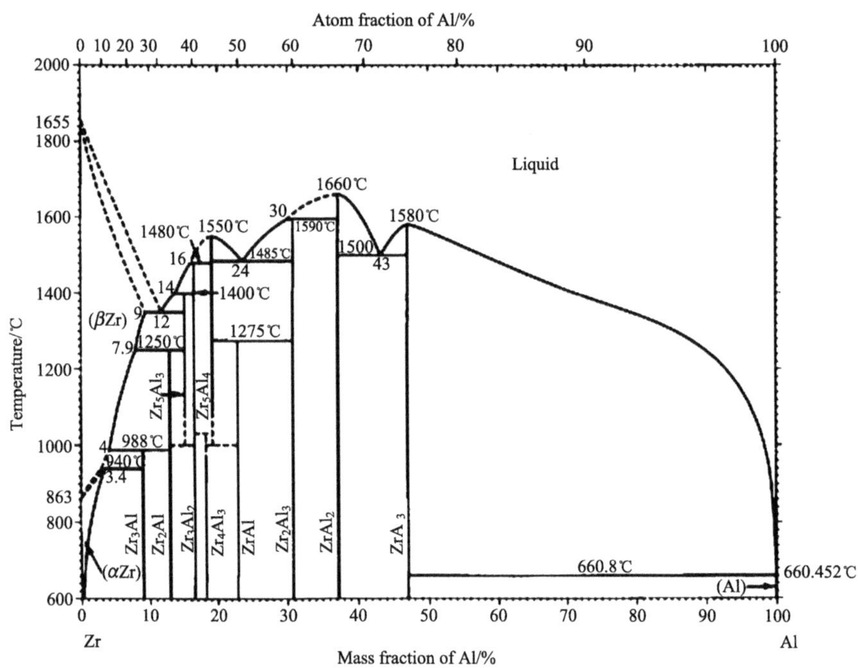

Fig. 6.4 Phase diagram of Al-Zr

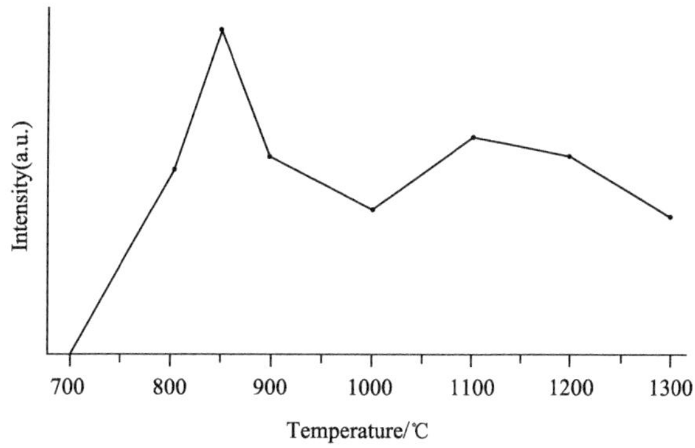

Fig. 6.5 Relationship between the formation trend of Al_3Zr and temperature

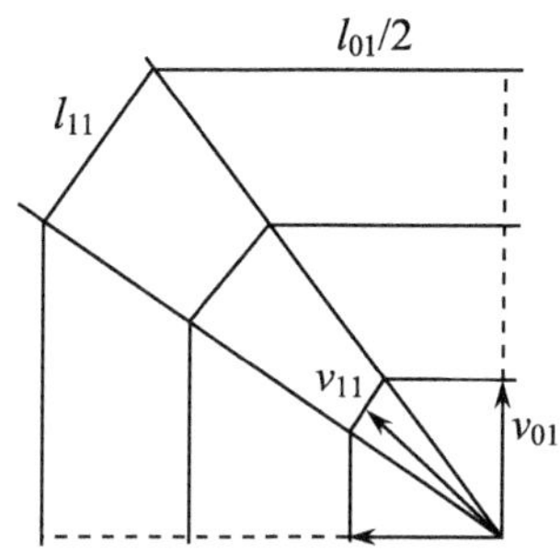

Fig. 6.6 Relation between the relative growth rate of crystal surface and crystal growth morphology

rate of crystalline planes and the crystalline growth morphology, which is shown in Fig. 6.6.

The l_{11} and l_{01} in Fig. 6.6 represent the size of (11) and (01) crystalline planes, and v_{11} and v_{01} represent the growth rate of (11) and (01) crystalline planes, respectively, in the two-dimensional model. It can be seen from Fig. 6.6 that the simple geometrical relation is:

$$v_{01} = l_{01}/2 + \sqrt{2}l_{11}/2 \tag{6.1}$$

$$l_{01}/2 = \sqrt{2}v_{11} - v_{01} \tag{6.2}$$

$$v_{11} = \sqrt{2}v_{01} - l_{11}/2 \tag{6.3}$$

$$l_{11}/2 = \sqrt{2}v_{01} - v_{11} \tag{6.4}$$

$$\frac{l_{01}}{l_{11}} = \frac{\sqrt{2}(v_{11}/v_{01})}{\sqrt{2} - v_{11}/v_{01}} \tag{6.5}$$

It can be observed from Eq. 6.5 that the two-dimension model has only single morphology of {11} when $\nu_{11}/\nu_{01} \geq \sqrt{2}$; when $\sqrt{2}/2 < \nu_{11}/\nu_{01} < \sqrt{2}$, it has an aggregate of two morphologies, i.e., {01} and {11}.

According to the crystalline growth theory [4], intermetallic compounds are usually grown in a way of small crystalline planes which have a zigzag solid/liquid interface on macro-scale and smooth atomic arrangement on micro-scale. The growth of this kind of interface has to rely on the steps formed on the interface which enable the atoms to diffuse from the liquid phase and deposit along the side direction thus extending it.

The research has proved that there are small facets on the surface of Al_3Zr crystal (as shown in Fig. 6.7b), and twin crystals also exist, as shown in Fig. 6.7a. The recognizable plane of twin crystal and the concave corners or grooves caused by twin crystal make the continuation of crystal growth in the direction parallel to the plane of twin crystal so that the crystal can grow continuously. Figure 6.8 shows the composite electron diffraction pattern of twin crystals. It can be concluded from the calibration of electron diffraction patterns (as shown in Fig. 6.8a) that the surface of twin crystals is $(\overline{1}\overline{1}4)$. The atomic arrangement on the $(\overline{1}\overline{1}4)$ plane along the direction of [110] is shown in Fig. 6.8b.

Figure 6.9 shows the twin model of Al_3Zr crystal along the $[2\overline{2}1]$ direction. Further analysis on the structure of Al_3Zr shows that the crystal has four close-packed planes

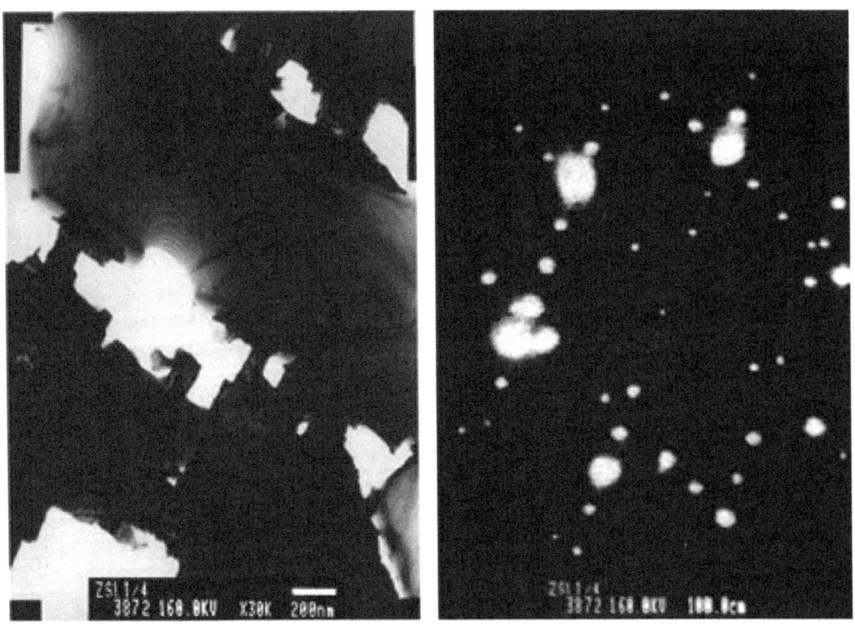

(a) (b)

Fig. 6.7 Twin crystal morphology of Al_3Zr crystal and composite electron diffraction pattern

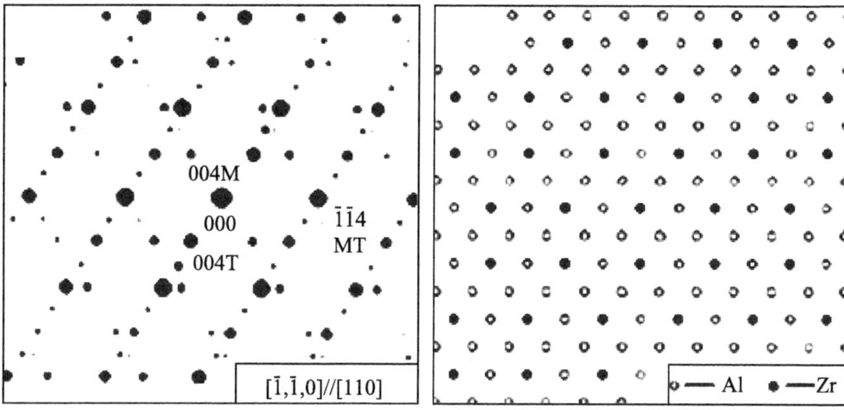

(a) (b)

Fig. 6.8 Diffraction pattern calibration of Fig. 6.7b and atomic arrangement on twin crystal plane $(\bar{1}14)$ along the direction of [110]

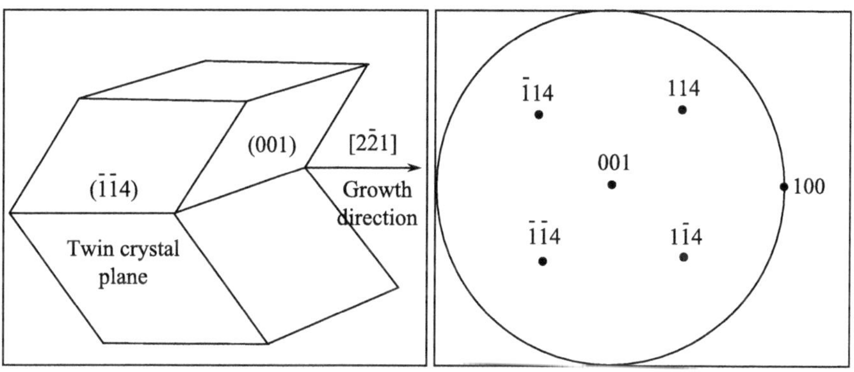

(a) (b)

Fig. 6.9 Al_3Zr crystal twin model diagram (**a**) and projection of Polar diagram of dense rows distributed in the direction of [001] (**b**)

(114), $(\bar{1}\bar{1}4)$, $(1\bar{1}4)$, and $(\bar{1}14)$, respectively, and can be seen on the projection of pole diagram along the [001] direction (as shown in Fig. 6.9b). Atomic dislocations can form on each close-packed plane to generate one or more twin crystals. The crystal usually grows in cuboid morphology when Al_3Zr generates single twin crystal and grows in polyhedron morphology when it generates several twin crystals. Thus, the acquisition of granular Al_3Zr is beneficial to control the generation and promote the formation of twin crystals.

The SEM micrographs of composites fabricated from Al-20%(K_2ZrF_6 + KBF_4) system under a pulsed magnetic field of 0.07 T and reaction time of 20 min with different initial reaction temperatures are shown in Fig. 6.10. It can be seen that

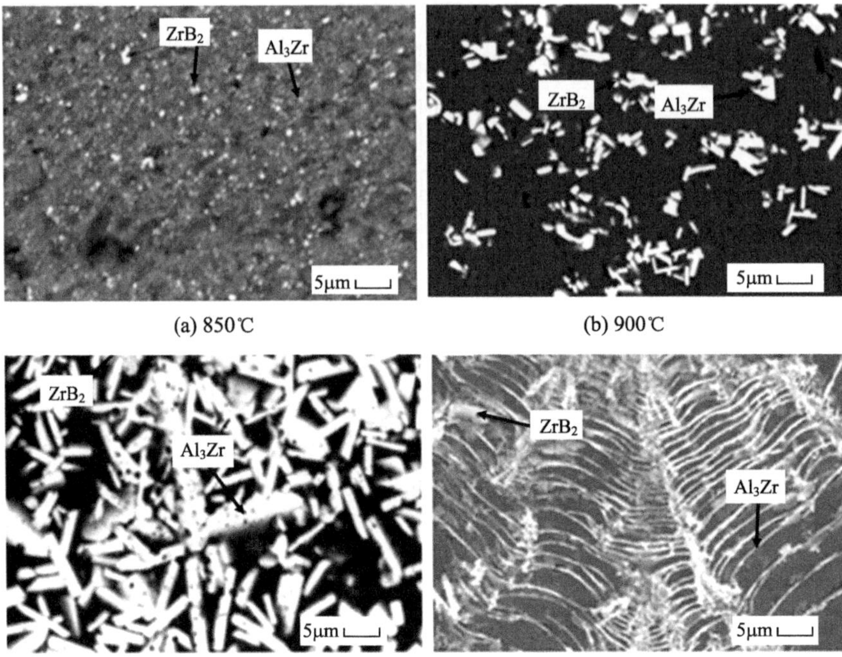

(a) 850℃ (b) 900℃ (c) 950℃ (d) 1000℃

Fig. 6.10 SEM micrographs of the composites fabricated at different initial reaction temperatures

the morphology of *in-situ* Al_3Zr changes by the variation of reaction temperature. When the initial temperature rises from 850 to 1,000 °C, the morphology of Al_3Zr varies from spherical to fibrous and that of ZrB_2 remains unchanged. When the initial reaction temperature is 850 °C, the Al_3Zr has spherical morphology and the ZrB_2 has tetragonal, both having a size of 0.2–0.4 μm, as shown in Fig. 6.10a. When the initial reaction temperature is 900 °C, most of the Al_3Zr is tetragonal with some short rods-like morphology, as shown in Fig. 6.10b. Figure 6.10c shows that the morphology of Al_3Zr is mainly rod-like at 950 °C and size is greater than that of at 850 °C. At 1,000 °C, the Al_3Zr has mainly fibrous morphology and scale-like distribution, as shown in Fig. 6.10d. It can be observed from the high magnification SEM micrographs (Fig. 6.11) of composite fabricated at 1,000 °C and the EPMA diagram of the white section from SEM image that the white substance with fibrous morphology is Al_3Zr and its size along diameter direction is about 180–200 nm. The main reason is that the growth morphology of Al_3Zr is greatly affected by temperature and the growth process is actually the process of Zr diffusion. According to the literature [5], intermetallic compounds are usually grown on small crystalline planes which have a zigzag solid/liquid interface on macro-scale and smooth atomic arrangement on micro-scale. The growth of this kind of interface has to rely on the steps formed on the interface which act as atom deposition sites for atoms from liquid

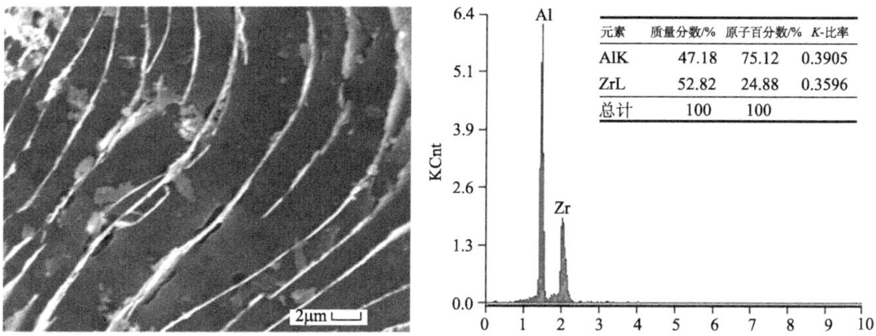

(a) The high-magnification SEM diagram of the composite fabricated at 1,000℃

(b) EPMA diagram of the white part

Fig. 6.11 **a** High magnification SEM micrographs of the composite fabricated at 1,000 °C, **b** EPMA diagram of the white section

phase. Al_3Zr crystal has four close-packed planes, and dislocations can form on each of the four close-packed planes to generate one or more twin crystals.

The twin crystal grows across solid/liquid interface forming concave corners and grooves where the atoms diffuse from the liquid phase and deposit, thus making possible the crystal growth in the direction parallel to the surface of the twin crystal. The crystal usually grows in cuboid morphology when Al_3Zr generates single twin crystal and grows in polyhedron morphology when it generates several twin crystals. It should be noted that the generation of the twin crystal is strongly related to the formation temperature and crystalline growth velocity of Al_3Zr, although more research should be conducted on the generation of multiple twin crystals and their mechanism. So, the initial reaction temperature is controlled at the range of (850 ± 5)°C to ensure the spherical morphology of generated Al_3Zr particles with small sizes and better strengthening effect on matrix properties in order to have the comprehensive properties of composites.

6.2 Morphology and Formation Mechanism of In-Situ Al_2O_3

6.2.1 Classification and Crystalline Structure of Al_2O_3

According to crystallographic data, the crystal structure of Al_2O_3 can be of three types: α-Al_2O_3, β-Al_2O_3, and γ-Al_2O_3. α-Al_2O_3 belongs to the hexagonal lattice, having the most stable structure with high hardness, high melting point, excellent wear-resistance, and corrosion-resistance properties. Research has found that Al_2O_3 generated from different reaction systems has different lattice types. For example, results of XRD and TEM have shown that the Al_2O_3 generated from Al-$Na_2B_4O_7$-K_2ZrF_6 in-situ reaction system is γ-Al_2O_3, while Al_2O_3 generated from Al-$Zr(CO_3)_2$ system is α-Al_2O_3.

6.2.2 Morphology and Growth Mechanism of Al_2O_3 Reinforcement Particles

It can be seen from the TEM image of composites in Fig. 6.12 that the in-situ generated particles have a spherical shape with uniform distribution and a small size of 20–100 nm. After calibration, it is found that the particles are those of the γ-Al_2O_3 phase. Analysis of HRTEM image of the interface between a particle and matrix (shown in Fig. 6.13a) indicates that the bright section is the lattice image of Al matrix while the dark section represents the projection of plane (021,113) of the γ-Al_2O_3 particle (which means the direction of electron beam is perpendicular to the crystalline plane). Fourier transform (acquired by digital micrograph) of Fig. 6.13a is shown in Fig. 6.13b. Fourier transform was conducted in order to get a detailed analysis of the diffraction pattern. The results show that points 1, 3, and 4 in Fig. 6.13b are the diffraction spots of γ-Al_2O_3 presented in Fig. 6.13a, while points 1 and 2 are the diffraction spot of Al matrix. Spot 1 belongs to γ-Al_2O_3 as well as to the Al matrix, which explains the higher brightness of spot 1 than the others. It can be seen from the image that the crystal plane (040) of γ-Al_2O_3 is parallel to crystal plane (200) of Al matrix. The two crystal planes have coherent correspondence with a distance of about 0.2024 nm, clear interface and without any reaction product.

From an interfacial energy perspective, the closer the resemblance of crystals is, atomic size, and distance of two atomic planes with each other, the smaller the interfacial energy between them is and the more easily one phase can nucleate and

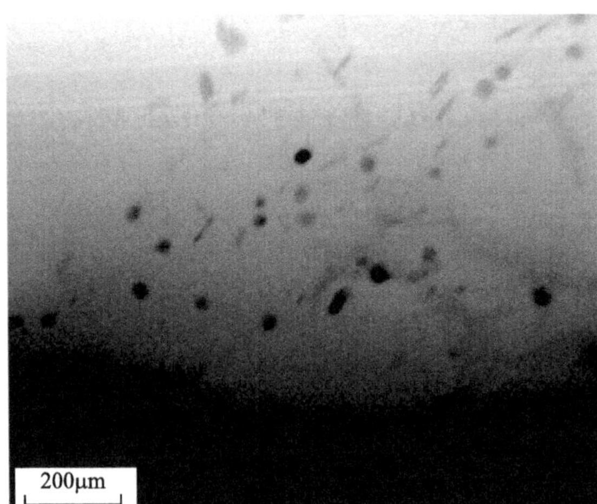

Fig. 6.12 TEM micrograph of $Al_2O_{3(np)}$/Al composites

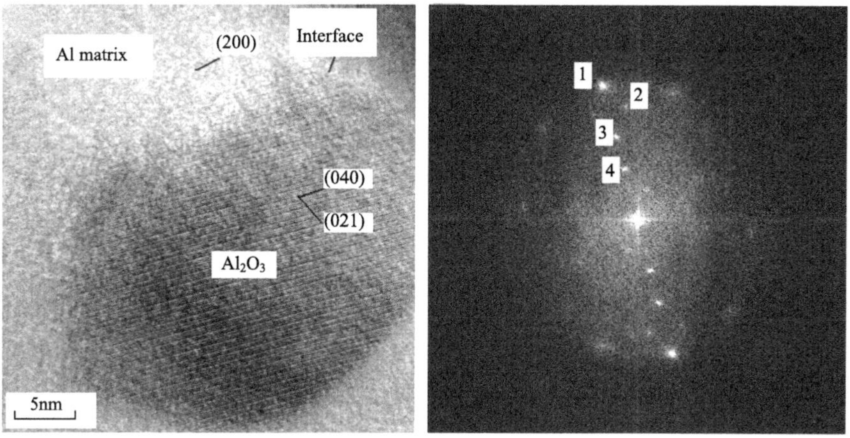

(a) (b)

Fig. 6.13 HRTEM image of interface between γ-Al_2O_3 particle and Al matrix

grow on the surface of the other phase. It can be concluded from the analysis that γ-Al_2O_3(040) // Al(200). The interplanar spacing of (040) and (200) is 0.1990 nm and 0.2024 nm, respectively. According to the formula of interfacial mismatch degree [6]:

$$\delta = \frac{2\left(d_{\mathrm{Al}} - d_{\gamma\text{ - }\mathrm{Al_2O_3}}\right)}{d_{\mathrm{Al}} + d_{\gamma-\mathrm{Al_2O_3}}} \times 100\% \tag{6.6}$$

where δ is the degree of mismatch; and d_{Al} and $d_{\gamma-\mathrm{Al_2O_3}}$ are the interplanar spacing of pure Al matrix and Al_2O_3. The planes of two phases have a coherent correspondence when $\delta \leq 5\%$ while incoherent correspondence happens when $\delta = 5$– 25%. Calculations reveal that the mismatch degree is only 0.8% for γ-Al_2O_3(040) // Al (200), so the condition of coherent correspondence is satisfied. It means that the primary Al phase can nucleate on the surface of γ-Al_2O_3 particles. The principle of energy of crystal growth indicates that the atomic array which first crystallizes should be the close packed orientation and the formed crystalline plane should be the close-packed plane. Al matrix has the lowest surface energy when it grows preferentially along the (200) plane. The Al matrix is attached to the γ-Al_2O_3 particles to crystallize and grow. Meanwhile, the surface energy of nano-scale γ-Al_2O_3 particles is also reduced due to the adsorption of extraneous atoms, which weakens the tendency of agglomeration of nano-scale particles having high surface energy and thus is conducive to the uniform distribution of γ-Al_2O_3 particles in the Al matrix.

Figure 6.14 shows the SEM micrograph and electron probe energy spectrum analysis of Al_2O_3 generated from Al-$Zr(CO_3)_2$ system under pulsed magnetic field. As shown in Fig. 6.14a, the morphology of in-situ Al_2O_3 particles is mainly olive-shaped and granular with a small size of 0.1–0.2 μm. Electron probe energy spectrum

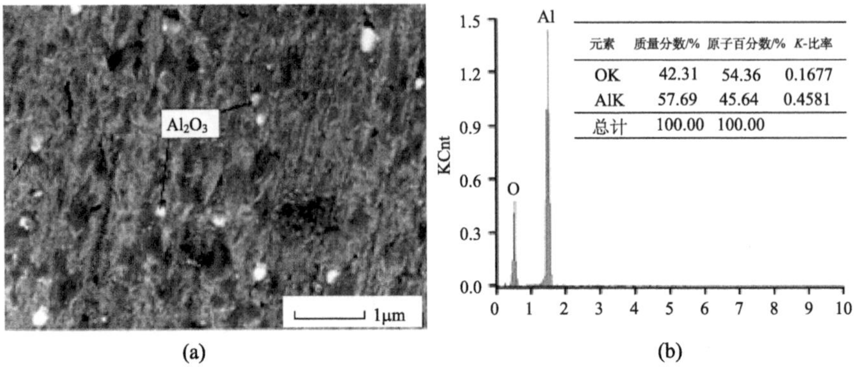

Fig. 6.14 SEM morphology and electron probe analysis of in-situ Al_2O_3

analysis shows that the atomic ratio of Al and O in in-situ particles is close to 2:3 and the particles are those of the Al_2O_3 phase.

Figure 6.15 shows the TEM morphology of in-situ Al_2O_3 particles. It can be seen that in-situ Al_2O_3 particles have small sizes, spherical morphology, and smooth interface.

From XRD and TEM analysis it is found that Al_2O_3 phase particles generated from Al-Zr$(CO_3)_2$ system under pulsed magnetic field are all α-Al_2O_3 phase particles.

Fig. 6.15 TEM morphology of in-situ Al_2O_3

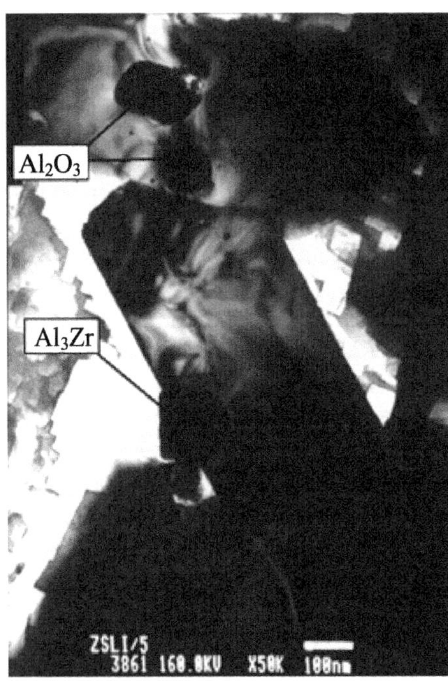

Figure 6.16 shows the electron diffraction pattern of α-Al_2O_3 crystal. The crystal structure model of α-Al_2O_3 is shown in Fig. 6.17 which belongs to the hexagonal lattice with $R\overline{3}c$ space group, the lattice constant $a_0 = 0.514$ nm, $\alpha = 55°17'$, $z = 2$, $CN_+ = 6$, $CN_- = 4$ and O^{2-} forms electrostatic bonds with four Al^{3+}.

O^{2-} arrays on the base of ABABAB… as hexagonal close-staked type with repetitive two-floor and Al^{3+} is filled in 2/3 of the octahedral void formed by O^{2-}. According to the Pauling's rule, the distance between Al^{3+} should be kept the furthest between layer and layer or in the same layer, showing uniform distribution on macro-scale, so as to reduce the electrostatic repulsion between Al^{3+} , which is beneficial to the stability of the structure where Al^{3+} is distributed in three forms: Al_D, Al_E, and Al_F, as shown in Fig. 6.18. They are arranged in order to satisfy the condition of the

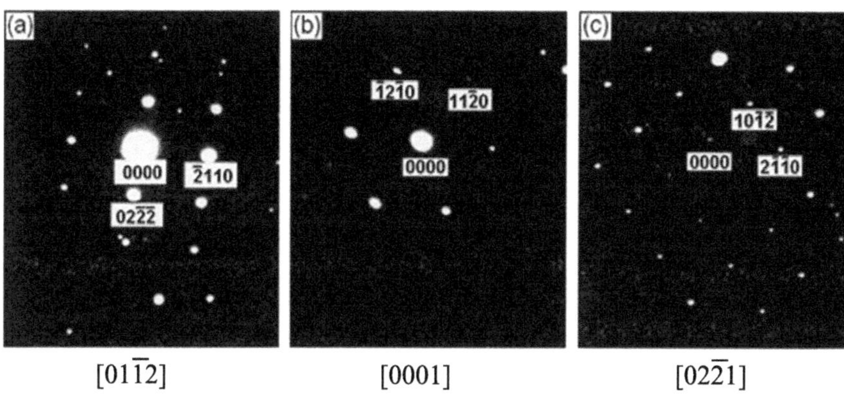

Fig. 6.16 Electron diffraction pattern of Al_2O_3 crystal

Fig. 6.17 Crystal structure model of α-Al_2O_3

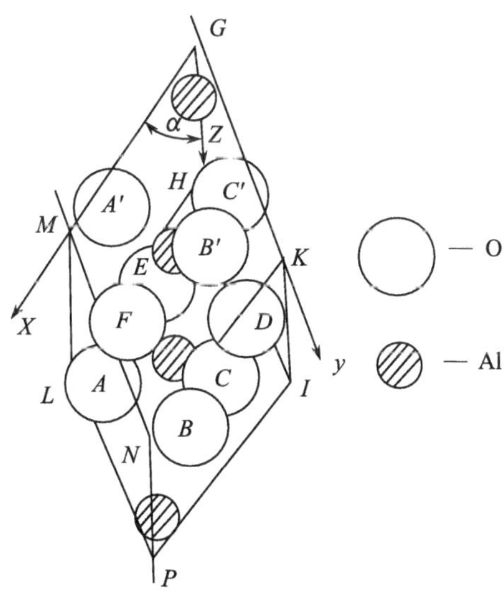

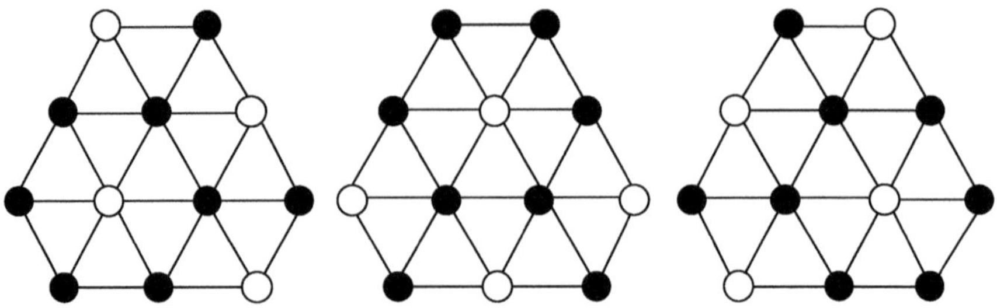

Fig. 6.18 Three forms of Al^{3+} distribution

longest distance between Al^{3+}. Considering O_A and O_B as the two arrangement types of O^{2-}, the arrangement order of O^{2-} and Al^{3+} in the α-Al_2O_3 crystal is as follows: $O_AAl_DO_BAl_EO_AAl_FO_BAl_DO_AAl_EO_BAl_F$. The above 12 arrays are considered as a unit, and the repetition of them constitutes the crystal structure of α-Al_2O_3.

6.2.3 Dislocation at the Particle/Matrix Interface

Figure 6.19 shows the dislocation morphology and distribution in pure Al and $Al_2O_{3(np)}$/Al composites. It can be seen from the figure that the dislocation density of pure Al is low, and it mainly distributes at the grain boundary (Fig. 6.19a). While the dislocation density is significantly increased, and the dislocation is small, dense, and irregularly distributed in the matrix of $Al_2O_{3(np)}$/Al composite (Fig. 6.19b). Al matrix is a plastic phase, and its thermal expansion coefficient is about 24×10^{-6}/°C, while Al_2O_3 particle is thermodynamically stable and has a low thermal expansion coefficient of only 7.92×10^{-6}/°C [6] due to the in-situ reaction.

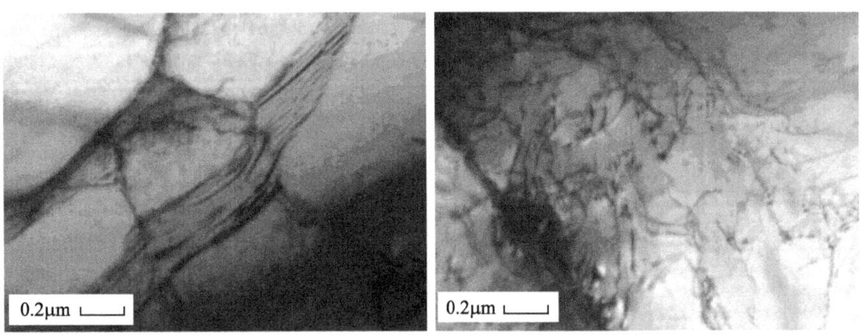

(a) Pure Al (b) $Al_2O_{3(np)}$/Al composites

Fig. 6.19 Dislocation distribution in Al matrix and composites

According to the literature [7], the density of dislocation can be calculated through the TEM morphology of composites by the following formula:

$$\rho = \frac{2\mathrm{N}}{LT} \tag{6.7}$$

where N is the total number of intersection points of dislocation line on a line segment L; T is the thickness of the sample, which is 0.3 μm at 120 kV; and ρ is the dislocation density. Calculation results indicate that the dislocation density in pure Al is $1.0 \times 10^{14}/m^2$, while the dislocation density in $Al_2O_{3(np)}/Al$ composite is $3.33 \times 10^{14}/m^2$.

6.2.4 Generation Mechanism of Dislocation

Dislocations usually form during solidification, fast cooling, high applied stress and are most likely to appear in two places: The first one is the grain boundary, which is regarded as the emission source or absorption source of dislocations due to the massive dislocation networks in the grain boundary; the second one is the surroundings of reinforcement phase particles. The following conditions may occur when the dislocation encounters reinforcement particle during the slip process: ① stop slipping, resulting in plugged dislocation; ② continue to slip through the particles, cutting the particle into half, which only happens when the strength of particle is low; ③ bypassing the particle, leaving a dislocation ring around the particle; ④ a variety of cross-slip might happen around the particles. The dislocation can originate or end at the grain boundary and can also originate or end at reinforced particles. The pinning and locking effect of particles on the dislocation is very obvious. Under the action of the dislocation locking, further slip results in bending of dislocation line, indicating the formation of mixed dislocation. As shown in Fig. 6.20, the dislocation in Fig. 6.20a terminates at the particle, and the particle has a pinning effect on the

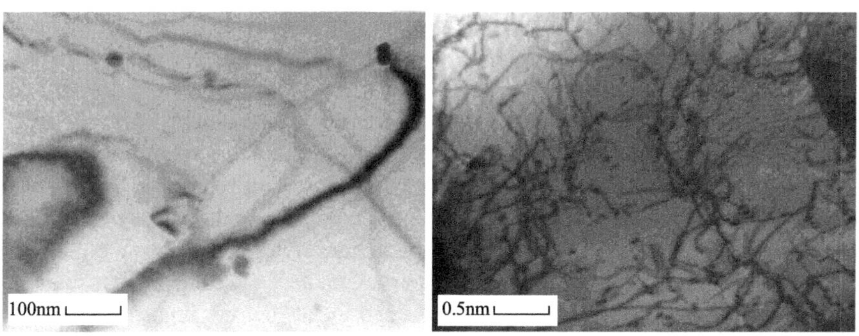

(a) Dislocation stops at particle (b) Dislocation plugged group stops at grain boundary

Fig. 6.20 Different dislocations in the matrix of $Al_2O_{3(np)}/Al$ composites dislocation. In Fig. 6.20b, the plugged group resulting from the pinning of dislocation slip terminates at the grain boundary [8].

Most of the particles formed by in-situ melt reaction are thermodynamically stable and have a low coefficient of thermal expansion. Strain caused by thermal mismatch and the residual stress between particle phase and matrix results in lattice distortion and induces an increase in the quantity of dislocations, forming the high density area, producing dislocation strengthening due to differences in thermal expansion coefficient when the high temperature composite is rapidly cooled and solidified. The dislocation density of this area can be expressed as [9]:

$$\rho \approx \frac{3f^{4/3}\left(\alpha_m - \alpha_p\right)\Delta T}{(1 - \nu)bD} \tag{6.8}$$

where α_p and α_m are the coefficient of thermal expansion of reinforcement phase particle and matrix, respectively; f is the volume fraction of reinforcement particles; b is the Burgers vector; D is the size of particle; ν is the Poisson's ratio; and ΔT is the variation of temperature. The dislocation density increases with the increase of thermal strain mismatch and particle volume fraction and with the decrease of grain size. The greater the difference of thermal expansion coefficient between the reinforcement particle and the matrix, the greater the dislocation density. The theoretical foundation of the property improvement of composites has been provided by the massive multiplication of dislocations in $Al_2O_{3(np)}$/Al composites.

6.3 Interface Structure of In-Situ ($Al_3Zr + Al_2O_3$)/A356 Composites

Research on the interface structure of composite materials is the frontier subject of current material science. Interfacial engineering of new materials with special properties can be developed by controlling and utilizing the physical and chemical properties of the material interface, which plays an important role in the development of high-performance new materials [10]. Interface structure and performance play a decisive role in the performance of metal matrix composites. In particles reinforced metal matrix composites, the load transfer from matrix to ceramic particles, the interaction between dislocation and particles, and the fracture behavior of composites are all related to the performance of the interface region [11]. Therefore, the key to obtain high-performance metal matrix composite materials is to study and master the law of interface reaction and its influence on performance and effectively control the structure and performance of particle/matrix interface. At present, the interfacial research of in-situ particle reinforced metal matrix composites at home and abroad is not thoronghly enough, such as whether there is a certain orientation relationship between particle and matrix, arrangement and matching of interfacial atoms, quantitative analysis of dislocations near the interfacial region. There have been no reports on the interfacial structure of Al_3Zr and Al_2O_3 particles reinforced Al matrix composites fabricated from Al-Zr-O system. This section mainly studies the interfacial structure of Al_3Zr/Al, Al_2O_3/Al, Al_2O_3/Si in ($Al_3Zr + Al_2O_3$)/A356 composites and factors influencing it, so as to optimize the interface structure and control the interfacial reaction of particles/matrix, forming effective load transfer, adjusting stress distribution to prevent crack propagation and obtain the interface with high bonding strength, and further improve the comprehensive performance of composites [12, 13].

6.3.1 Interfacial Structure of Al_3Zr/Al and Al_2O_3/Al

Figures 6.21 and 6.22 show the TEM morphology and electron diffraction patterns of the interface between particles and matrix in ($Al_3Zr + Al_2O_3$)/A356 composite. The electron diffraction analysis of region A in Fig. 6.21 is shown in Fig. 6.22a, and the diffraction pattern is calibrated as Al_3Zr. The electron diffraction analysis of region B in Fig. 6.21 is shown in Fig. 6.22b, and the diffraction pattern is calibrated as Al. Therefore, the interface

in region A and region B in Fig. 6.21 shows the morphology of Al_3Zr/Al interface. Through observation of a lot of Al_3Zr/Al interfaces in the composite, the results show that the Al_3Zr/Al interface is clear, smooth, and round with no reactants.

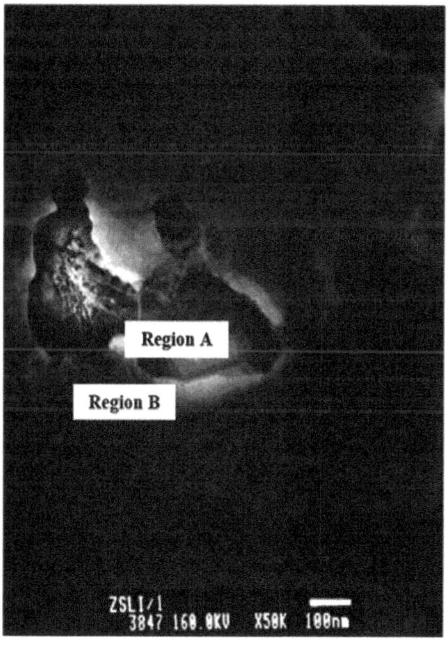

Fig. 6.21 TEM morphology of Al_3Zr/Al interface

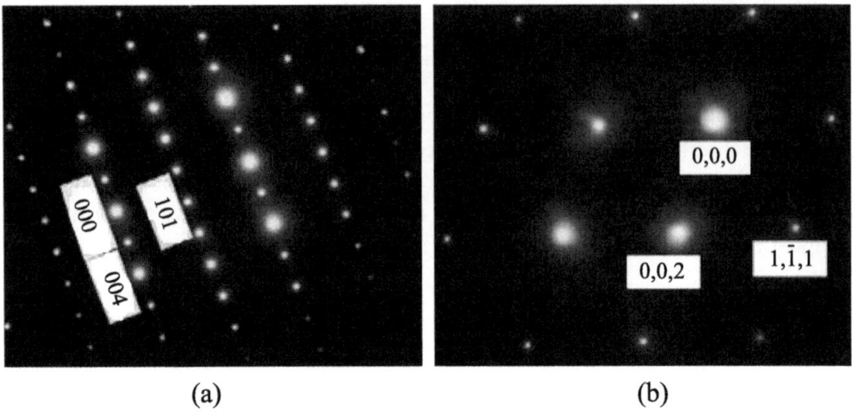

Fig. 6.22 Electron diffraction pattern of regions A and B in Fig. 6.21

Figure 6.23 shows another TEM morphology of particles/matrix interface in (Al_3Zr + Al_2O_3)/A356 composites. Electron diffraction patterns of regions C and D in Fig. 6.23 are shown in Fig. 6.24. Analysis of electron diffraction patterns shows that region C belongs to α-Al_2O_3 and region D belongs to Al matrix, so these regions together form the α-Al_2O_3/Al interface. It can be seen from Fig. 6.23 that the interface

Fig. 6.23 TEM morphology of Al_2O_3/Alinterface

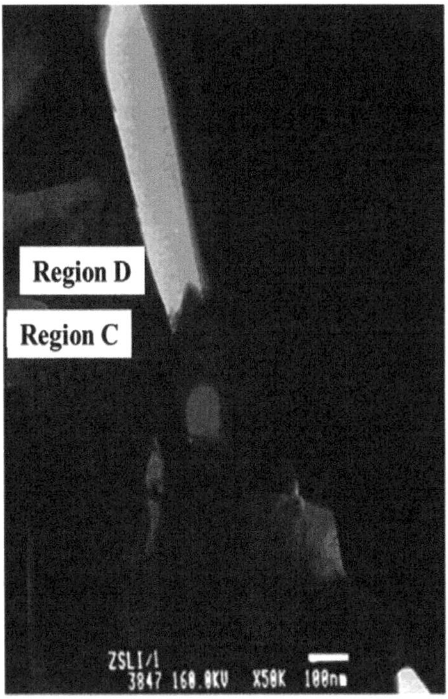

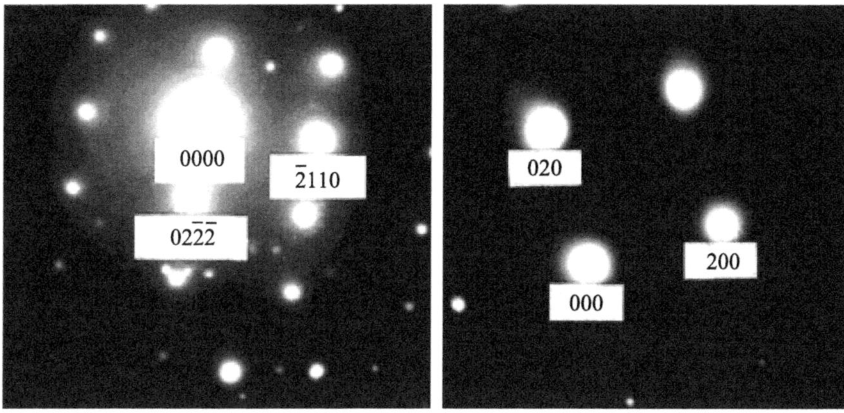

(a) (b)

Fig. 6.24 Electron diffraction pattern of regions C and D in Fig. 6.23

between Al_2O_3 particle and Al matrix is clean, without the formation of by-products, and the interface is smooth and round.

6.3.2 *Orientation Relationship of Al_3Zr/Al Interface and Atomic Arrangement*

Figure 6.25 shows the simulated composite diffraction pattern and indexed diffraction pattern of Al_3Zr/Al interface. Results indicate that the correlation between the Al_3Zr particle and the α-Al phase is: $[\overline{22}1]Al_3Zr\|[001]_{Al}$, $(012)Al_3Zr\|(1\bar{1}0)_{Al}$. Figure 6.26 shows the atomic arrangement of the Al_3Zr/Al interface. As can be seen from Fig. 6.26, atoms on the plane (001) of Al_3Zr along the [100] direction almost coincide with atoms on the plane (001) of Al at the interface, but they are only slightly deviated due to the difference in lattice constants between them. Therefore, the two interfaces have an excellent coherent correspondence.

6.3.3 *The Interfacial Structure of α-Al_2O_3/Si*

Figure 6.27 shows the TEM morphology of the interface between the Si particles and matrix in the ($Al_3Zr + Al_2O_3$)/A356 composite, and Fig. 6.28 shows the electron diffraction pattern of regions E and F of Fig. 6.27. It can be seen from the analysis of diffraction pattern, in Fig. 6.28, that regions E and F are α-Al_2O_3 and Si, and Si phase nucleates and grows on the surface of α-Al_2O_3 particle in the way of twin

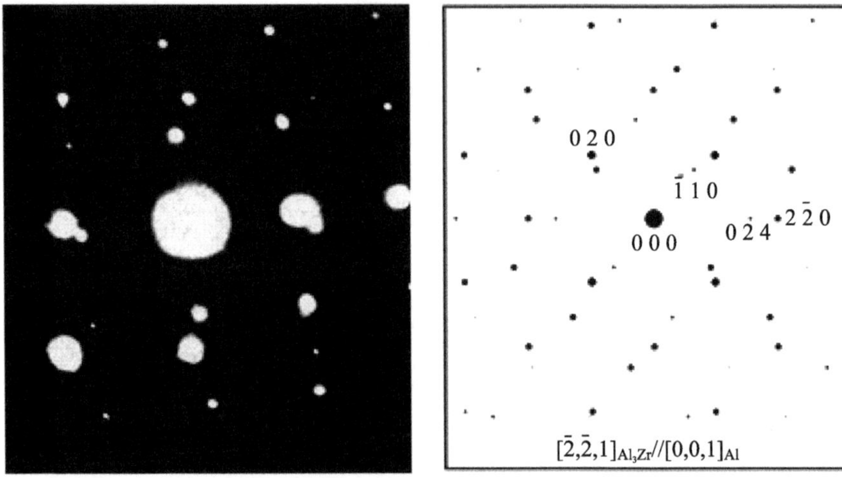

(a) Diffraction composite diagram (b) Exponent calibration of Fig. 6-25(a)

Fig. 6.25 Microregion electron diffraction pattern of Al_3Zr/Al interface

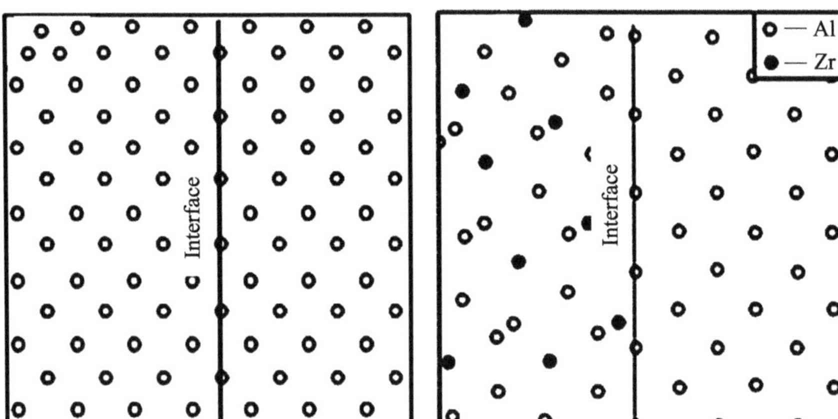

(a) Al_3Zr(100)//Al(100) (b) Al_3Zr(012)//Al(110)

Fig. 6.26 Schematic diagram of atomic arrangement and matching of Al_3Zr/Al interface

crystals. Observation of α-Al_2O_3/Si interface indicates that the interface is clean with no reactants and is a dense interfacial combination.

Figure 6.29 shows the atomic arrangement at the interface of α-Al_2O_3/Si. It can be seen that the arrangement of Si atoms and internal atoms at the interface of Al_2O_3 have a one-to-one correspondence, which indicates that α-Al_2O_3/Si interface

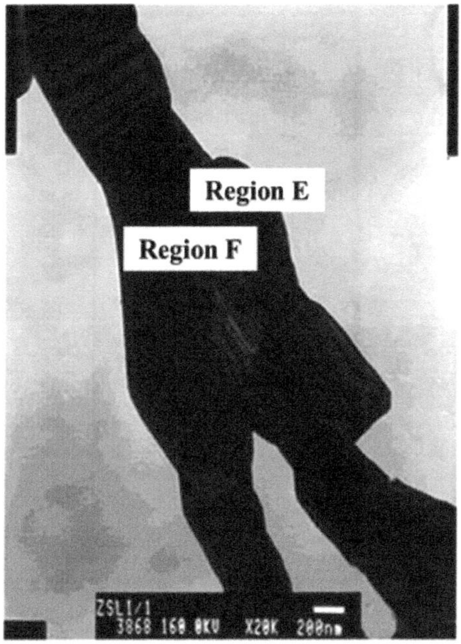

Fig. 6.27 TEM morphology of Al_2O_3/Si interface

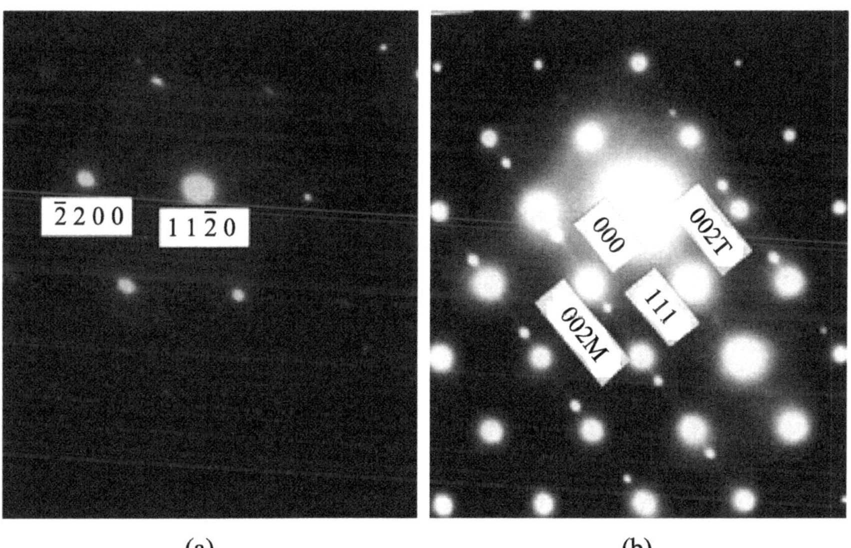

Fig. 6.28 Electron diffraction pattern of regions E and F in Fig. 6.27

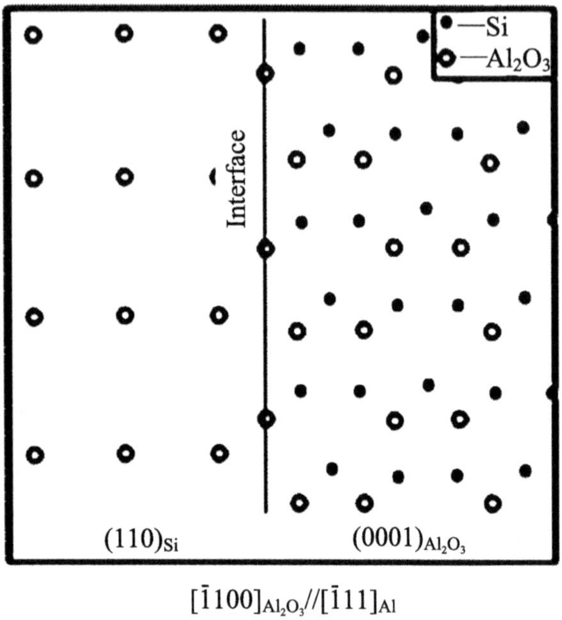

Fig. 6.29 Schematic diagram of atomic arrangement and matching at α-Al_2O_{3p}/Si interface

has a good orientation relationship. According to the literature [14], its orientation relationship is as follows:

$$[\bar{1}100]_{\alpha\text{-}Al_2O_3}//[\bar{1}11]_{Si}(0001_{\alpha\text{-}Al_2O_3})//(110)_{Si}$$

According to the theory of generation of interfacial energy [15], the more closely the geometry of atomic arrangement, atomic size, and distance of two crystal planes contact with each other, the smaller interfacial energy between the two crystal planes, and the more easily one phase can get nucleation and growth on the surface of the other phase. The degree of mismatch δ is usually used to express the matching condition of interface crystal nucleus and basal atoms:

$$\delta = \frac{a_S - a_N}{a_N} \tag{6.9}$$

where a_S and a_N are the atomic distance of substrate and crystal nucleus, respectively.

The smaller the δ value, the better the match between those two, and the lower the interface tension. Therefore, the supercooling degree of the nonspontaneous nucleation is smaller. According to the literature, when $\delta \leq 5\%$, the corresponding plane of two phases can achieve coherent correspondence; when $\delta = 5$–25%, the coherent correspondence is bad.

Bramfitt [16] proposed that the calculation of mismatch should make the low exponent plane of the crystal nucleus coincide with the low exponent plane of the

Table 6.1 Matching parameter of plane (0001) in α-Al_2O_3 particle and planes (110) and (001) of Si

	$(0001)_{Al_2O_3}//(110)_{Si}$			$(0001)_{Al_2O_3}//(001)_{Si}$		
$[uvw]_{Al_2O_3}$	$[\bar{1}100]$	$[\bar{2}110]$	$[0\bar{1}10]$	$[\bar{1}2\bar{1}0]$	$[\bar{1}100]$	$[\bar{2}110]$
$[uvw]_{Si}$	$[\bar{1}11]$	[001]	$[1\bar{1}1]$	[100]	[110]	[010]
$d[uvw]_{Al_2O_3}$	0.963	0.556	0.963	0.556	0.963	0.556
$d[uvw]_{Si}$	0.940	0.543	0.940	0.543	0.768	0.543
θ	5.26	30	5.26	0	15	30
$\delta/\%$	5.08			11.61		

substrate, and then the mismatch model of the two-dimensional lattice is established. The formula is as follows:

$$\delta^{(hkl)_s}_{(hkl)_n} = \frac{1}{3}\sum_{i=1}^{3}\frac{\left|d[uvw]^i_s\cos\theta - d[uvw]^i_n\right|}{d[uvw]^i_n} \times 100\% \tag{6.10}$$

where $(hkl)_s$ and $(hkl)_n$ are the low exponent surface of the substrate and crystal nucleus, respectively; $[uvw]_s$ and $[uvw]_n$ are the low exponent orientation of $(hkl)_s$ and $(hkl)_n$, respectively; $d[uvw]_s$ and $d[uvw]_n$ are the atomic distances along the $[uvw]_s$ and $[uvw]_n$ crystal orientation, respectively; and θ is the included angle between $[uvw]_s$ and $[uvw]_n$.

Table 6.1 gives the matching parameter of plane (0001) in α-Al_2O_3 phase and plane (110), (001) of Si calculated according to formulas (6.6–6.24). Figure 6.29 and Table 6.1 show that the plane (0001) of α-Al_2O_3 particle and crystal plane (110) of Si phase have excellent interfacial matching, the mismatch degree is less than 6%, which is 5.08%, as Al_2O_3 particles basically correspond coherently with Si phase and Si can nucleate on the surface of α-Al_2O_3 particle. Therefore, volume fraction of Al_2O_3 particles increases and the eutectic in the matrix is refined with the introduction of pulse magnetic field to the (Al_3Zr + Al_2O_3)/A356 composites. Theoretical analysis is consistent with the microstructural observation results of the fabricated composites.

6.4 Distribution of Dislocations and Micro-hardness Near the Interface

According to the literature [17], the main strengthening mechanism of discontinuously reinforced metal matrix composites (DRMMCs) is the increase of dislocation density. Therefore, it is helpful to understand the strengthening mechanism of composites by analyzing and characterizing the distribution of dislocation near the matrix in the interface region. This section mainly studies the morphology and distribution of dislocations, and their relationship with microhardness in Al matrix composites reinforced by Al_3Zr and Al_2O_3 particles through in-situ reaction of melt under pulsed magnetic field.

6.4.1 Dislocations at Particle/Matrix Interface

Figure 6.30 shows the TEM image of dislocation morphology and distribution in the matrix of (Al_3Zr + Al_2O_3)/Al composites fabricated by melt reaction method under pulsed magnetic field through Al-$Zr(CO_3)_2$ system. It can be seen from Fig. 6.30 that the density of dislocations near the particle/matrix interface in the matrix of the composite was significantly increased after applying the pulsed magnetic field, and the dislocation was small, dense, and has a reticular distribution.

It is predicted in the literature [18] that the increment of dislocation density is:

$$\Delta\rho = \frac{\Delta\alpha \cdot \Delta T N A}{b} \tag{6.11}$$

where $\Delta\alpha$ is the difference between the coefficient of matrix and reinforcement; ΔT is the variation of temperature; N is the number of particles; b is the Burgers vector;

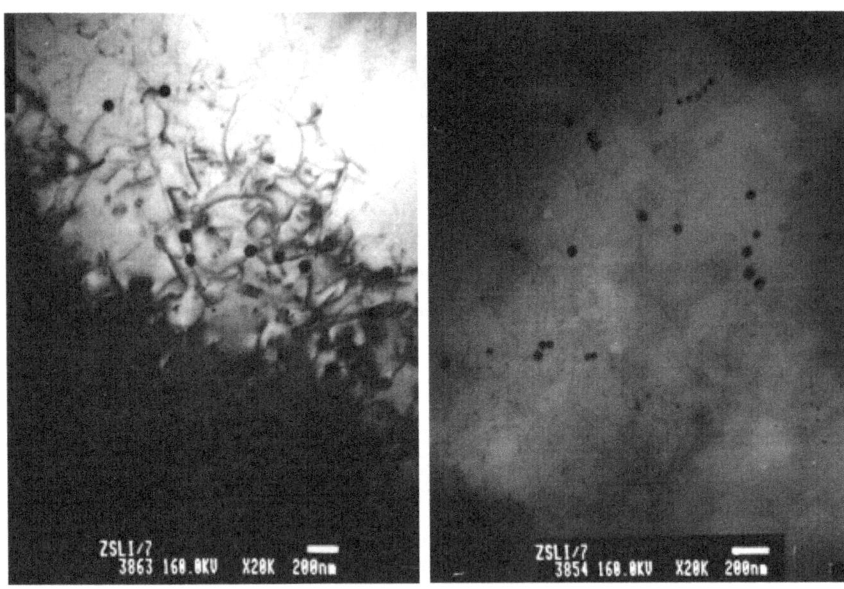

(a) Dislocation near interface of particle/matrix (b) Distribution of dislocation

Fig. 6.30 TEM micrograph of dislocation morphology and distribution in (Al_3Zr + Al_2O_3)/Al composites and A is the total superficial area of single particle. Miller and Humphrey obtained a similar expression by assuming the particle to be cubic [19]:

$$\Delta\rho = 12\frac{\Delta\alpha \cdot \Delta T V_f}{bd} \tag{6.12}$$

where d is the size of particles.

From Eqs. 6.11 and 6.12, the contribution of dislocation strengthening in the strengthening mechanism of in-situ Al matrix composites is increased by the generation of more dislocations induced by the increase of in-situ reinforcement particles in composite and the refinement of grains in the matrix.

6.4.2 Micro-hardness of Particle/Matrix Interface

Figure 6.31 shows a comparison of microhardness distribution near the particle/matrix interface in the $(Al_3Zr + Al_2O_3)$/Al composites synthesized with and without pulsed magnetic field. The load of Vickers hardness test is 0.24 N, and it proceeds along the interface to the deep of the matrix. It can be seen from Fig. 6.31 that the microhardness of composites fabricated under pulsed magnetic field is much higher than that fabricated without pulsed magnetic field at the same near interfacial distance, especially the region near the interface. Also, the microhardness in the matrix of the fabricated composites renders gradient distribution for composites with or without pulsed magnetic field. The region near the interface shows high microhardness while the region far from the interface shows low mircohardness, which is mainly attributed to the high dislocation density in the near interface region.

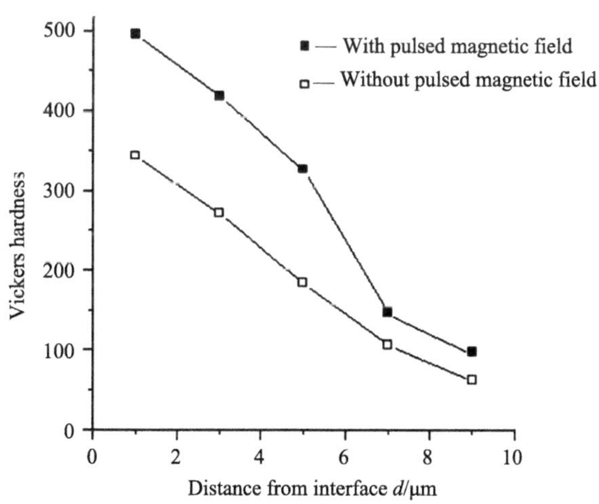

Fig. 6.31 Microhardness distribution near particle/matrix interface of $(Al_3Zr + Al_2O_3)_P$/Al composite fabricated under pulsed magnetic field

References

1. Zhang SL. The microstructure and properties of aluminum matrix composites synthesized under physical field. Zhenjiang: JiangSu University; 2007. (in Chinese).
2. Brandes EA, Brook GB. Smithells metal reference book. Oxforcl: Butter worth-Heinemcnnltd., 1992.
3. Hashimoto S, Yamaguchi A, Yasuda M. Fabrication and properties of novel composites in the system Al-Zr-C. J Mater Sci. 1998;33(19):4835–42.
4. Hu HQ. Principle of metal solidification. Beijing: China Machine Press; 1991. (in Chinese).
5. Kühn G, Kurz W, Fisher DJ. Fundamentals of solidification. Pfaffikon: Trans Tech Publications; 1986.
6. Bai PC, Pei J, Dai XJ, et al. HRTEM study on interface structure of Al_2O_3/2024Al composites. Rare Metal Mater Eng. 2009;38(s1):1–5 (in Chinese).
7. Wang H, Li G, Zhao Y, et al. In situ fabrication and microstructure of Al_2O_3, particles reinforced aluminum matrix composites. Mater Sci Eng A. 2010;527(12):2881–5.
8. Chen K, Liu H, Zhang Z, et al. The improvement of constituent dissolution and mechanical properties of 7055 aluminum alloy by stepped heat treatments. J Mater Process Technol. 2003;142(1):190–6.
9. Zuo Y, Cui J, Dong J, et al. Effect of low frequency electromagnetic field on the constituents of a new super high strength aluminum alloy. J Alloy Compd. 2005;402(1–2):149–55.
10. Tao J. Introduction to New Techniques for Fabrication of Metal Matrix Composites. Beijing: Chemical Industry Press; 2007. (in Chinese).
11. Wang J. MMCp and its micromechanical behavior fabricated by high-energy ultrasonic method. Nanjing: Southeast University; 1997. (in Chinese).
12. Zhao YT. Study on microstructure and properties of Al_3Zr and Al_2O_3 reinforced aluminum matrix composites produced by melt reaction of Al-Zr-O system. Nanjing: Southeast University; 2001. (in Chinese).
13. Zheng M. Microstructure and properties of aluminum matrix composites synthesized bymelt reaction under ultrasonic field. Zhenjiang: Jiangsu University; 2011. (in Chinese).
14. Arsenault RJ, Shi N. Dislocation generation due to differences between the coefficients of thermal expansion. Mater Sci Eng. 1986;81(1–2):175–87.
15. Surappa MK, Rohatgi PK. Matrix/reinforcement development for aluminum-based composites. Mater Lett. 1986;16:562–4.
16. Bramfitt BL. The effect of carbide and nitride additions on the heterogeneous nucleation behavior of liquid iron. Metall Trans. 1970;1(7):1987–95.
17. Nutting J. Metallography of phase transformations. Int Mater Rev. 1973;18:37.
18. Fan H, Aubry S, Arsenlis A, et al. Grain size effects on dislocation and twinning mediated plasticity in magnesium. Scripta Mater. 2016;112:50–3.
19. Christman T, Needleman A, Nutt S, et al. On microstructural evolution and micromechanical modelling of deformation of a whisker-reinforced metal-matrix composite. Mater Sci Eng, A. 1989;107:49–61.

Chapter 7
Mechanical Properties of In-Situ Aluminum Matrix Composites

The endogenous reinforcement particles possess many advantages, such as small size, uniform distribution, and interfacial bonding well with the matrix. These features endow the in-situ aluminum matrix composites with impressive performances that are different from traditional composites, e.g., mechanical, thermal, electromagnetic, and optoelectronic properties. In this section, the mechanical properties of (Al_2O_3 + Al_3Zr)/Al, (ZrB_2 + Al_3Zr)/A356, ZrB_2/2024Al, and Al_2O_3/6063Al composites prepared under different physical fields (pulsed magnetic field, ultrasonic field, acoustic-magnetic coupling field) are introduced, and the corresponding fracture and strengthening mechanisms are discussed.

7.1 Mechanical Properties of In-Situ Aluminum Matrix Composites at Room Temperature

7.1.1 Mechanical Properties of Aluminum Matrix Composites Synthesized Under Pulsed Magnetic Field

Figure 7.1 shows the mechanical properties of (Al_2O_3 + Al_3Zr)/Al composites synthesized by the melt reaction of Al–$Zr(CO_3)_2$ system under an initial temperature of 850°C, the pulsed magnetic field intensity of 0–0.07 T, and $Zr(CO_3)_2$ addition of 5–25% for 20 min [1]. It can be seen from Fig. 7.1a that as the $Zr(CO_3)_2$ amount improves, the tensile strength σ_b and yield strength σ_s of the composites increase significantly, but the elongation δ decreases. When $Zr(CO_3)_2$ is added up to 20%, the σ_b and σ_s are 167.8 and 138.5 MPa, which are about 2.042 and 3.169 times pure aluminum matrix, respectively. However, once the addition level of $Zr(CO_3)_2$ exceeds 20%, the σ_b and σ_s fall. Figure 7.1b reveals that as the pulsed magnetic field intensity enhances, the tensile strength of the composite increases but δ decreases. For example, the intensity reaches 0.07 *T*, the σ_b and σ_s of the composite are 169.75

Y. Zhao, *In-Situ Synthesis of Aluminum Matrix Composites*,
https://doi.org/10.1007/978-981-16-9120-1_7

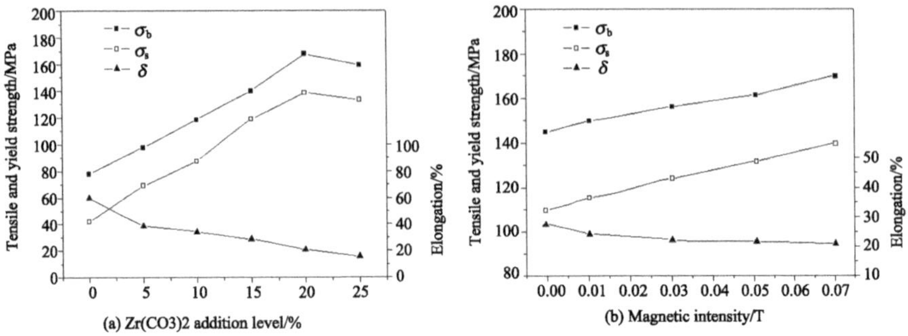

Fig. 7.1 Tensile properties of $(Al_2O_3 + Al_3Zr)/Al$ composite fabricated by Al–$Zr(CO_3)_2$ system

and 139.41 MPa, which are 18.02 and 28.33% higher than those of the composite without magnetic field application.

Figure 7.2 shows the fractography of $(Al_2O_3 + Al_3Zr)/Al$ composite after tensile tests at room temperature. There are many dimples and tearing edges on the fracture

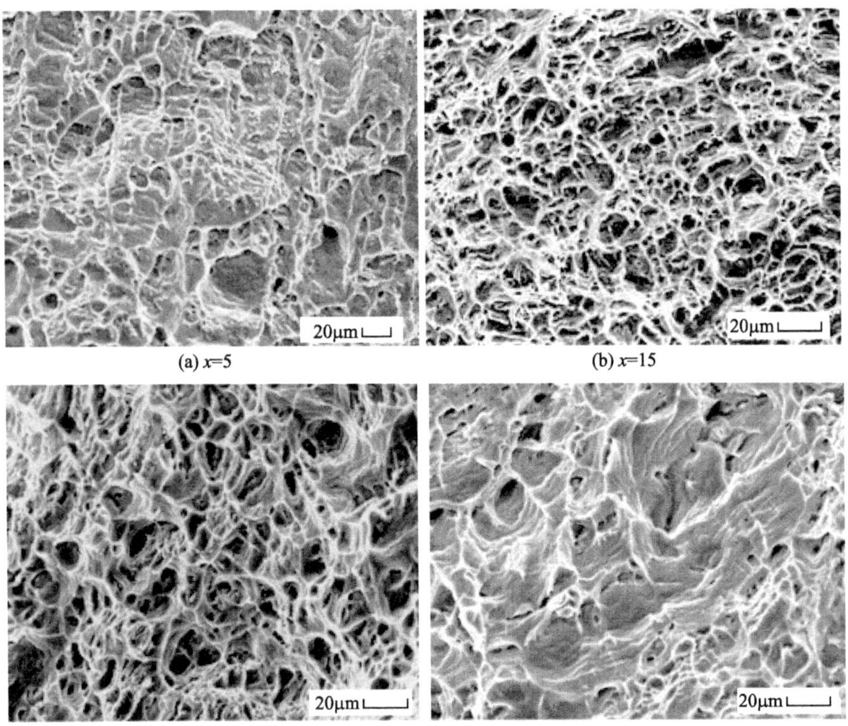

(a) x=5 (b) x=15 (c) x=20 (d) x=25

Fig. 7.2 SEM images of the tensile fracture surface of the as-prepared specimens fabricated from Al-x% $Zr(CO_3)_2$ system with $x = 5, 15, 20, 25$

surface. Fine particles are located within the dimples, and the tearing edges are dense, curved, and entangled with each other, which indicates a ductility fracture pattern. With the increase in $Zr(CO_3)_2$ amount, the number of dimples and tear edges on the surface are significantly reduced, and the dimple size becomes larger. In particular, when 25% $Zr(CO_3)_2$ is added, although the fracture pattern is not changed, the resulting elongation is relatively low since most dimples become shallow, and some flattened ones are connected (Fig. 7.2d). It is well known that the fracture process of metals includes two stages, i.e., microcrack initiation and microcrack,

propagation. However, for particle-reinforced metal matrix composites, the fracture behaviors generally relate to the particle size, distribution, and the interface bonding between the enhancement and matrix. In this experiment, although the reinforced particles are in-situ formed inside the matrix, however, they still belong to the external components for the Al matrix. The distorted crystal lattice and extra intrinsic energy induced by discontinuous ceramic particles in the matrix can effectively hinder the dislocation movement, explaining why the strength of composite improves. On the other hand, owing to the difference in the thermal expansion coefficient between ceramic particles and the matrix, thermal stress generates around the particles during the solidification process, which inevitably reduces the interfacial bonding during tensile tests. Therefore, although ceramic particles have two effects, i.e., strengthening and damaging, on the matrix, it seems that the strengthening effect plays an essential role in our works.

7.1.2 Mechanical Properties of Aluminum Matrix Composites Synthesized Under Ultrasonic Field

Figure 7.3 shows the mechanical properties of $(Al_3Zr + ZrB_2)$/A356 composites prepared from A356-20%$(K_2ZrF_6 + KBF_4)$ system under an initial temperature of 850°Cand the different ultrasonic power for 10 min [2]. The mass ratio of K_2ZrF_6 to KBF_4 is about 2.25:1. It is evident that as the ultrasonic power improves, the tensile strength σ_b, yield strength σ_s, and elongation δ of the composites increase. In particular, σ_s enhances remarkably. When the ultrasonic power reaches 1.2 kW, the σ_b and σ_s are 403.61 and 343.98 MPa, which are 16.09 and 12.9% higher than those of the composites without ultrasonic field application.

Figure 7.4 shows the fractography of as-fabricated composites after tensile tests at room temperature. As seen from Fig. 7.4a, there are many brittle and flat areas and fine tearing edges, accompanying locally small dimples on the surface, indicating a mixture model of quasi-cleavage and ductility fractures. With the increase of ultrasonic power, the dimple numbers increase gradually, and corresponding fracture characteristics tend to ductility pattern. As shown in Fig. 7.4d, many subtle dimples occupy the fracture surface, which is a typical ductility fracture model. The effect of ultrasonic power on fracture morphology of the composites is consistent with the variations in tensile properties at room temperature.

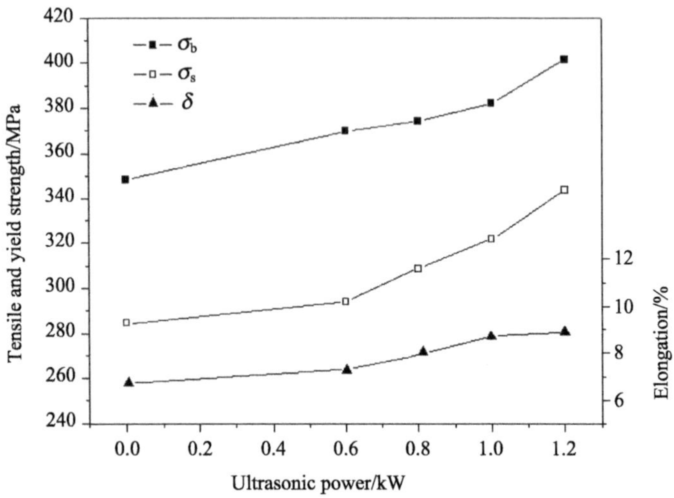

Fig. 7.3 Tensile properties of $(Al_3Zr + ZrB_2)$/A356 composites fabricated from Al-20wt.% $(K_2ZrF_6 + KBF_4)$ system under different ultrasonic power

7.1.3 Mechanical Properties of Aluminum Matrix Composites Synthesized Under Ultrasonic-Magnetic Coupling Field

Figure 7.5 shows the mechanical properties of 15vol.%$(Al_2O_3 + ZrB_2)$/A356 composites prepared from A356–K_2ZrF_6–KBF_4–$Na_2B_4O_7$ system under an initial temperature of 850 °C, and under ultrasonic field (1.0 kW, 3 min, intermittent mode), high-frequency magnetic field (0.10 T, 8 min, intermittent mode), and coupling fields for 20 min [3]. It is evident that when the melt reaction aided with ultrasonic-magnetic coupling fields, the tensile strength σ_b and elongation δ of the as-cast composites are as high as 302 MPa and 9.1%, which are 14.8 and 48.5% higher than those only under ultrasonic treatment, as well as 34.22 and 100% higher than those without physical field. Besides, due to the agglomeration of ZrB_2 particles, the fracture ductility of the composites is lower than the A356 alloy. However, the nano-sized ZrB_2 particles are ready to exfoliate from these aggregates and dispersed uniformly in the matrix under ultrasonic-magnetic coupling fields. Meanwhile, the primary α-Al grains are refined, and the eutectic silicon changes from needle-like to granular. The above microstructure modifications make a higher ductility for the composites than that of A356 alloy.

The effect of the ultrasonic field on melt contains two effects, i.e., acoustic flow and cavitation. When the ultrasonic field introduces intense oscillation in the melt, agglomerated particles are broken and dispersed well, during which more TiB_2 can act as heterogeneous nuclear. Thus, the significantly refined matrix grains lead to higher

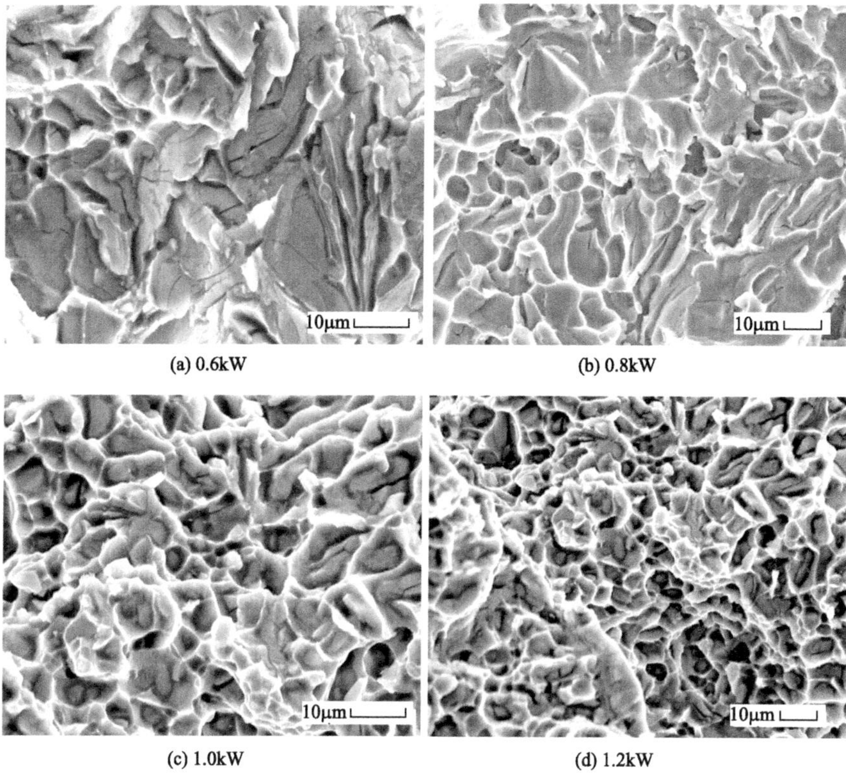

Fig. 7.4 SEM images of the tensile fracture surface of $(Al_3Zr + ZrB_2)/A356$ under different ultrasonic power

tensile strength and elongation [4]. The low-frequency magnetic field is mainly manifested in electromagnetic stirring. With the enhanced magnetic field frequency, the increasing electromagnetic stirring enforces more particles to take part in melt action. Simultaneously, the oscillation and forced convection can accelerate the diffusion rate of solute atoms and promote aggregates decomposition, which facilitates the reaction process. With the combination of acoustic flow, cavitation, and electromagnetic effects, the generated enhancement phases decompose into fine particles and rapidly diffuse out of reaction regions, offering nucleation core for the subsequent formed reinforcement phases. Such impacts also accelerate the in-situ melt reaction, where more Ti and B atoms react to generate enhancement particles that are distributed uniformly in the melt [5].

As is well known, the fracture process of metals includes two stages, i.e., microcrack initiation and microcrack propagation. However, for particle-reinforced metal matrix composites, the fracture behaviors generally relate to the particle size, distribution, and the interface bonding between the enhancement and matrix. Figure 7.6

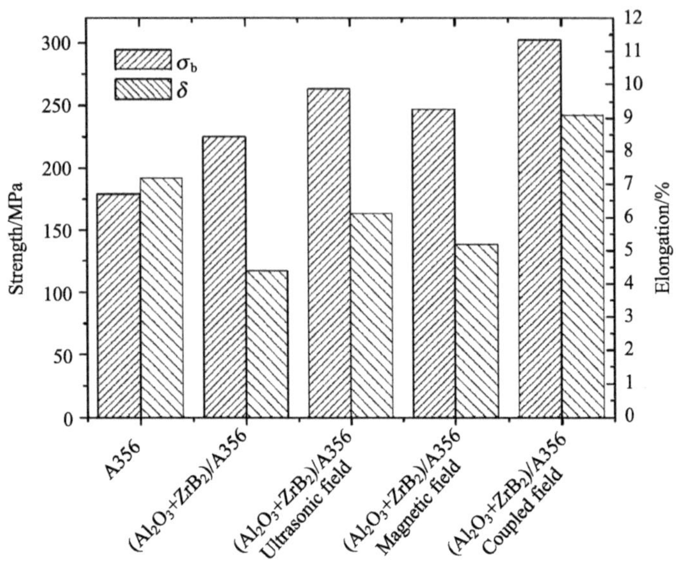

Fig. 7.5 Tensile properties of $(Al_2O_3 + ZrB_2)$/A356 composites synthesized under different situations

shows the tensile fractography of 15vol.% $(Al_2O_3 + ZrB_2)$/A356 composites synthesized under different physical fields. As seen in Fig. 7.6a, for A356 alloys, there are brittle fracture zones over a large area, accompanying dimples and tear edges locally, indicating a brittle–ductility fracture mechanism, and the corresponding elongation is 7.2%. Figure 7.6b gives the fractography of the composite synthesized by direct melt reaction. It is evident that the number of dimples on the fracture surface decreases, and more particles, brittle flat zones, and tearing edges appear. Many dimples become deep and large, within which some fine particles locate as aggregates. Therefore, the longation remarkably reduces from 7.2% for the matrix to 4.4% for the composites, with approximately 38.9% drop. The fracture mechanism is a typical mixture of the brittle and plastic patterns. When the ultrasonic field with 1.0 kW power is applied to the melt, more and subtle dimples replacing brittle flat zones dominate the fracture surface, meaning a transformation trend from cleavage fracture to ductility mechanism (Fig. 7.6c). Figure 7.6d shows the Tensile fractography of the composite under a high-frequency magnetic field. The fracture surface exhibits ductility characteristics, where brittle flat area decreases, and small and shallow dimples increase. Although the high-frequency magnetic field can improve the composites' mechanical properties to a certain extent, the agglomerated particles still occur within the dimples. When two physical fields couple together in the best way during melt reaction, it can be seen that more fine and uniform dimples dominate the whole fracture surfaces of the composites. That is a typical ductile fracture pattern, and the elongation is as high as 9.1% (as shown in Fig. 7.6e).

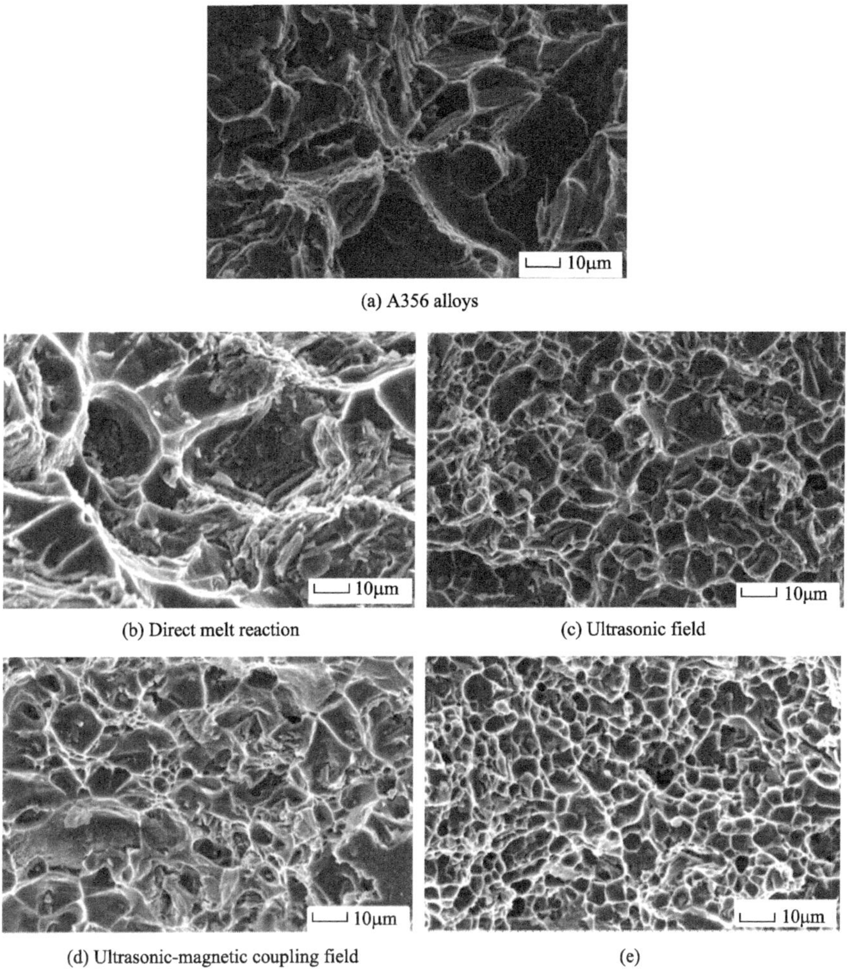

Fig. 7.6 Tensile fractography morphologies of composites synthesized under different situation

7.1.4 Mechanical Properties of In-Situ Aluminum Matrix Nanocomposites

Figure 7.7 shows the mechanical properties of ZrB_2/2024Al composites prepared from Al-K_2ZrF_6–KBF_4 system under an initial temperature of 870 °C and a pulsed magnetic field of 0.07 T for 30 min [6]. As seen in Fig. 7.7b, the tensile strengths increase as the ZrB_2 amount improves. Besides, it is worthy to note that the elongation is significantly enhanced when the ZrB_2 addition levels increase, which transgresses the previous research. When ZrB_2 reaches 5.4%, the tensile strength, yield stress,

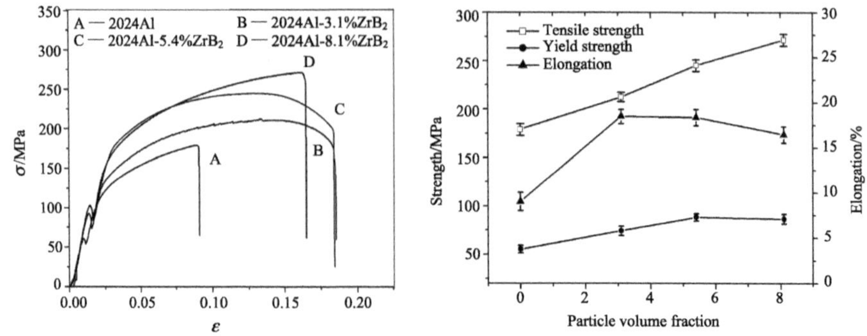

(a) Engineering stress-strain curves (b) Relationship between properties and particle volume fraction

Fig. 7.7 Tensile properties of the ZrB_2/2024Al nanocomposites

and elongation are 245 MPa, 88 MPa, and 18.3%, respectively, which are 1.36, 1.57, and 2.03 times greater than those of 2024Al alloy.

Figure 7.8 compares the fractography of 2024Al matrix and ZrB_2/2024Al composites after tensile test at room temperature. From Fig. 7.8a, the fracture surface of the 2024Al alloy exhibits brittle flat zones and tearing edges, with little and shallow dimples located along the grain boundary, which are characterized as cleavage fracture. Compared with the matrix alloy, the number of brittle flat zones and tearing edges remarkably reduces in the composites, corresponding to a quasi-cleavage pattern. In addition, it can be from Fig. 7.8d that loss of dimples are observed at the grain boundary, and the observed dimples are shallower and smaller in size. No residual particle is kept in the fracture place, indicating that the interfaces between in-situ particles and the aluminum matrix are well. The fracture types are intergranular and apparent ductile fracture, contributing to the increase of the tensile properties for the composites.

In the present reaction system, the composites' strengthening mechanisms mainly include load transfer, grain refinement and crack bridging, branching, and deflection [7]. Based on the load transfer mechanism, the reinforcements can bear the matrix's stress if their interface is well. In this work, no intermediate products are generated in the interface between the matrix and ZrB_2, meaning their relatively strong atom bonds. It seems that the well-bonded interface can delay particles' detachment from the aluminum matrix, resulting in an effective load transfer and strength enhancement. On the other hand, the refined grain size by adding particles increases the resistance for the dislocations' motion across grain boundaries. Hence, the tensile strengths and yield strengths of composites are improved by ZrB_2 nanoparticles. The elongation variation is a result of the cooperation of grain refinement and particle reinforcement. Firstly, the deformation uniformity increases due to the activated deformation in more crystal grains associated with grain refinement behavior. Besides, the smaller the grain size is, the more tortuous the grain boundary is. It is believed that tortuous boundaries are not conducive to the crack propagation. Thus, the ability to

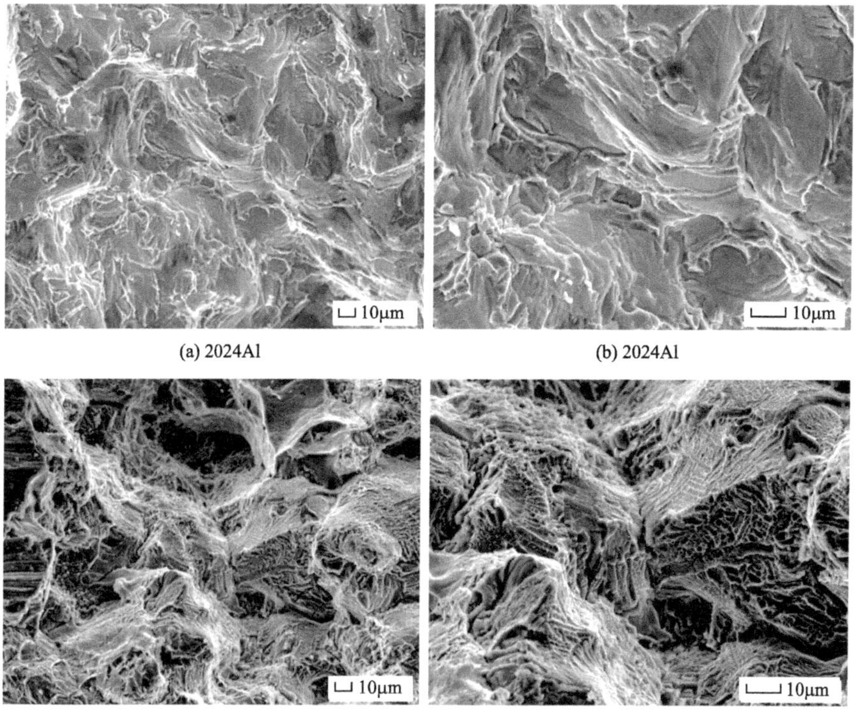

(a) 2024Al (b) 2024Al

(c) ZrB_2/2024Al nanocomposites (d) ZrB_2/2024Al nanocomposites

Fig. 7.8 Images of the tensile surface of the matrix and composites

withstand deformation before the fracture is enhanced. Secondly, the grain boundaries act as microcracks nucleation sites because of residual stresses induced by thermal elastic anisotropy of ZrB_2 particles during the solidification process. Then the microcrack develops locally at grain boundaries due to deformation subsequently. In this case, the ZrB_2 nanoparticles located at the grain boundaries can alter the crack growth path, i.e., crack bridging, branching, and deflection. These processes consume more energy and lead to an increase in crack propagation resistance, giving rise to the improvement of fracture toughness. It should be noted that these particles may act as barriers for dislocation migration, which leads to the elongation reduction, although this weakening effect is insignificant compared to the strengthening effect. Therefore, the elongation of the composites increases significantly when ZrB_2 nanoparticles are introduced. The above effect is considered to be more profound with the raises of the ZrB_2 volume percentage. However, the strengthening effect sensitively depends on the distribution of particles throughout the matrix. With the increase of ZrB_2 volume fraction, the extent of particle aggregation increases largely. During the tensile deformation, the particle aggregation will lead to early interface debonding and void formation in the ductile matrix. Also, the crack will be triggered and propagate easily in the particle aggregation region. Therefore, the elongation declines with the increase of particle volume fraction.

7.2 High-Temperature Mechanical Properties of In-Situ Aluminum Matrix Composites

7.2.1 *High-Temperature Tensile Properties*

7.2.1.1 Particle Content

The high-temperature tensile properties of 2024Al matrix and ZrB_2/2024Al nanocomposites with different volume fractions were tested at a temperature of 480°C and the strain rate $1 \times 10^{-3\ s^{-1}}$. Table 7.1 lists the elongation δ and flow stress σ of different materials. Figure 7.9 gives the variations in elongation and flow stress with respect to ZrB_2 nanoparticle content. It is evident that δ and σ of the 2024Al matrix alloy are 59.65% and 27.55 MPa, respectively, at conditions of 480°C and $1 \times 10^{-3\ s^{-1}}$ strain rate. In contrast, for the composites, δ and σ vary from 84.3 to 89.2%, 17.1 to 19.4 MPa, respectively. As the particle content increases, the elongation initially increases and decreases, but the flow stress decreases and then increases slightly. When the ZrB_2 addition level reaches 5.4%, the max elongation (89.2%) and relatively low stress (17.1 $_{\mathrm{MPa}}$) are achieved. Compared with the 2024Al matrix alloy, the composites' high-temperature properties are significantly enhanced, especially when ZrB_2 content is 5.4%, the elongation increases by 49.5%, and the flow stress reduces by 37.9%.

Table 7.1 Elongations and flow stresses of ZrB_2/2024Al nanocomposites under high-temperature tensile

Materials	Elongation/%	Mean elongation/%	Flow stress/MPa	Mean flow stress/MPa
2024Al	50.8	59.65	28.3	27.55
	48.5		26.8	
2024Al-3.1% ZrB_2	88.5	87.05	18.9	19.35
	85.6		19.8	
2024Al-5.4% ZrB_2	90.2	89.2	16.7	17.1
	88.2		17.5	
2024Al-8.1% ZrB_2	82.5	84.3	20.2	19.4
	86.1		18.6	

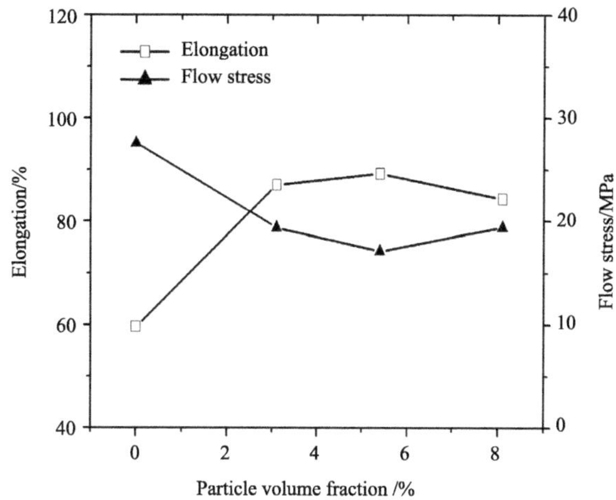

Fig. 7.9 Effect of particle volume fraction on high-temperature tensile properties

7.2.1.2 Deformation Temperature

The tensile tests of 5.4vol.%ZrB_2/2024Al nanocomposites were conducted at 1×10^{-3} s^{-1} and 420–510 °C to assess the influence of deformation temperature. Table 7.2 lists the high-temperature elongation δ and flow stress σ of composites. Figure 7.10 gives the relationship between deformation temperature and properties. The δ and σ vary remarkably as the temperature increases from 420 to 510 °C. As shown in Fig. 7.10a, the δ continuously improves as the temperature increases from 420 °C, and a max value of 92.3% occurs at 480 °C. Hereafter, the δ decreases as the temperature increases and finally drops to 73% at 510 °C. From Fig. 7.10b, the σ decreases with the temperature rise and reaches the minimum value of 16.6 *MPa* at 510 °C. Comprehensively comparison shows the composites exhibit the highest δ and the lower σ at 480 °C. Thus, the optimal deformation temperature is 480 °C. Generally,

Table 7.2 Elongations and flow stresses of samples tested under different deformation temperatures

Number	Deformation temperature/°C	Tensile properties			
		Elongation/%	Mean elongation/%	Flow stress/MPa	Mean flow stress/MPa
1	420	70.8	69.65	32.2	34
2	420	68.5		35.8	
3	450	78.2	77	17.4	19.5
4	450	75.8		21.5	
5	480	89.3	90.8	20.8	19.7
6	480	92.3		18.6	
7	510	72.5	73	16.4	16.6
8	510	73.5		16.8	

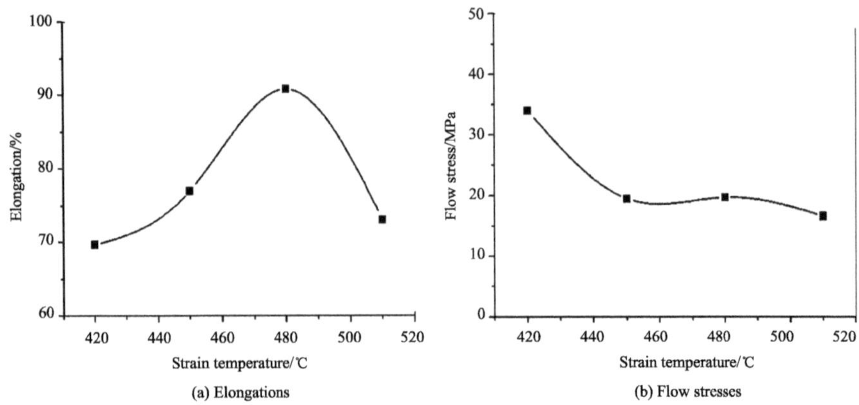

Fig. 7.10 Effect of deformation temperature on elongations and flow stress of nanocomposites

as the deformation temperature increases in a given range, the composites' flow stress decreases but elongation increases. This can be attributed to the high atom activation induced by high temperature and low critical shear stress, promoting grain boundary slip and improving grain boundary diffusion. On the other hand, high temperature leads to the grain growth. If the temperature is too high, e.g., close to or exceeds the over-burning point of Al alloys, the severe growth grains and reduced grain boundaries will restrict grain boundary slip, resulting in a sharp drop in the elongation of the composites.

7.2.1.3 Strain Rate

To determine the optimal strain rate, 5.4vol.%ZrB_2/2024Al nanocomposites were tested at the optimal deformation temperature of 480 °C and different strain rates. Table 7.3 lists the δ and σ of the composites, and Fig. 7.11 gives the relationship between the strain rate and tensile properties. It can be seen from Fig. 7.11a that δ periodically varies as the strain rate increases. When the strain rate equals $1 \times 10^{-3\ s-1}$, δ reaches 89.4%. From Fig. 7.11b, the σ of the composites exhibits a similar variation. When the strain rate equals $1 \times 10^{-3\ s-1}$, σ is relatively low, only 18.03 MPa. The results indicate that flow stress improves with the increase in strain rate, which can be attributed to the enhanced resistance of dislocation movement within grain and grain boundary slip, thereby leading to stress concentration. Therefore, it can be inferred from the curves that the highest elongation and the lowest flow stress of the composites can be achieved at a strain rate of $6 \times 10^{-4\ s-1}$.

Table 7.3 Elongation and flow stress of samples tested under different train rate

Number	Strain rate/s^{-1}	Mechanical properties			
		Elongation/%	Mean elongation/%	Flow stress/MPa	Mean flow stress/MPa
1	1×10^{-1}	65.8	68.15	48.56	49.42
2	1×10^{-1}	70.5		50.28	
3	1×10^{-2}	60.8	64.65	39.55	38.54
4	1×10^{-2}	68.5		37.52	
5	1×10^{-3}	88.2	89.4	17.46	18.03
6	1×10^{-3}	90.6		18.6	
7	1×10^{-4}	74.5	75.35	24.53	23.18
8	1×10^{-4}	76.2		21.82	

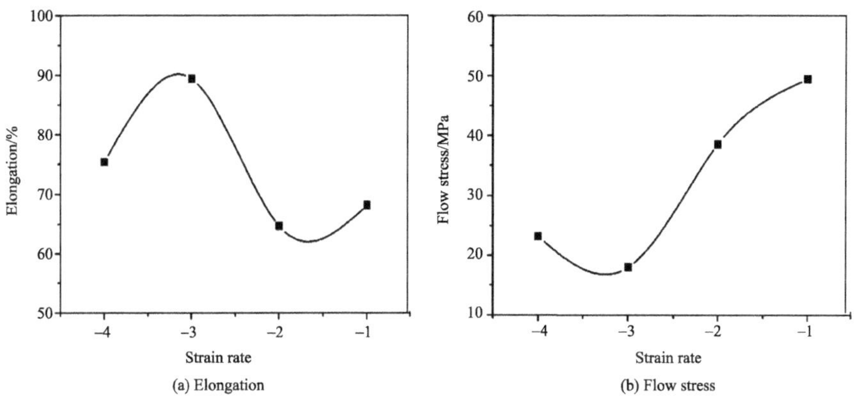

Fig. 7.11 Effect of strain rate on elongations and flow stresses of nanocomposites

7.2.1.4 High-Temperature Plastic Deformation Mechanism

Figure 7.12 shows the optical microstructure of the 2024Al matrix and 5.4vol.% ZrB_2/2024Al nanocomposite near fracture zone after testing at $1 \times 10^{-3\ s-1}$ strain rate and 480°C temperature. As seen in Fig. 7.12a, b, there are a large number of second-phase particles with a size of 5–10 μm in the 2024Al alloys. However, as shown in Fig. 7.12c, d, due to the addition of ZrB_2 nanoparticles, the presence of second-phase particles in the composites become fine and uniform, i.e., their size reduces from about 5 to 1 μm.

Figure 7.13 shows SEM images of the fractography, where the void growth mechanism controlled by plastic deformation dominates the fracture behavior. For the 2024Al matrix (Fig. 7.13a, b), the fracture surface macroscopically fluctuates, and some voids with different sizes can be observed. Figure 7.13b reveals that a specific horizontal connection occurs during voids vertically growing. Therefore, the fracture

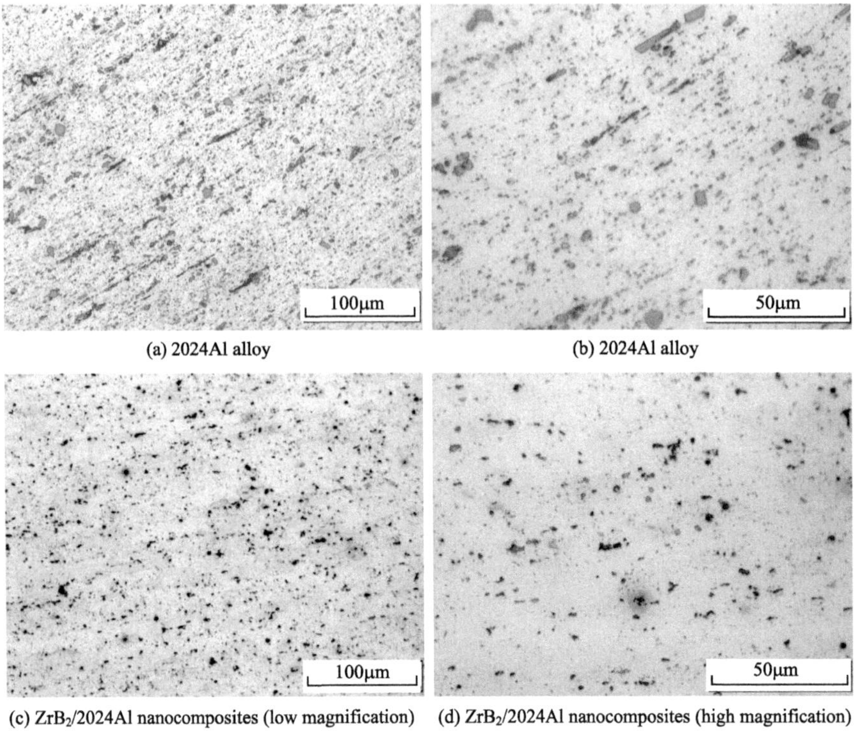

Fig. 7.12 Optical images of the area near the fracture of 2024Alalloys and ZrB_2/2024Al nanocomposites

process of the 2024Al matrix can be divided into void nucleation, growth, and connection, which leads to the final failure. Figure 7.13c–e exhibits the fracture morphology of 5.4vol.%ZrB_2/2024Al nanocomposite. The uniform and fine dimples appear on the flat surface, which is characterized as the void pattern of intergranular fracture. A magnification of the L zone located in Fig. 7.13 shows that there are many long metal whiskers, i.e., the trace of grain boundary slip, indicating that the grain boundary sliding has begun (Fig. 7.13e).

Many studies conclude that particle-reinforced metal matrix composites mainly contain three strengthening mechanisms, i.e., grain refinement, dislocation strengthening, and Orowan strengthening. At lower temperatures, these mechanisms can significantly enhance composites' strength. However, these strengthening mechanisms may fail at high temperature and low strain rate, resulting in lower flow stress for ZrB_2/2024Al composites than the 2024Al matrix at high temperature. For the grain refinement mechanism, high-temperature plastic deformation may be caused by the grain boundary slip. The presence of nanoparticles at the grain boundary can hinder grain growth due to Zener pinning effect. The finer the matrix grain, the lower the strength; Thus, easy grain boundary slip leads to the high elongation. Based on the dislocation strengthening mechanism, dislocation slip and climb can be readily

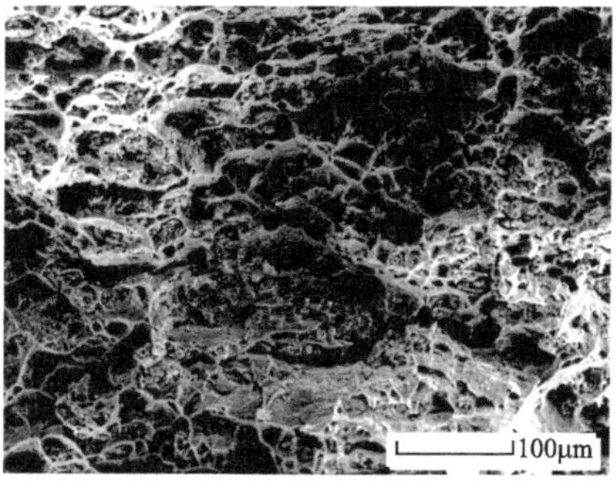

(a) 2024Al

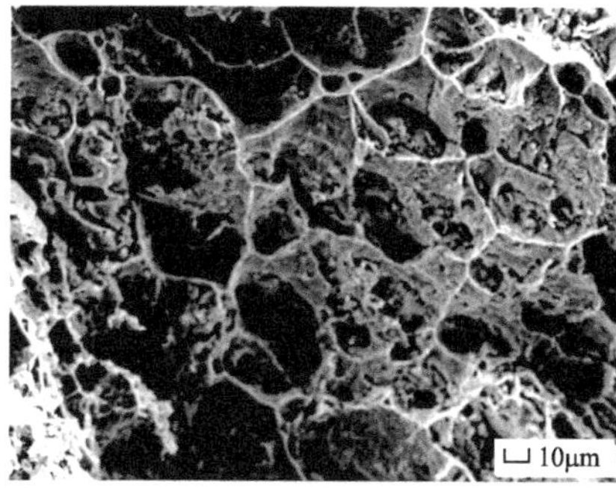

(b) 2024Al

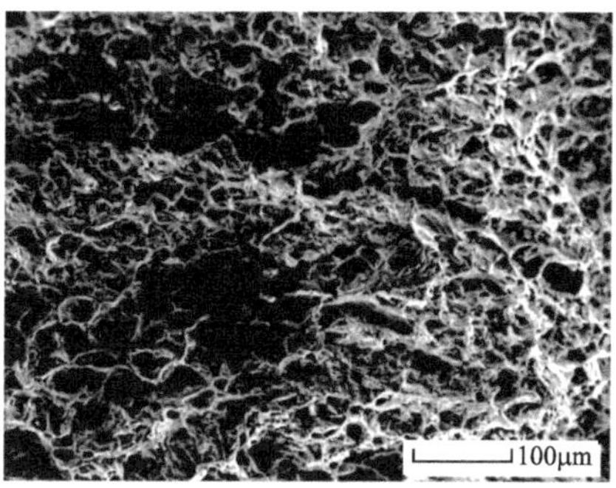

(c) ZrB2/2024Al nanocomposites

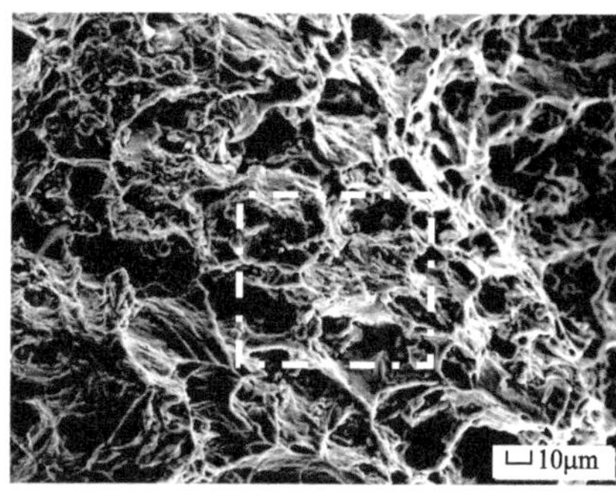

(d) ZrB2/2024Al nanocomposites

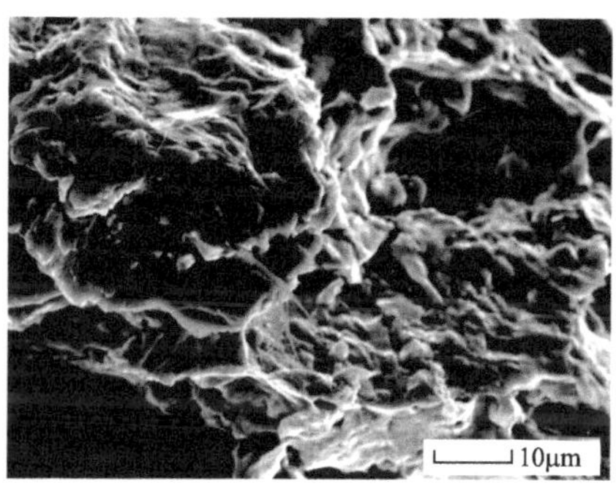

(c) ZrB2/2024Al nanocomposites

Fig. 7.13 SEM images of the high temperature tensile surface of 2024Al and ZrB_2/2024Al nanocomposites

activated at high temperatures, which leads to a low dislocation density. Moreover, low strain rates usually lead to sufficient recovery that is associated with dislocation contribution. Thus, the flow stress of composites decreases. For Orowan strengthening, since a high temperature will activate dislocation movement and reduce the shear elastic modulus, the dislocation can easily cross the reinforcement particles, and the corresponding Orowan strengthening effect is weakened.

7.2.2 *High-Temperature Creep Properties*

Creep means that a material slowly deforms under constant temperature and load over time. The accumulated damage finally leads to fracture, expressing different fracture patterns. Here, the creep behaviors of Al_2O_3/6063Al in-situ composites under different temperatures, stresses, and particle contents are introduced. The related fracture mechanisms are discussed based on the fracture characteristics.

7.2.2.1 Influencing Factors

Figure 7.14a shows the creep curves of 5vol.%Al_2O_3/6063Al composites tested at 525 K and different stresses. Although these curves contain three stages, the entire process shortens as the applied stress increases. It is evident that the applied stress has a remarkable influence on the second stage. When 50 MPa is used, the second stage lasts longer, and the curve keeps relatively stable. With the enhancement in the applied stress, this stage shortens, and the corresponding steep curve occurs. As the stress increases up to 90 MPa, the creep curve is quickly unstable and enters the third stage [8].

Figure 7.14b shows the creep curves of composites tested at 50 MPa and different test temperatures. The creep process is prolonged as the temperature decreases. Under 623 K, the second stage only keeps for a short time, and brittle fracture happens. By contrast, when the temperature equals 523 K, the curve becomes stable and lasts for tens of hours.

Figure 7.14c gives the relationship between steady-state creep rate and applied stress. It can be seen from Fig. 7.14 that as the temperature increases, the steady-state creep rate of the composites significantly improves. Under the same applied stress, the steady creep rate increases by 1–2 orders of magnitude for every 50 K increase, indicating that the temperature substantially influences the creep resistance. The long stress period lasts before failure at high temperature, which results in a lower elongation. That means that under high temperature and constant load conditions, slow and continuous plastic deformation occurs in the composites even if the applied stress is far low than the yield strength. Since the service performance at high temperatures deteriorates, it is not reasonable to use the composites' short-time high-temperature strength as a service criterion.

Figure 7.14d gives the relationship between the steady-state creep rate and applied stress of 5–7vol.%Al_2O_3/6063Al in-situ composites at a temperature of 523 K. At this temperature, the variation tendency of creep rate remains unchanged. As the particle addition level increases, the steady-state creep rate decreases by about one order of magnitude, indicating that the reinforcement content has a greater impact on the composite's creep resistance.

Figure 7.14e shows the creep fracture life of Al_2O_3/6063Al composites tested under different applied stress and 523–623 K. At the same temperature, the higher the particle volume fraction, the longer the composite's creep life. Besides, under

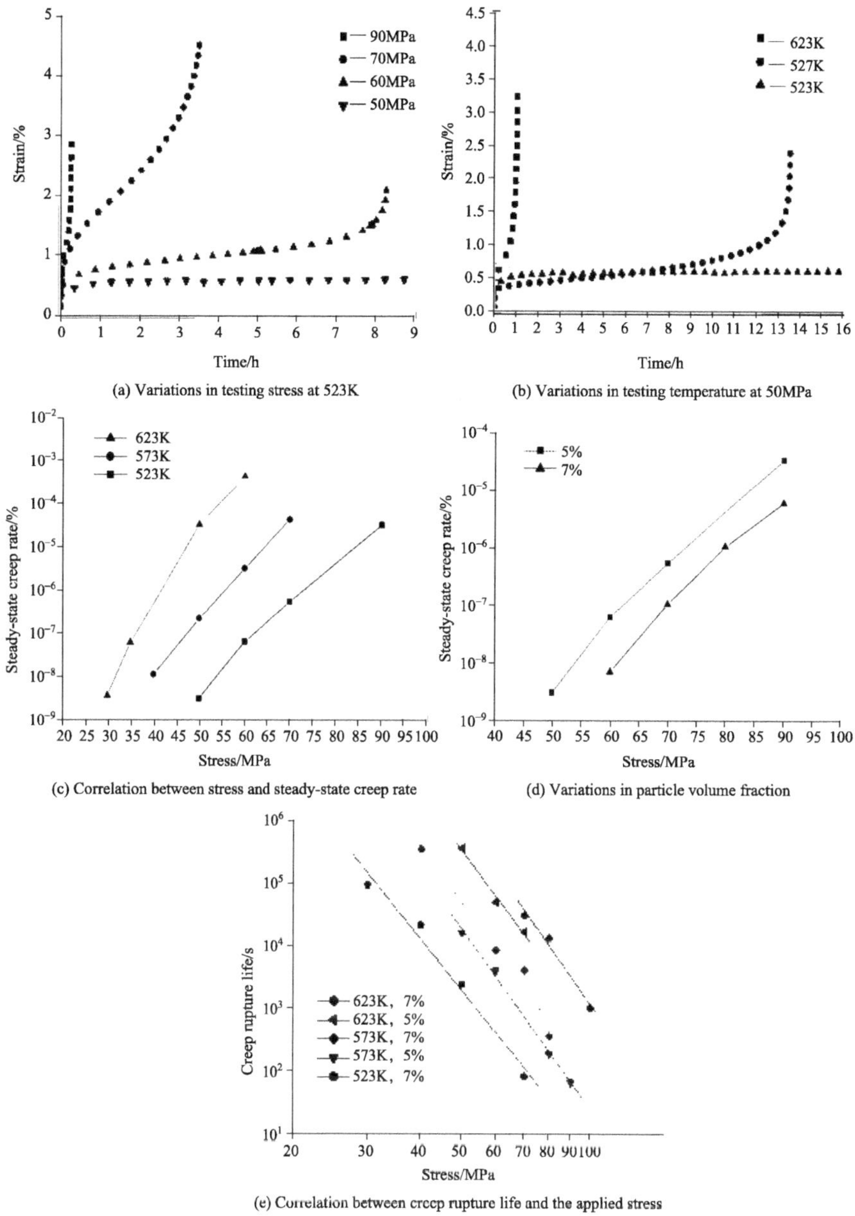

Fig. 7.14 Creep properties of Al_2O_3/6063Al composites under different conditions

the same stress, the improved particle content can also significantly increase the composites' creep life. It means that the incorporation of reinforcement particles is beneficial to improve the creep resistance of the composites.

7.2.2.2 Constitutive Equation of High-Temperature Creep

For Al_2O_3/6063Al composites, the steady-state creep rate can be represented as a power law of the applied stress as:

$$\dot{\varepsilon} = A\sigma^n \exp\left(-\frac{Q}{RT}\right) \tag{7.1}$$

where $\dot{\varepsilon}$ is the steady-state creep rate; A is a constant related to material properties; σ is the applied stress; n is the nominal creep stress index; Q is the apparent creep activation energy, R is the gas constant; and T is the absolute temperature. Figure 7.15 shows the curves treated using a double logarithm of the applied stresses and steady-state creep rates obtained from the experiment. Table 7.4 lists the calculated values of the apparent stress exponent of the Al_2O_3/6063Al composites. The composites' nominal stress index is much larger than that of pure aluminum (n = 3–5, Q = 142 kJ/mol). It is noted that there is no large difference in the nominal stress exponents

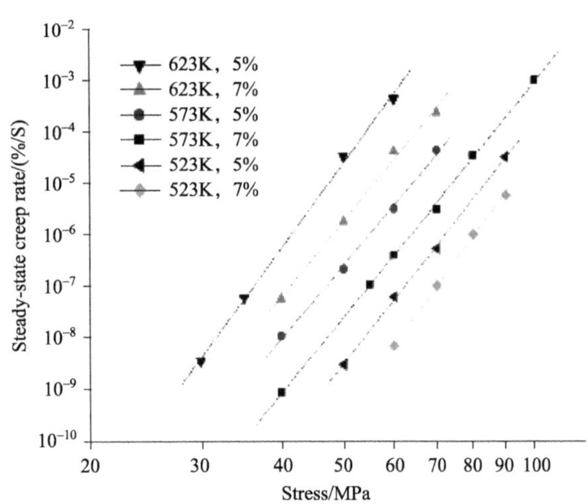

Fig. 7.15 Relation between the applied stress and the steady-state creep rate of in-situ Al_2O_3/6063Al composites

Table 7.4 Apparent stress exponent of the Al_2O_3/6063Al composite at different temperatures

Materials	Apparent stress		
	523 K	573 K	623 K
5vol.%Al_2O_3/6063Al	15.57493	14.64718	16.94162
7vol.%Al_2O_3/6063Al	16.58836	15.08671	15.05654

of Al_2O_3/6063Al composites with different contents at high temperatures. This is because of these reinforcement particles with similar content and size, and the same high-temperature creep mechanisms for the composites.

Assuming that the steady-state creep rate is only sensitive to the temperature, Q and A can set as constants at different temperatures. Based on the relationship between strain rate and temperature under the same stress, the apparent activation energies of 5vol.% and 7vol.% Al_2O_3/6063Al composites are calculated to be 194 and 201 kJ/mol, respectively. Generally, Q depends on the creep mechanism. Herein, Q of the composites is greater than that of pure Al ($Q = 142$ kJ/mol) and improves as the particle content increases. The results indicate that the Al_2O_3/6063Al composites possess higher creep resistance than matrix alloys, and increasing particle content is an effective means to improve the creep resistance of the composite. A similar conclusion has been drawn in the works of the creep behavior of Al, 6061Al, and 2024Al matrix composites [9]. The high creep stress index and creep activation energy in Al_2O_3/6063Al composites can be related to effective stress, σ-σ_0. That means that creep occurs when the applied stress is larger than the critical value. Therefore, Eq. (7.1) can be modified as:

$$\dot{\varepsilon} = B\left(\frac{\sigma - \sigma_0}{E}\right)^N \exp\left(-\frac{Q_0}{RT}\right) \tag{7.2}$$

where B is a constant associated with material properties; σ_0 is the threshold stress; E is the elastic modulus; N is the true stress exponent; and Q_0 is the true activation energy. Taking a double linear fitting of $\dot{\varepsilon}^{1/N}$ and applied stress, the threshold stress can be achieved by a linear regression method at $\dot{\varepsilon} = 0$ if a linear relationship occurs. For Al_2O_3/6063Al composites, the values of N are taken as 3, 5, and 8 for the creep processes of dislocation slide, high-temperature dislocation climb controlled by lattice self-diffusion, and a constant structure model, respectively. In this case, the N is equal to 8, giving the best linearity of the creep data, which indicates that the creep mechanism of Al_2O_3/6063Al composite is controlled by a constant structure model (Fig. 7.16).

Table 7.5 lists the threshold stress values achieved by linear extrapolation at $N =$ 8. The fabricated composites exhibit improved threshold stress with the increase of volume fraction of in-situ nanoparticles. Subsequently, the creep data are normalized treated with an aim to describe the creep mechanism. The curves of normalized effective stress, $(\sigma - \sigma_0)/E$ vs the creep rate, $\dot{\varepsilon}/DL$ are plotted in Fig. 7.17 using a double logarithmic method. The fitting slope from creep data of Al_2O_3/6063Al composites at different temperatures closes to 8, indicating that the creep mechanism is dominated by a constant structure model controlled by lattice self-diffusion [10]. The above results are the same as other reports for TiB_2/Al, SiC_w/Al, (Al_2O_3 + TiB_2)/Al, (Al_2O_3 + TiB_2)/Al-Cu composites, and so on.

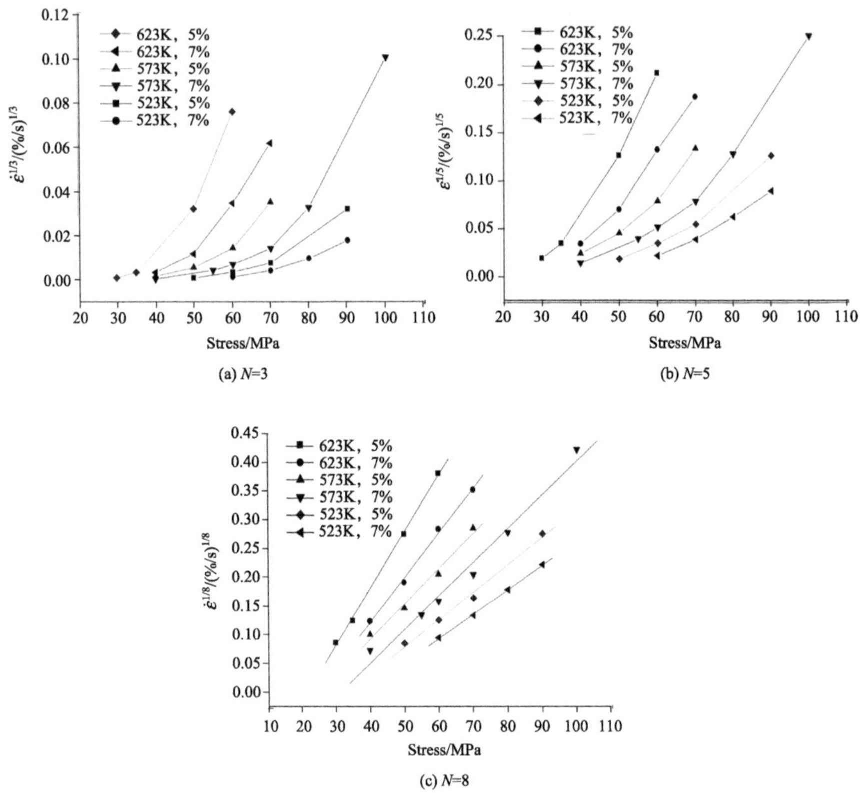

Fig. 7.16 Plots of 1/3, 1/5, and 1/8 power of the steady-state creep rate vs applied stress for the Al_2O_3/6063Al composites

Table 7.5 Threshold stresses of the Al_2O_{3p}/6063Al composite at different temperatures (MPa)

Materials	Threshold stresses		
	523 K	573 K	623 K
5vol.%Al_2O_3/6063Al	31.24553	23.02841	21.08301
7vol.%Al_2O_3/6063Al	37.5934	27.20822	23.24075

7.2.2.3 High-Temperature Creep Fracture Mechanism

Figure 7.18 shows that the fracture surface of Al_2O_3/6063Al composites is dominated by a rock-candy pattern in microscopical, which belongs to intergranular fracture. The rapid fracture zone exhibits transgranular fracture morphology, but dimple-like holes have appeared on the grain boundary, as shown in Fig. 7.18b. As is well known, depending on stress level, temperature, and strain rate, the creep failure can be divided into the brittle pattern along grain boundaries or ductile pattern across matrix grains.

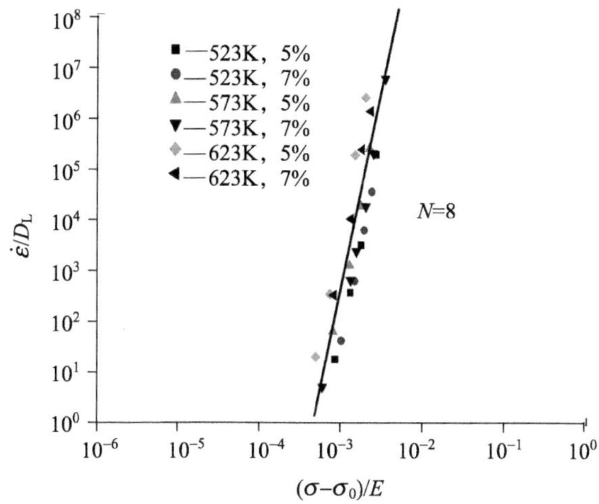

Fig. 7.17 Relationship of effective stress compensated to the modulus of Al_2O_3/6063Al nanocomposites and steady-state creep rate normalized to the diffusion coefficient

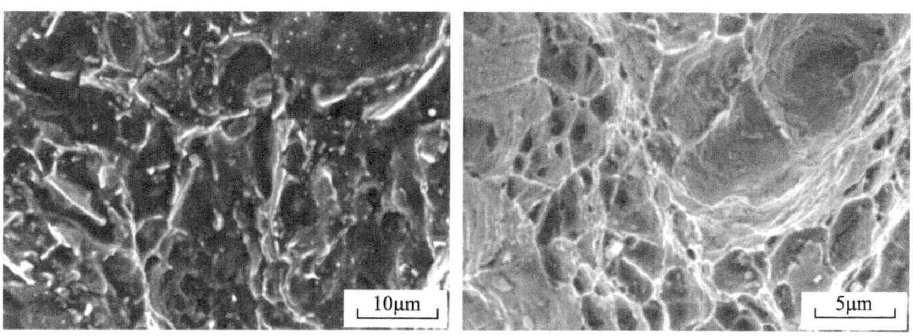

(a) Overall morphology (b) Microporous grain boundaries

Fig. 7.18 SEM images of creep fracture of Al_2O_3/6063Al composite

The former usually happens under low stress, low temperature, or strain rate, while the latter only occurs under high stress, high temperature, or strain rate.

Creep fracture occurs via intergranular crack growth that develops through two routes. One is the initiation of wedge-shaped cracks at the trigeminal grain boundary, and the other is the grain boundary cavity. As shown in Fig. 7.19, Al_2O_3/6063Al composites generally exhibit wedge-shaped cracks under high stress and low temperature. Since triple grains' junction acts as a hinder for grain boundary slip, some holes are inevitably formed around stress concentration and further torn into cracks. As slip continuously proceeds, the expansion and connection of wedge-shaped cracks at grain boundaries finally lead to material intergranular damage. When the Al_2O_3/6063Al composites undergo creep at low stress and high temperature, grain boundary cavities are preferentially formed at the protrusions or around the fine

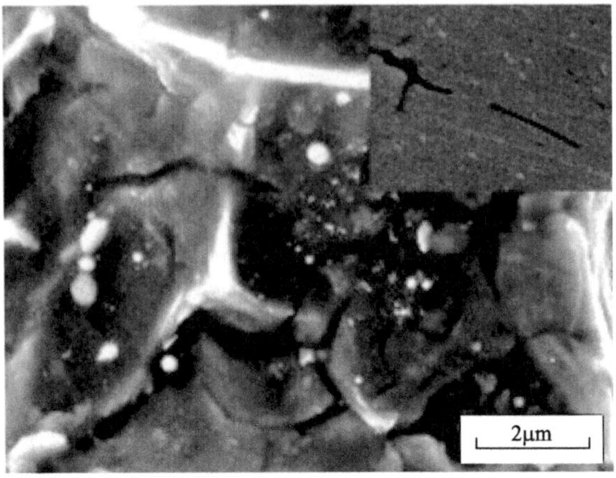

Fig. 7.19 Wedge-shaped crack on the grain boundaries of the Al_2O_3/6063Al composite formed under stress of 70 MPa at 523 K

precipitation phase, as shown in Fig. 7.20. It is well known that many non-equilibrium vacancies generated under high temperatures tend to aggregate into small holes. If applying constant stress, these holes will grow up due to long time diffusion of atoms, resulting in microhole aggregation fracture. Such behavior may be the initial crack nucleation way during high-temperature creep of the composites. Although the microholes are similar to conventional dimples in morphology, there are some differences in nature. The appearance of the microholes along the grain boundaries is a result of atom diffusion, while dimples are mainly caused by the multi-slip, corresponding to the grain destruction.

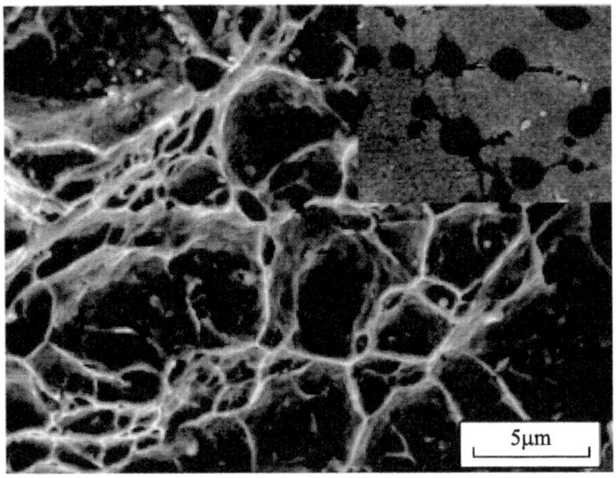

Fig. 7.20 Holes on the grain boundaries of the Al_2O_3/6063Al composite formed under the stress of 70 MPa at 623 K

7.3 Tensile Failure Behavior of In-Situ Aluminum Matrix Composites

For in-situ aluminum matrix composite synthesized with physical fields, fracture observation is a post-analysis, then the actual damage process and affect factors are difficult to be determined. Thus, SEM equipped with a tensile accessory is used to help to illuminate the initiation and growth for the cracks and rupture behaviors under dynamic applying load for the as-prepared composites. Based on the observed phenomena, the factors that influence the composites' fracture behavior are discussed, and the relevant fracture model is established.

7.3.1 In-Situ Tensile

Figure 7.21 shows the SEM image of the ($Al_3Zr + Al_2O_3$)/Al composite in-situ tensile process [11]. At the beginning of loading, the gray–white fine slip zone appears in Fig. 7.21a. As the load increased, the slip zone appeared staggered, as shown in Fig. 7.21b. At the same time, microscopic slip zones were observed in dense areas of slip zones or around particles. The appearance of cracks and rips, and the rips are short and thin. Figure 7.21f shows the high magnification structure of the staggered slip zone shown in Fig. 7.21b. When the load continues, these rips become thicker, longer, and connected. The cracks' growth and connection tend to be along the intersection line of the slip zone, some along the particle's edge, and some through the particle, causing the particle to break and forming a crack as shown in Fig. 7.21c. As the stretching continues, the main cracks after initiation further expand and extend, as shown in Fig. 7.21d. At the same time, some small microcracks continue to be produced near the main cracks. The propagation direction of the main crack is along with the interface between the particle-enriched area and the depleted area of the crack, or when it encounters a high strain area close to the direction of crack propagation during the propagation process, the crack tip becomes blunt and the expansion of the crack is hindered. As the load continues to increase, the main crack widens and new cracks expand upward. When the load reaches a certain level, the main crack connects with the new crack, as shown in Fig. 7.21e, and the composite finally fractures completely.

According to the above in-situ tensile investigations and the theory, the crack initiating mechanisms for the in-situ ($Al_3Zr + Al_2O_3$)/Al composites are discussed below.

When the composites are subject to pull-force, the obvious interleaved slipping bandings are observed during the in-situ tension for the in-situ ($Al_3Zr + Al_2O_3$)/Al. The dislocations are accumulated while meeting the in-situ particles or the crystal boundaries in the expanding course. These result in stress concentration. Besides, the stress concentrations are generated at the converging point for two dislocation bandings. The stress concentrations lead to the crack breach initiations and growths.

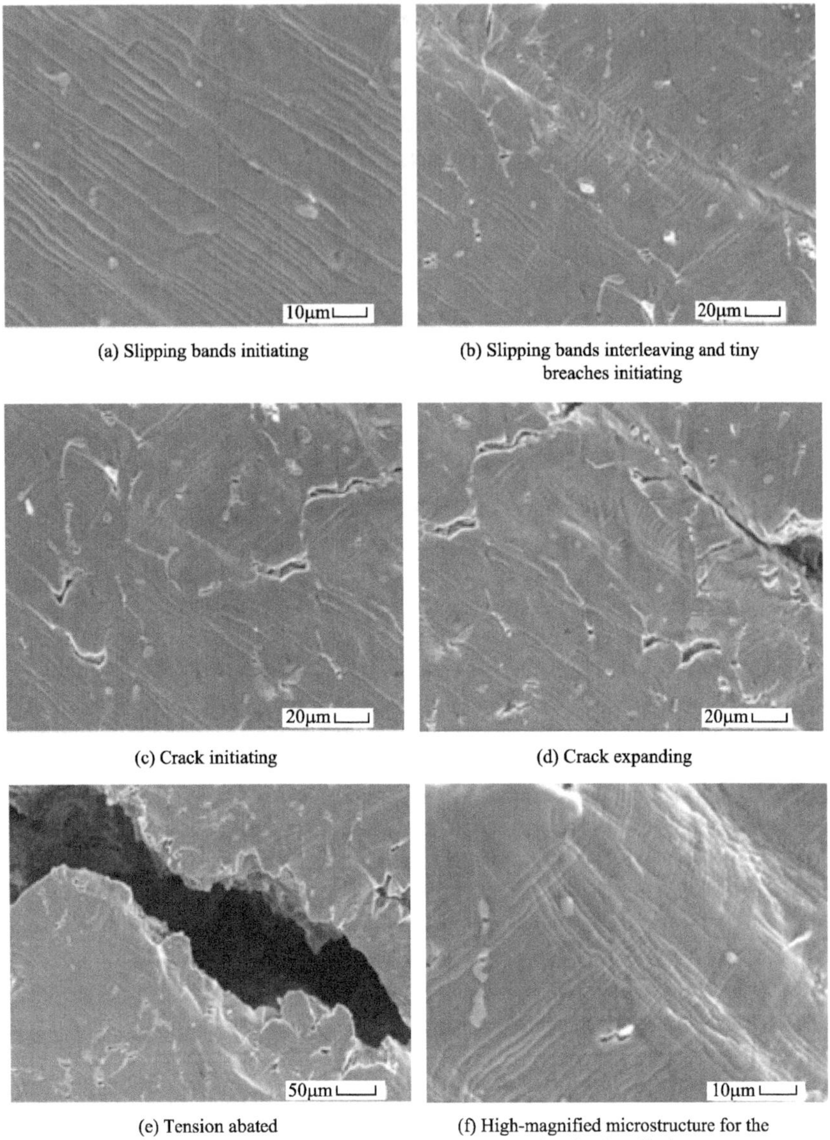

(a) Slipping bands initiating

(b) Slipping bands interleaving and tiny breaches initiating

(c) Crack initiating

(d) Crack expanding

(e) Tension abated

(f) High-magnified microstructure for the interleaving slipping

Fig. 7.21 SEM micrographs for the in-situ tension of the ($Al_3Zr + Al_2O_3$)/Al composites

The model for the crack initiation caused by the dislocation actions during the slipping course is shown in Fig. 7.22.

According to the dislocation accumulation theory [12], the model for the particles standing force during the pull-off course is established as shown in Fig. 7.23. $\overline{\sigma}_x$ is

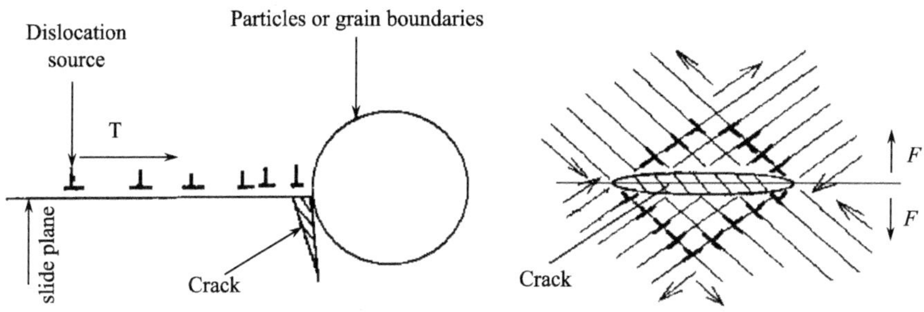

(a) Dislocation accumulation for the slipping band

(b) Dislocations converge for two different slipping bands

Fig. 7.22 Model of crack initiation due to the dislocation accumulation

Fig. 7.23 Model for the particles standing force during the pull-off course

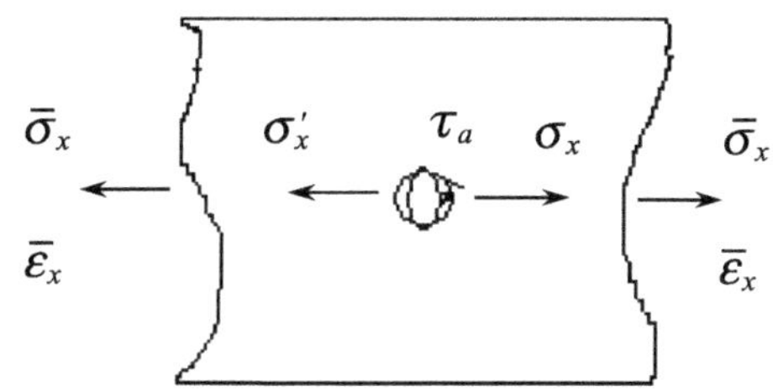

the tensile stress applied on the ends, and τ_a is the component shearing stress of the $\overline{\sigma}_x$ in the dislocation accumulation slippage surface.

The criterion for the interface between the in-situ particles and the aluminum matrix disengaging is following:

$$\sigma_x' + \xi \cdot n\overline{\sigma}_x \geq \sigma_{th} \tag{7.3}$$

where σ_{th} is the interface bond strength of the in-situ particles and the aluminum matrix; σ_x' is the tensile stress applied on the end of the particles due to the deformations of the in-situ particles and the aluminum matrix; $\xi \cdot n\overline{\sigma}_x$ is the stress concentration generated by the dislocation accumulation; n is the minimal dislocation numerical measurement for the interface disengaging; and ξ is the coefficient related to the structures of the materials and the deforming conditions, $0 < \xi < 0.5$.

In order to explain the model, the in-situ particles are simplified as an ellipsoid and further as a short cylinder with a height a and a diameter b. The stress $sigma_x'$ applied on the end of the in-situ particles is expressed as:

$$\sigma_x' = \kappa\varepsilon_x = \left(1 + \frac{\sqrt{2}}{2} \cdot \frac{a}{b}\right)^{\theta} \cdot \overline{\sigma}_x \tag{7.4}$$

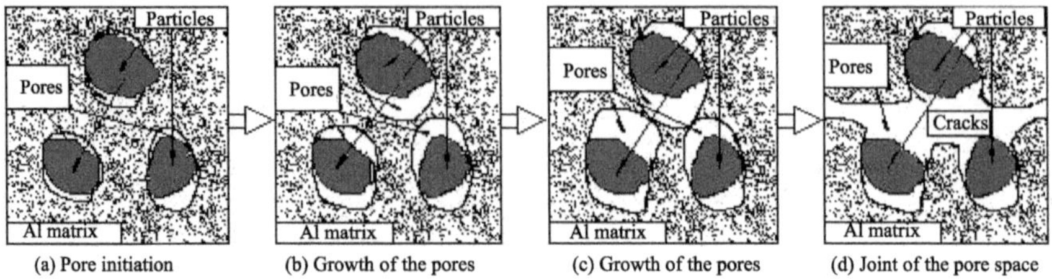

Fig. 7.24 Crack initiation caused by the joint of the pore space near the particles

where k is the coefficient for the strain hardening and θ the exponent. Using Eqs. (7.3) and (7.4), we arrive at:

$$n \geq \frac{\overline{\sigma}_{th} - \left(1 + \frac{\sqrt{2}}{2} \cdot \frac{a}{b}\right)^{\theta} \cdot \overline{\sigma}_x}{\xi \cdot \overline{\sigma}_x} \tag{7.5}$$

From Eq. (7.5), we can know that the minimal dislocation numerical measurement for the interface disengaging is depended on the interfacial bonding strength, applied load, and the properties of the matrix materials.

During the initial stage of the tensile course, the (Al_3Zr + Al_2O_3)/Al composites show obvious plastic deformation due to the in-situ particles' cooperation motions and the aluminum matrix, and the sharp slipping bandings are observed. When the plastic deformation is developed to some degree, the cooperation motions are impeded because of the dislocation accumulations and the crystal boundaries. Thus, enormous stress concentrations are generated in the in-situ particles' interfaces and the aluminum matrix. When the slipping shearing stress is larger than the interface bonding strength and the in-situ particles' shearing strength, the in-situ particles are disengaged from the aluminum matrix and broken, respectively, and the cavities are formed near the particles. The growth and connections of the cavities induce the initiation of the cracks. The model for the crack initiation caused by the cavities due to the particles disengaging is shown in Fig. 7.24.

In the fabrication course, the aluminum is degassed and refined with slag removed. However, there are a small number of gas voids and enclosed slags as shown in Fig. 7.25. When the aluminum matrix is tensiled and deformed, the stress concentration is generated near the gas voids or the enclosed slags, and the cracks are initiated.

7.3.2 Strengthening Mechanisms

According to the research and analysis of mechanical properties and in-situ tensile properties of the aluminum matrix composites reinforced by endogenous particles,

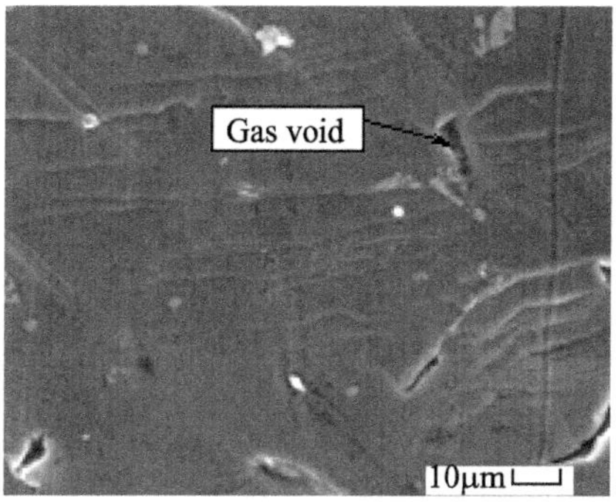

(a) Gas void in the aluminum matrix

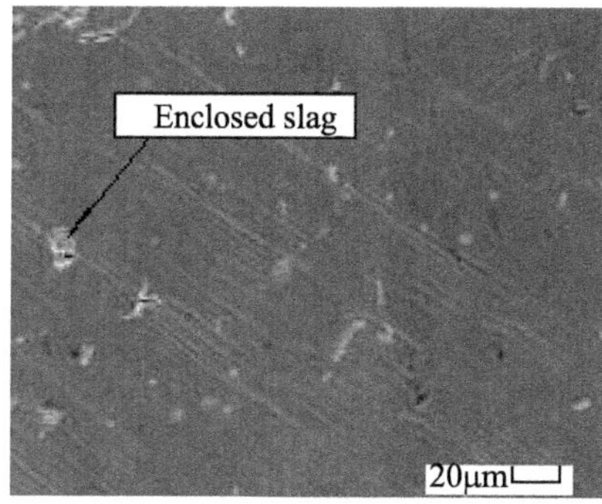

(b) Enclosed slag in the aluminum matrix

Fig. 7.25 Crack initiation caused by defects in the matrix

the strength of the composites at room temperature increases obviously with the increase of the total integral number of endogenous particles. The in-situ tensile test of the composite confirmed that the growth of the crack is selective. Due to the uneven distribution of the reinforced particles, when the main crack propagates in front of the particle-intensive area, the direction of expansion will be deflected, and the particles will be depleted bypassing the dense particle area, and expand and extend forward along with the interface between the dense particle area and the depleted area. The strengthening mechanisms of (Al_3Zr + Al_2O_3)/A356 composites mainly include Orowan strengthening, dispersion strengthening, fine-grain strengthening, solid solution strengthening, and dislocation strengthening. These strengthening mechanisms can be expressed by the following linear relationship:

$$\sigma_{\text{composite}} = \sigma_{\text{Orowan}} + \sigma_{\text{grain}} + \sigma_{\text{solution}} + \sigma_{\text{dislocation}} \tag{7.6}$$

7.3.2.1 Orowan Strengthening

Orowan strengthening is caused by the hindrance of dislocations when they pass through fine and hard particles in close distances. In the endogenous Al_3Zr and Al_2O_3 particle-reinforced aluminum matrix composites, due to the small endogenous particles and high particle volume fraction, most of the particles are distributed in the grain boundaries. From Orowan's dislocation hindrance theory, it can be known that the smaller the particle spacing, the greater the curvature of the dislocation when it passes around the particle, the more excellent the resistance to the movement of the dislocation, and the higher the strength of the material. The particle spacing λ can be expressed by the following formula:

$$\lambda = D\left(\frac{\pi}{6V_f} - \frac{2}{3}\right)^{1/2} \tag{7.7}$$

where D is the particle diameter; and V_f is the particle volume. According to the Orowan calculation formula [13]:

$$\sigma_{\text{Orowan}} = \frac{2Gb}{2\pi(1-\nu)^{1/2}}\frac{1}{\lambda}\ln(D/b) \tag{7.8}$$

where G is the shear modulus of the matrix; b is the Burgers vector; ν is Poisson's ratio; D is the average diameter of the particles; and λ is the particle spacing. Based on Eqs. (7.7) and (7.8), we can obtain

$$\sigma_{\text{Orowan}} = \frac{2Gb}{2\pi(1-\nu)^{1/2}}\frac{1}{D\left(\frac{\pi}{6V_f} - \frac{2}{3}\right)^{1/2}}\ln(D/b) \tag{7.9}$$

It can be seen from Eq. (7.9) that the higher particle volume fraction V_f leads to small particle spacing λ. Under certain conditions of the matrix, the ability of the composite to resist deformation is improved. For aluminum matrix composites, according to the literature, $G = 26900$ MPa, $b = 2.8 \times 10^{-10}$ m, $\nu = 0.34$. Under our experimental conditions, the composite particle volume fraction after applying the pulsed magnetic field is significantly improved compared to those without the pulsed magnetic field. The average particle diameter is reduced considerably with the size of 0.08–0.15 μm. The Orowan contribution is evident after applying a pulsed magnetic field. Set V_{f1} and D_1 to be the particle volume fraction and average particle size of the composite prepared under a pulsed magnetic field, respectively. Set V_{f2} and D_2 to be the particle volume fraction and average particle size of the composite without a pulsed magnetic field. For Fig. 3.7, particle volume fraction of the composite prepared by applying a pulsed magnetic field is $V_{f1} = 12\%$, and the average particle size is $D_1 = 0.12$ μm. By contrast, the particle volume fraction of composite prepared without a pulsed magnetic field is $V_{f2} = 8\%$, and the average particle size is $D_2 = 1.5$ μm. According to Eq. (7.9), the Orowan strengthening contribution after applying the pulsed magnetic field can be calculated as:

$$\begin{aligned}\Delta\sigma_{\text{Orowan}} &= \Delta\sigma_{1\text{Orowan}} - \Delta\sigma_{2\text{Orowan}} \\ &= \frac{2Gb}{2\pi(1-\nu)^{1/2}}\left(\frac{\ln(D_1/b)}{D_1\left(\frac{\pi}{6V_{1f}} - \frac{2}{3}\right)^{1/2}} - \frac{\ln(D_2/b)}{D_2\left(\frac{\pi}{6V_{2f}} - \frac{2}{3}\right)^{1/2}}\right) = 21.26\,\text{MPa}\end{aligned}$$

Therefore, the Orowan effect in the in-situ Al matrix composite prepared under the pulsed magnetic field is significant and cannot be ignored.

On the other hand, for (Al_3Zr + Al_2O_3)/A356 composites, since A356 is an age-hardening matrix, the influence of Al_3Zr and Al_2O_3 reinforcements on the distribution and size of the Mg_2Si precipitates in the matrix A356 alloy is significant. The (Al_3Zr + Al_2O_3)/A356 composite matrix synthesized under a pulsed magnetic field has dense dislocations distributed in a network and is connected to the aging precipitation. This indicates that the introduction of in-situ particles significantly increases the matrix's dislocation density, and the dislocations are beneficial to the precipitation of the aging precipitates and make the precipitates fine and uniform. Therefore, the particles in the composite material play a role in promoting and strengthening the precipitated phases.

7.3.2.2 Fine-Grain Strengthening

The grain structure is one of the main factors that affect the mechanical and physical properties of as-casted metal matrix composites. When the dislocations in the grains move to the vicinity of the grain boundaries, they are hindered by the grain boundaries, resulting in dislocation plugging. If the deformation continues to increase, external forces must be increased, and the more refined the crystal grains, the greater the proportion of grain boundaries, and the more hindrance of dislocation movement. Besides, due to the orientation difference between adjacent crystal grains, each crystal grain's slip tendency is different. When a crystal grain undergoes plastic deformation, it will be restrained and hindered by the surrounding crystal grains and coordinated with elastic deformation. This elastic deformation becomes the resistance to the continued plastic deformation, promoting the opposition to plastic deformation of the polycrystal. There are more impurity atoms in the grain boundary, which hinder the plastic deformation and enhance the above influence. The refinement of the grains can also improve the plasticity and toughness of the metal. This is because the more refined the crystal grains, the greater the number of crystal grains in a specific volume. With the same amount of deformation borne by more crystal grains, the deformation is more uniform and does not cause serious stress concentration, so the metal can produce greater plastic deformation. Without breaking, the metal shows good plasticity. Since the grain boundaries of refined grains are twisted and bent, it helps restrain the propagation of cracks. If the cracks pass through the grain boundaries, more energy is consumed, so the toughness is improved.

At present, there are two main mechanisms for the grain refinement of the composite material matrix: one is the heterogeneous nucleation mechanism on the surface of the reinforcement in the matrix; the other is the heat exchange mechanism between the reinforcements and the matrix alloy. For Al_3Zr and Al_2O_3 reinforced aluminum matrix composites, since the reinforced phases are generated from the melt through a chemical reaction, Al_3Zr and Al_2O_3 can offer heterogeneous nucleated sits for α-Al phase and Si phase, respectively. Due to the presence of Al_3Zr and Al_2O_3 particles, the α-Al dendrites are refined into small equiaxed grains, while

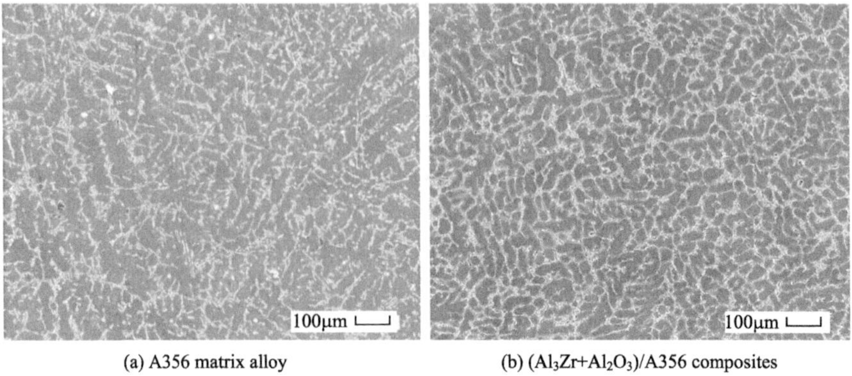

Fig. 7.26 SEM microstructures of A356 matrix alloy and ($Al_3Zr + Al_2O_3$)/A356 composites

the eutectic silicon changes from long needles to short rods or granules. Therefore, the grain refinement mechanism of the in-situ ($Al_3Zr + Al_2O_3$)/A356 composite originates from the matrix's heterogeneous nucleation.

According to the Hall–Petch relationship, the grain refinement contribution to the strength of the composite can be expressed as [14]:

$$\sigma_{grain} = \sigma_0 + kd^{-1/2} \tag{7.10}$$

where σ_{grain} is the yield strength related to the grain size; σ_0 is the frictional resistance of dislocations within the grains; d is the average grain diameter; and k is a constant related to the material. At room temperature, for face-centered cubic metals, k is about 0.05–0.1 MPa·m$^{1/2}$. For body-centered cubic metals, k is about 0.3–1.8 MPa·m$^{1/2}$.

For ($Al_3Zr + Al_2O_3$)/A356 composites synthesized after applying a pulsed magnetic field, the matrix grain size (d_1) is significantly smaller than that of the A356 matrix alloy (d_2), as shown in Fig. 7.26. Therefore, the yield strength increase due to grain refinement can be expressed as:

$$\Delta\sigma_{grain} = k(d_1^{-1/2} - d_2^{-1/2}) \tag{7.11}$$

If $k = 0.1$ MPa·m$^{1/2}$, $d_1 = 4$ μm, $d_2 = 10$ μm, $\Delta\sigma_{grain} = 18.45$ MPa can be achieved.

7.3.2.3 Solid Solution Strengthening

When foreign atoms are dissolved in the matrix, it can hinder the movement of dislocations. On the other hand, the foreign atoms and the metal atoms of the matrix have different sizes, it will produce lattice distortion, generate strain fields, and interact with the dislocations. The solute atoms act as an obstacle to the movement of dislocations and increase the resistance to plastic deformation. The reasons can be summarized in two aspects. First, the solute atoms cause lattice distortion and increase the dislocation density. The degree and solubility of lattice distortion caused by solute atoms differ between solute atoms and solvent atoms and the way of dissolution. The greater the number of solute atoms dissolve, the more serious the lattice distortion and corresponding larger strengthening contribution. Second, the interactions between solute atoms and dislocations make the dislocations stable. Solute atoms tend to distribute around dislocations. In replacement solid solution, solute atoms with sizes smaller than the solvent atoms tend to diffuse to the pressure position above the edge dislocation. In contrast, solute atoms that are larger than the solvent atom or solute atoms in interstitial solid solution tend to diffuse to the tensile position below the edge dislocation, forming a "Cottaell gas mass." The Korotkoff gas mass formation reduces the dislocation energy and stays in a stable state, restraining the dislocation movement. If dislocations break from the Korotkoff gas mass constraints and move, a greater external force must be exerted, which expresses the increase in deformation resistance.

Whether it is a displacement solid solution or an interstitial solid solution, there is a symmetric strain field. According to the literature [15], the interaction between this strain field and dislocations can be given as follows.

$$\sigma_{\text{solute}} = G\varepsilon\sqrt{x_f/4} \tag{7.12}$$

where G is the shear elastic modulus; ε is the change rate between the foreign atom's diameter and the diameter of the matrix; and x_f is the concentration of the foreign atoms.

It can be seen from Eq. (7.12) that the greater the change rate of the diameter difference between the foreign atoms and the matrix atoms and the higher the concentration of foreign atoms in the matrix, the stronger the solid solution strengthening ability. For the $(Al_3Zr_+ Al_2O_3)$/A356 composites synthesized under a pulsed magnetic field, Zr atoms are dissolved in α-Al. It is known that the atomic volume of Zr is 0.023272nm^3, and the atomic volume of Al is 0.016603nm^3. Since their size difference will produce lattice distortion and strain field, solid solution strengthening can play an important role in the composites. According to the literature [16], $G = 26900$ MPa, $\varepsilon = 0.119$, $x_f = 0.1\%$. From Eq. (7.12), the strengthening contribution is:

$$\sigma_{\text{solute}} = G\varepsilon\sqrt{x_f/4} = 16\,\text{MPa}.$$

It is evident that in the $(Al_3Zr_+ Al_2O_3)$/Al composites, the solid solution strengthening induced by Zr has a specific contribution to the strength.

7.3.2.4 Dislocation Strengthening

Based on predictions of Arsenault and Shi [17], an increase in dislocation density caused by granular particles can be obtained.

$$\overline{p} = \frac{\Delta\alpha\Delta T N A}{b} \tag{7.13}$$

where $\Delta\alpha\Delta T$ is the thermal mismatch strain; N is the number of particles; b is the Burgers vector; and A is the total surface area of each particle.

While Miller and Humphrey assumed that the shape of the particle is a cuboid, they obtained similar expressions:

$$\overline{p} = 12\frac{\Delta\alpha\Delta T V_f}{bd} \tag{7.14}$$

where d is the particle size; and V_f is the volume fraction of the particle.

From the above equations, it is clear that the dislocation density increases as the particle size decreases; meanwhile as the particle volume fraction improves, the dislocation density also increases, and then the strength increases. In the ($Al_3Zr_+ Al_2O_3$)/A356 composites, since reinforcement particles with small size and high volume fraction possess large thermal expansion coefficients from the matrix, many dislocations are generated around these particles, thereby increasing the strength of the composites.

There are always dislocations that occur in metal. Calculations from crystal integrity and strength theory show that dislocation elimination is a way to enhance strength in the crystal. In contrast, the crystal defect theory points out that an increase in the dislocation density can also increase strength effectively. In particle-reinforced metal matrix composites, the vast difference in the thermal expansion coefficient between the reinforcement and the matrix will result in large thermal stress. Then the plastic deformation, in turn, increases the dislocation density, which can be calculated as [18]:

$$\Delta\rho = \frac{\Delta\alpha\Delta T N A}{b} \tag{7.15}$$

where $\Delta\alpha = \alpha_1 - \alpha_2$, α_1 and α_2 are the thermal expansion coefficients of the matrix and reinforcement, respectively; $\Delta T = T - T_0$, T_0 is the initial temperature, T is the instantaneous temperature; N is the number of particles; b is the Burgers vector; and A is the total surface area of each particle. Assuming that the particle shape is spherical, a similar expression is obtained:

$$\Delta\rho = \frac{2\Delta\alpha\Delta T V_f}{bd} \tag{7.16}$$

where V_f is the particle volume fraction; and d is the particle size.

It can be seen from Eq. (7.16) that as the particle volume fraction increases and the particle size decreases, the more dislocation density is generated. The contribution of dislocation density to the yield strength of the matrix is usually expressed by the following formula [19]:

$$\Delta\sigma_{\text{dislocation}} = AGb\sqrt{\Delta\rho} \tag{7.17}$$

combination of Eqs. (7.16) and (7.17), we can obtain:

$$\Delta\sigma_{\text{dislocation}} = AGb\sqrt{\frac{2\Delta\alpha\Delta T V_f}{bd}} \tag{7.18}$$

It is assumed that the particle content and the particle diameter generated under a pulsed magnetic field are V_{f1} and d_1, while without a pulsed magnetic field, the particle content and the particle diameter are V_{f2} and d_2. The strengthening contribution of dislocation strengthening after applying the pulsed magnetic field is:

$$\Delta'\sigma_{\text{dislocation}} = AGb\left(\sqrt{\frac{2\Delta\alpha\Delta T V_{f1}}{bd_1}} - \sqrt{\frac{2\Delta\alpha\Delta T V_{f2}}{bd_2}}\right) \tag{7.19}$$

For the 12vol.%(Al_3Zr + Al_2O_3)/A356 composite with the aid of a pulsed magnetic field, the particle size is about 0.12×10^{-6} m, $\alpha_{Al_3Zr} = 12.1 \times 10^{-6}$ m/°C, $\alpha_{Al_2O_3} = 9 \times 10^{-6}$ m/°C, $\alpha_{Al} = 22.41 \times 10^{-6}$/°C, $b = 2.8 \times 10^{-10}$ m, if the temperature variation $\Delta T = 150$°C, according to Eq. (7.19), $\Delta'\sigma_{\text{dislocation}} = 24.85$ MPa.

Therefore, for the in-situ particle-reinforced aluminum matrix composites synthesized under a pulsed magnetic field, due to the small particle size and high volume fraction, the large thermal mismatch strain leads to high dislocation density, which greatly increases the mechanical properties of composites. Therefore, based on Eq. (7.6), $\sigma_{\text{composite}} = 80.56$ MPa.

The experimentally measured yield strength σ_{s} of the composite material prepared with and without a pulsed magnetic field are 339.74 MPa and 278.95 MPa, respectively. Then the enhanced strength $\Delta\sigma_{\text{b}} = 60.79$ MPa, showing that the predicted calculated value is close to the experimental value although there is also a certain gap. This is mainly due to many complicated factors during the quantification of various strengthening mechanisms, such as particle size and particle volume fraction. According to the above analysis, it can be concluded that the strengthening mechanisms of in-situ Al matrix composites mainly contain Orowan strengthening, fine-grain strengthening, solid solution strengthening, and dislocation strengthening.

References

1. Zhang S, Zhao Y, Chen G, et al. Microstructures and mechanical properties of aluminum matrix composites fabricated from Al–*x* wt.% $Zr(CO_3)_2$ ($x = 5, 10, 15, 20, 25$) systems. J Alloys Compd. 2007;429:198–203.
2. Zhang S, Zhao Y, Chen G, et al. Microstructures and dry sliding wear properties of in situ ($Al_3Zr + ZrB_2$)/Al composites. J Mater Process Technol. 2007;184:201–8.
3. Yang H, Zhao Y, Chen G, et al. Preparation and microstructure of in-situ $(ZrB_2+Al_2O3+Al_3Zr)_p$/A356 f7composite synthesized by melt direct reaction. Trans Nonferrous Met Soc China. 2012;22:571–6.
4. Zhang S, Dong X, Zhao Y, et al. Preparation and wear properties of TiB_2/Al–30Si composites via in-situ melt reactions under high-energy ultrasonic field. Trans Nonferrous Met Soc China. 2014;24:3894–900.
5. Zhao Y, Zhang S, Chen G, et al. Dynamics responses of the in situ magnetochemistry reaction for Al-$Zr(CO_3)_2$ system. J Alloy Compd. 2008;457:164–70.
6. Tian K, Zhao Y, Jiao L, et al. Effects of in situ generated ZrB_2 nano-particles on microstructure and tensile properties of 2024Al matrix composites. J Alloy Compd. 2014;594:1–6.
7. Kueck AM, Ramasse QM, de Jonghe LC, et al. Atomic-scale imaging and the effect of yttrium on the fracture toughness of silicon carbide ceramics. Acta Mater. 2010;58:2999–3005.
8. Yutao Z, Weili L, Xizhou J, et al. Microstructure and high temperature creep properties of nano Al_2O_3/6063Al composites synthesized in situ by sonochemical. J Compos Mat. 2015;32(5):1399G1407.
9. Tjong SC, Ma ZY. High-temperature creep behaviour of powder-metallurgy aluminium composites reinforced with SiC particles of various sizes. Compos Sci Technol. 1999;59:1117–25.
10. Ma ZY, Tjong SC. Creep deformation characteristics of discontinuously reinforced aluminium-matrix composites. Compos Sci Technol. 2001;61:771–786.
11. Zhang S, Zhao Y, Chen G, et al. ($Al_2O_3+Al_3Zr$)/A356 nanocomposites fabricated by magnetochemistry in situ reaction. J Alloy Compd. 2009;475:261–7.
12. Zuhan L. Crystal defects and mechanical properties of metals. Beijing: Metallurgical Industry Press; 1988. (in Chinese).
13. Zhang Q, Xiao BL, Wang WG, et al. Reactive mechanism and mechanical properties of in situ composites fabricated from an Al-TiO_2 system by friction stir processing. Acta Mater. 2012;20:7090–103.
14. Hsu CJ, Chang CY, Kao PW, et al. Al-Al_3Ti nanocomposites produced in situ by friction stir processing. Acta Mater. 2006;54:5241–9.
15. Verhoeven JD. Fundamentals of Physical metallurgy. New York, NY: John Wiley & Sons Inc.; 1975. p. 518–24.
16. Bai M, Xue Q, Guo H. Reciprocal sliding wear of SiC particle reinforced Al-Cu aluminum matrix composites against stainless steel, high speed tool steel and ceramics. I. Tribological properties and construction of tribo-maps. Wear, 1996;195:152–161.
17. Arsenault RJ, Shi N. Dislocation generation due to differences between the coefficients of thermal expansion. Mater Sci Eng. 1986;81:175–87.
18. Alizadeh M. Strengthening mechanisms in particulate Al/B_4C composites produced by repeated roll bonding process. J Alloy Compd. 2011;509:2243–7.
19. Liu ZY, Xiao BL, Wang WG, et al. Analysis of carbon nanotube shortening and composite strengthening in carbon nanotube/aluminum composites fabricated by multi-pass friction stir processing. Carbon. 2014;69:264–74.

Chapter 8
Plastic Forming of In-Situ Aluminum Matrix Composites

In order to make practical metal matrix composite material components, the metal matrix composite material needs to be subjected to secondary plastic forming processing [1]. Since the addition of reinforcements brings great difficulties to the secondary processing of metal matrix composites and the reinforcements have high hardness and wear resistance, which makes the plastic forming processing of this composite material very difficult [2, 3]. The processing requirements and difficulties of different types of metal matrix composite components are very different. For continuous fiber-reinforced metal matrix composite components, the plastic forming process is generally completed in the composite process, supplemented by a small amount of cutting processing and connection. Short fibers, whiskers, and particle-reinforced metal matrix composite materials can be made into practical components by secondary processing such as casting, plastic forming, and welding.

8.1 Hot Extrusion of In-Situ Aluminum Matrix Composites

Hot extrusion is extrusion by heating the metal material to the hot forging forming temperature, i.e., extruding by heating the billet to a temperature above the recrystallization temperature of the metal before extrusion. Hot extrusion is the earliest used extrusion technology. It carries out various extrusion forming of metals at the temperature of hot extrusion bits good plasticity [4, 5]. At present, hot extrusion is mainly used to manufacture equal cross sections long parts, profiles, pipes, bars, and various machine parts. Hot extrusion can not only form non-ferrous metals and their alloys with good plasticity and relatively low strength, low and medium carbon steels, etc., but also form high-carbon and high-alloy steels with higher specific strength, such as special structural, stainless steel, high-speed tool steel and heat-resistant steel, etc. The hot extrusion plastic processing technology is used in in-situ aluminum-based composite materials to improve the microstructure of the as-cast composite material, reduce the as-cast defects, and improve the performance.

Y. Zhao, *In-Situ Synthesis of Aluminum Matrix Composites*,
https://doi.org/10.1007/978-981-16-9120-1_8

8.1.1 *The Effect of Hot Extrusion on the Microstructure of $Al_2O_{3(p)}$/6063Al Composites*

Figure 8.1 shows the optical images of the $Al_2O_{3(}p_{)}$/6063Al composite ingot before and after extrusion. Comparing of the microstructure of the $Al_2O_{3(}p_{)}$/6063Al composite before and after extrusion, there is no obvious plasticity processing defects such as cracks after extrusion. The structure of the extruded $Al_2O_{3(}p_{)}$/6063Al composite is fibrous and the as-cast structure disappears completely, the gray part has a fine subcrystalline structure. This is because the composite material grains are extruded in the *X* and *Z* axes directions. During the pressing process, it is squashed by the pressing force, and the overall direction of the crystal grains tends to be parallel to the *Y* axis.

Figure 8.2 is the SEM images of the $Al_2O_{3(}p_{)}$/6063Al composite before and after extrusion. The Al_2O_3 particles in the 6063Al matrix are elliptical; while part of Al_2O_3 particles before extrusion have sharp corners, and there are many inclusions in the

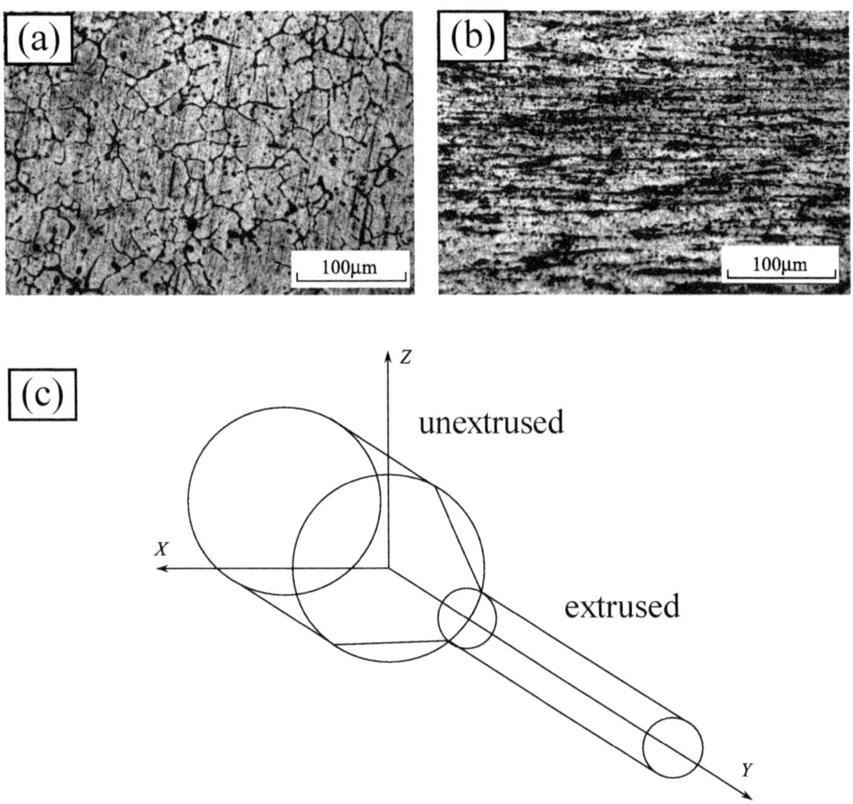

Fig. 8.1 Optical micrographs of composite materials before and after extrusion **a** as-cast; **b** hot extrusion; **c** direction of thermal extrusion

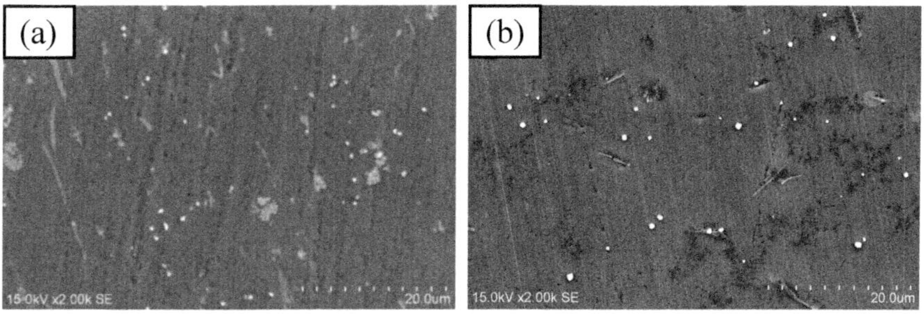

Fig. 8.2 SEM images of $Al_2O_{3(p)}$/6063Al composite: **a** as-cast; **b** hot extrusion

system [6, 7], the particle distribution is not uniform (Fig. 8.2a). In the structure of the extruded composite material, the shape of Al_2O_3 particles is relatively rounded, and the dispersion is more uniform. The inclusions in the structure are broken, broken, and dispersed in the matrix.

Due to the influence of the extrusion die during the hot extrusion process, the force on the *X* and *Z* directions of the sample is greater than the force in the *Y* direction. Therefore, the grains are squeezed in the *Y* direction and squashed in the *X* and *Z* directions. As the composite material has undergone significant plastic deformation, the relative position of the Al_2O_3 particles has changed significantly, thus greatly improving the distribution of reinforcement particles in the composite material [8].

8.1.2 *The Effect of Thermal Extrusion on Structure of ZrB_2/6063Al Composites*

Figure 8.3 shows the SEM images of ZrB_2/6063Al composite before and after extrusion. It can be seen from Fig. 8.3 that the ZrB_2 reinforcement particles in the composite material are dispersed and the agglomeration phenomenon is significantly

Fig. 8.3 SEM images of ZrB_2/6063Al composite before and after extrusion: **a** before extrusion; **b** after squeezing

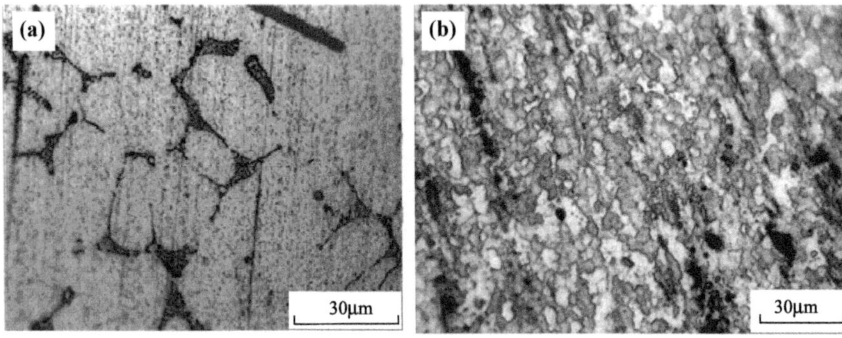

Fig. 8.4 Comparative optical micrographs of cross section of ZrB_2/6063Al: **a** before extrusion; **b** after extrusion

reduced. During the extrusion process, the composite material is subjected to extrusion forces in the transverse and longitudinal directions in the vacuum extrusion die, and the matrix grains undergo plastic flow. The reinforcement ZrB_2 particles gradually move with the flow of the matrix grains. When a certain amount of deformation has taken place, the matrix and reinforcement are in contact with each other, and the friction and extrusion between the two are broken. Compared with the original as-cast sample in Fig. 8.3a, the distribution of reinforcement particles is more uniform and fine, all below 10 μm, and dispersed in the 6063Al matrix. In addition, compared with the directional arrangement and distribution of the reinforcement in the forged composite material, the reinforcement particles after extrusion have no obvious order in the extrusion direction and tend to be messy and dispersed.

The optical structure of the composite material after corrosion before and after extrusion is analyzed, as shown in Fig. 8.4. From Fig. 8.4b, it can be observed that the matrix grains of the composite material after extrusion processing are regular bundled fiber grains with a grain diameter of about 5–15 μm, which is similar to Fig. 8.4a of the original as-cast composite material. Compared with the matrix crystal grains are obviously thinner, the matrix structure is more compact and dense. Since the extrusion processing temperature of the composite material is close to the recrystallization temperature of the matrix aluminum, and it can be observed from the figure that the matrix grains of the composite material become fine and uniform after extrusion, it is speculated that the ZrB_2/6063Al composite matrix Al during the extrusion process Recrystallization occurred during the extrusion process.

8.1.3 The Effect of Hot Extrusion on the Microstructure of ZrB_2/2024Al Composite

Figure 8.5 shows the metallographic structure of the hot extruded bar of ZrB_2/2024Al composite with a particle volume fraction of 5.4%. Figure 8.5a is the cross section of

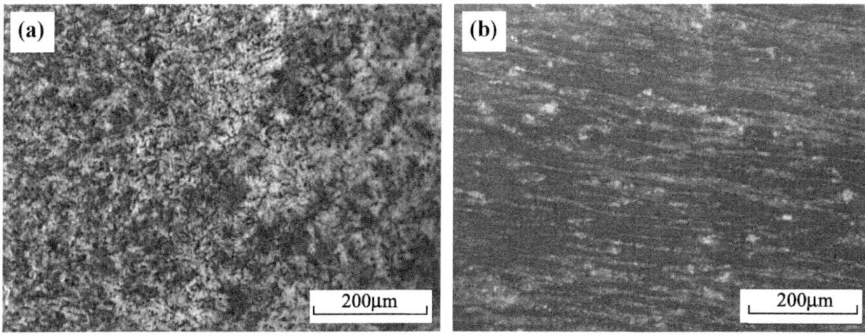

Fig. 8.5 Optical micrographs of 5.4 vol% ZrB_2/2024Al composites after hot extrusion: **a** cross section of sample; **b** longitudinal direction

the bar, and Fig. 8.5b is the longitudinal section of the bar. It can be seen from Fig. 8.5 that the matrix grains of the ZrB_2/2024Al composite material are elongated after hot extrusion, most of the nano-ZrB_2 particles are distributed along the grain boundaries, a few particles are distributed in the matrix grains, the coarse second phase and nano-ZrB_2 particles at the grain boundary are streamlined along the extrusion direction. In addition, hot extrusion eliminates the casting defects of composite materials and makes the structure more dense [9, 10].

8.2 Forging and Rolling of In-Situ Aluminum Matrix Composites

Both forging and rolling belong to plastic processing, or pressure processing. Forging is a forming and processing method that uses the hammer or punch of a forging machine or applies pressure to the blank through a die to cause it to be plastically deformed to obtain a part of the required shape and size. It is mainly used to produce metal parts. Rolling is a pressure processing method in which metal blanks pass through a pair of rotating rolls (various shapes). The cross section of the material is reduced and the length is increased due to the compression of the rolls. This is the most commonly used production method for producing steel. Production of profiles, plates, pipes.

8.2.1 *The Influence of Forging and Rolling on Microstructure of $Al_2O_{3(p)}$/6063Al Composites*

8.2.1.1 The Effect of Forging on the Structure of $Al_2O_{3(p)}$/6063Al Composites

Figure 8.6 shows the optical images of $Al_2O_{3(p)}$/6063Al composite after forging. The grain size after forging is finer than before forging, and the structure of the composite material is dense. Comparing the metallographic microstructure of the extruded sample, the study shows that there is a certain orientation difference of the grains after forging, which is mainly due to the continuous change of the plastic deformation direction of multi-directional forging (see schematic diagram, Fig. 8.6c, resulting in the microstructure of the composite material after plastic deformation as shown in the figure. By observation of grain structure after forging in Fig. 8.6b, it can be found that the grain size is reduced.

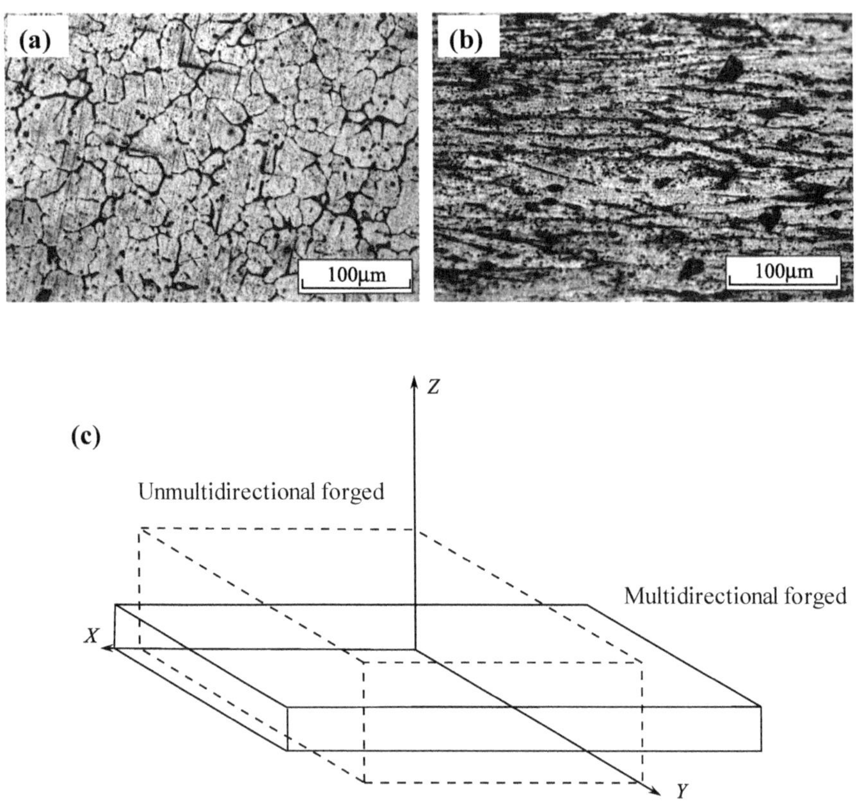

Fig. 8.6 Optical images of forged composite: **a** as-cast; **b** forged state; **c** forging direction

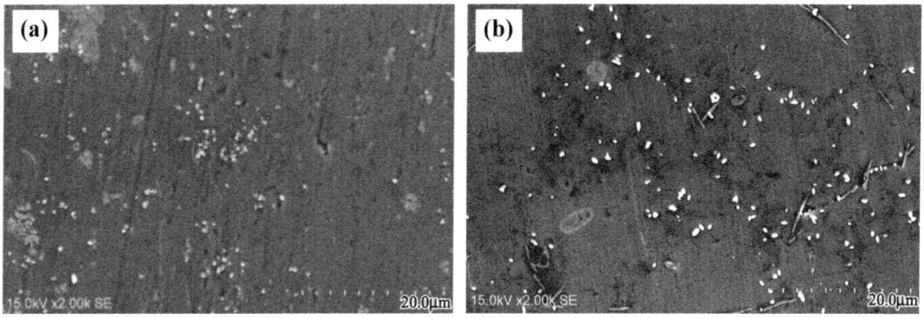

Fig. 8.7 SEM images of $Al_2O_{3(p)}$/6063Al composite: **a** as-cast; **b** forged state

Figure 8.7 shows SEM images of $Al_2O_{3(p)}$/6063Al composite after forging. It can be seen that the distribution of Al_2O_3 particles is more uniform than before forging, and the particle size has not been significantly reduced. In the structure of the composite material after multi-directional forging, the shape of Al_2O_3 particles is relatively rounded, and the dispersion is more uniform. The inclusions in the structure are squeezed, broken, and dispersed in the matrix.

8.2.1.2 The Effect of Rolling on the Structure of $Al_2O_{3(p)}$/6063Al Composite

Figure 8.8 shows the optical picture of the composite sample before and after cold rolling. It can be seen from Fig. 8.8a that the metallographic structure of $Al_2O_{3(p)}$/6063Al composites has a large grain size, mainly composed of nearly square and nearly spherical grains. In addition, there are some rose-like grains with fine Mg_2Si phases distributed between the grains. It can be seen from Fig. 8.8b that the as-cast structure is basically eliminated after rolling, and the crystal grains have changed significantly compared to before rolling, showing a flat shape mainly along the rolling direction. In addition, it can be seen that after rolling, the rose crystals in the 6063 aluminum alloy matrix show a crushed morphology, and some of the rose-shaped boundaries are bent and deformed or even broken, forming broken rose crystals. There is a relatively high orientation difference in the grains, which is mainly due to the deformation and uneven stress distribution caused by cold rolling [11], resulting in the microstructure of the composite material after cold rolling showing the morphology shown in Fig. 8.8b.

Figure 8.9 shows the SEM microstructure of composite before and after cold rolling. Before rolling, the distribution of Al_2O_3 reinforcement particles in the composite material is relatively uneven, and there are some acute angles, but after the sample is rolled, it can be found that the distribution of Al_2O_3 particles has been improved to a certain extent, and the morphology has rounded corners. There is no obvious change in the size of Al_2O_3. The plastic deformation caused by rolling

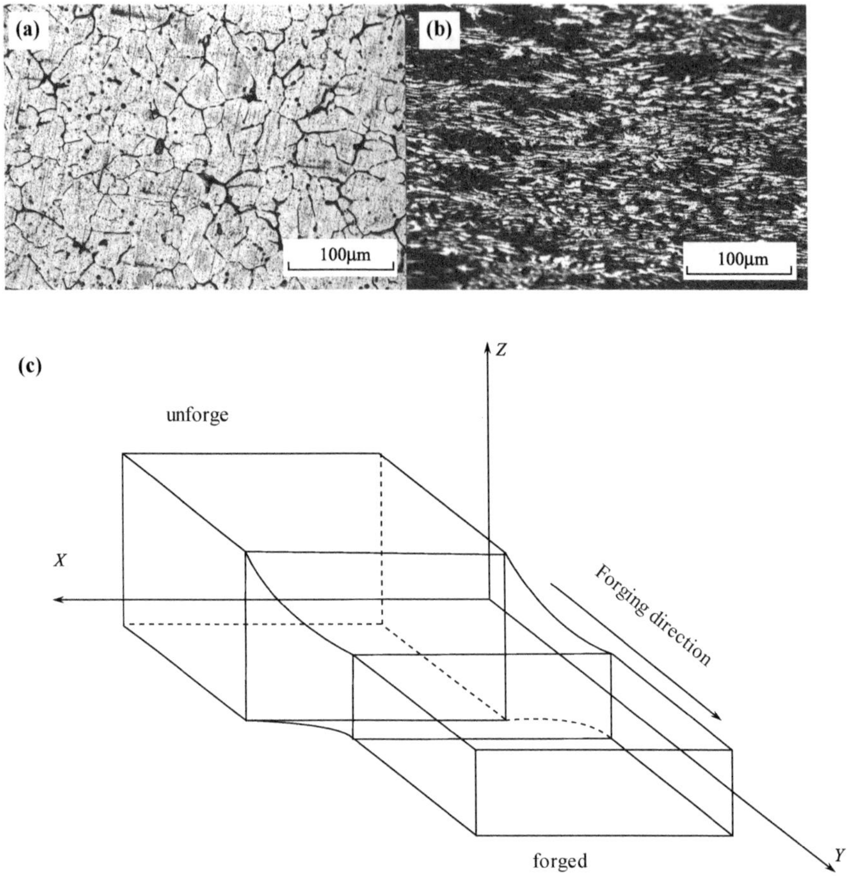

Fig. 8.8 Optical micrographs of composites before and after rolling (200×): **a** as-cast; **b** rolled; **c**) rolling direction

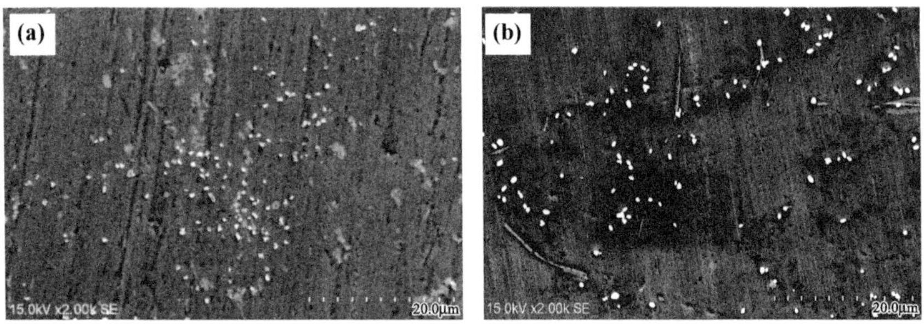

Fig. 8.9 SEM images of $Al_2O_{3(p)}$/6063Al composite: **a** as-cast; **b** rolled

causes the reinforcing particles to move under force, rub, and collide with each other, so that the irregular sharp corners are continuously spheroidized.

8.2.2 The Effect of Forging on the Structure of Al-Zr-B Composites

8.2.2.1 The Effect of Forging on the Microstructure of ZrB_2/6063Al Matrix Composites

Figure 8.10a, b shows the comparison images of the ultra-depth-of-field morphology before and after forging processing. Figure 8.10a is the ultra-depth-of-field map of the metallographic structure of the original casting sample after corrosion with Keller reagent. As shown in Fig. 8.10, the matrix has a grain size of about 100–150 μm and they are arranged in a coarse dendritic structure. Figure 8.10b is the sample after forging processing. There are obvious forging streamlines, the grain direction is perpendicular to the forging force direction, and the grains are obviously elongated in this direction. Figure 8.10ac, d shows the electron scanning microscope pictures before and after forging and plastic deformation. Before forging, most of the ZrB_2

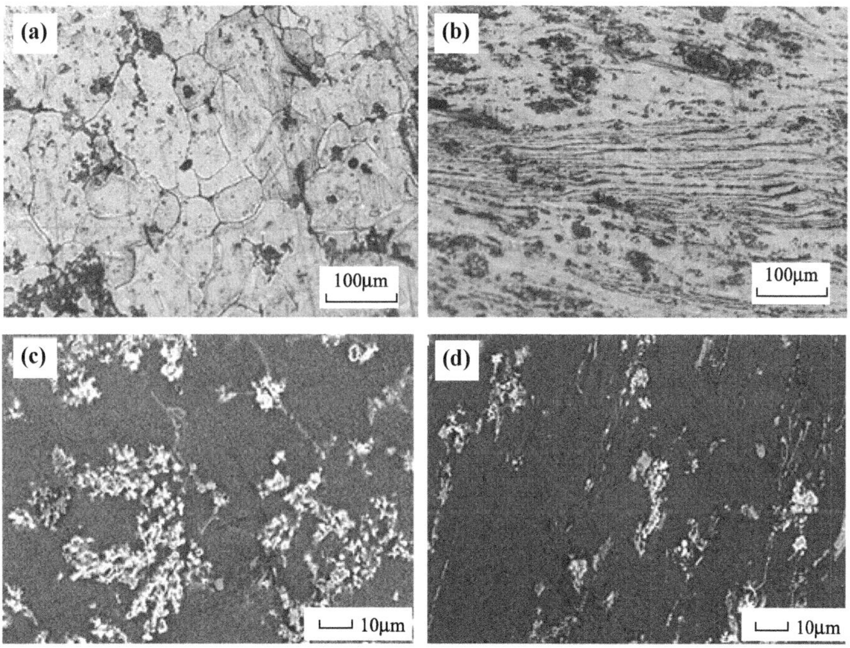

Fig. 8.10 Ultra-depth of field optical images and SEM images of ZrB_2/6063Al composite: **a**, **c** before forging; **b**, **d** after forging

particles (about 65–70%) have obvious segregation, and a small part (about 30–35%) small particles are scattered along the grain boundary and are distributed end to end in strips. After forging, the structure showed a clear river-like pattern, and the ZrB_2 reinforcement particles were also distributed in chains along with the directional flow of the matrix, and the segregation phenomenon was significantly improved. Larger particles may be forged. The forged particles are distributed along the flow direction of the matrix in the form of chain links; ZrB_2 particles with smaller sizes distributed along the grain boundary are distributed in a curved band with the flow of the grain boundary, casting defects such as porosity and shrinkage are significantly improved, and the internal structure of the material is more uniform.

8.2.2.2 The Effect of Forging on Microstructure Changes of Al_3Zr/6063Al Composites

Figure 8.11 shows the as-cast and forged SEM images of 5 wt.%Al_3Zr/6063Al composites. In Fig. 8.11a, before forging deformation of Al_3Zr/6063Al composite material, part of Al_3Zr reinforcement is slab-like or long needle-like, with a size of about 100–200 μm, a small part is short rod-shaped particles, with a size of about 10–20 μm having agglomerated tendency. It can be seen from Fig. 8.11b that after the forging process, the lath and needle-like Al_3Zr particles in the composite material have obvious fracture marks. During the forging deformation process, the reinforcement particles Al_3Zr and the metal matrix are deformed in an incompatible manner resulting in stress concentration of reinforcement particles and fracture. The long slab-shaped Al_3Zr particles are forged and basically disappeared, and the particle size is significantly reduced. After the fracture, the particle size is about 10–30 μm. There are a small amount of Al_3Zr particles with smaller size in the as-cast state. The shape and size of the particles are basically unchanged after forging. However, due to the flow pressure perpendicular to the forging direction, the particles slip around, which improves the segregation phenomenon and the distribution, which tends to be uniform and has a certain streamlined distribution.

In addition to fracture, the reinforcing particles will also move and rotate. As a result, when the forging deformation is large enough, the particles will be distributed

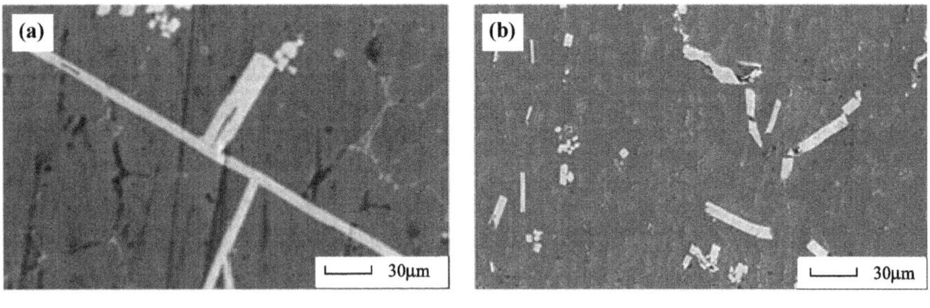

Fig. 8.11 SEM images of Al_3Zr/6063Al composite: **a** before forging: **b** after forging

in an orderly orientation perpendicular to the forging direction. The movement and rotation of the particles are caused by the plastic deformation of the matrix. Dislocation slip and climbing play a leading role in the rotation of the particles. During forging, the matrix grains deform and move, and the reinforced particles receive a certain torque. It rotates under the action, and the reinforcing particles also move along with the movement of the matrix crystal grains and recrystallization, so that the fractured reinforcing particles will not form void defects but be filled by the matrix.

In addition, there are some void defects in the as-cast composite material before forging. During hot forging, the base alloy undergoes dynamic recrystallization to produce fine recrystallized grains, and the grains deform and move when the aluminum alloy matrix is subjected to forging force, welding the internal defects of the raw material (such as microcracks, holes, etc.), which makes the organization more dense, and the defects are obviously reduced.

8.2.2.3 The Influence of Forging Plastic Deformation on the Matrix Grain

Figure 8.12 is the optical diagram of composite samples before and after forging at 850 °C 5 wt.%Al_3Zr/6063Al, comparing the as-cast and forged microstructures. Before forging, the matrix grains are irregularly shaped dendrites with a size of about 80–100 μm. After forging, the morphology of the matrix grains is obviously deformed, showing a streamlined long fibrous structure with a width of 5–15 μm and a length of 90–110 μm, which is consistent with the forging direction. During the forging process, the speed of the air hammer can reach about 10–30 m/s. The particles inside the material are forged under the action of huge impulsive force, and the inside of the forging is subjected to three-dimensional stress, which makes the crystal grains XY plane under compressive stress. In the direction of extension, the grains deform into a fibrous structure. The grains are streamlined as a whole and become smaller. At the same time, part of the dendrites are forged, making the number of grains larger

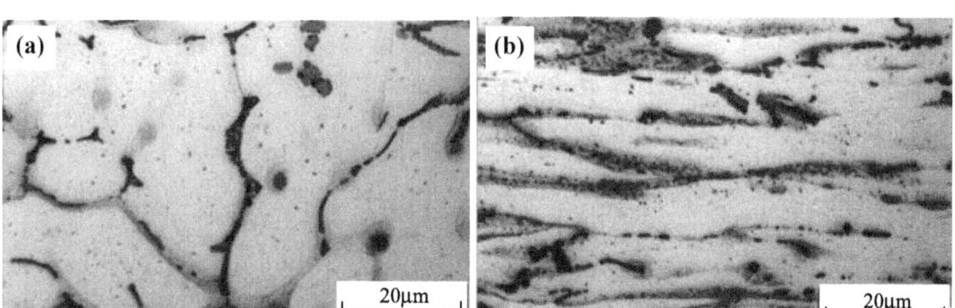

Fig. 8.12 Optical microstructure of Al_3Zr/6063Al composites: **a** before forging: **b** after forging

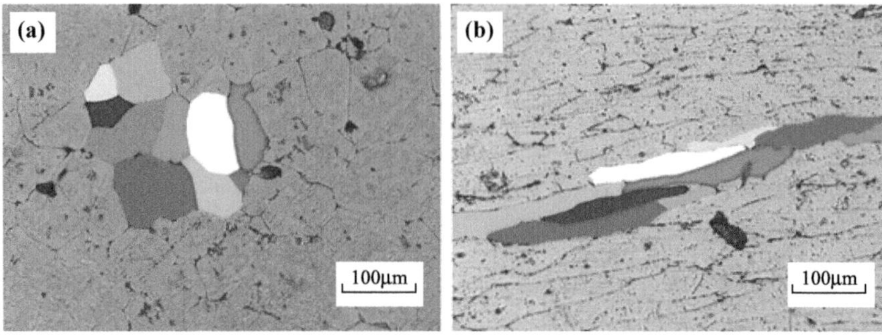

Fig. 8.13 Microstructure diagram and grain color marking diagram of composite: **a** as-cast: **b** forged state

and smaller. Therefore, the overall forging plastic deformation affects the matrix. The grains are refined to a certain extent in the forging direction.

Figure 8.13 is the corrosion superdepth of field map before and after forging of the composite material after the additive color treatment. Figure 8.13a is a typical as-cast structure. The grains are coarse dendritic grains. There are defects such as holes at the junction of some three grains. Figure 8.13b shows the forged microstructure. It can be seen that the original equiaxed crystals are elongated to form a fibrous structure. The organization is denser and defects are significantly reduced.

8.2.2.4 XRD Phase and Matrix Lattice Orientation Analysis of Forged Composite Materials

Figure 8.14 shows the X-ray diffraction curve of 15 wt.% Al_3Zr/6063Al composite material at 850 °C. XRD shows that the phase of Al_3Zr/6063Al composite material has not changed before and after forging. In Fig. 8.14, comparing the X-ray diffraction patterns of 15 wt.%Al_3Zr/6063Al composite with the forged sample, it is found that after Al_3Zr/6063Al composite material is forged, the matrix grain orientation has twisted, and distortion of α-Al phase occurs.

Further analyze shows the lattice distortion of matrix after forging, as shown in Fig. 8.15. It can be seen from Fig. 8.15 that the diffraction peaks in the (111), (220), (200), (311) directions of the α-Al phase of Al_3Zr/6063Al composite after forging are all shifted to the right, (114), (The intensity of diffraction in the direction of 220) and (311) increases significantly, while the intensity of diffraction in the direction of (200) decreases. According to the Bragg equation $2d\sin\theta = n\lambda$, the diffraction peak shifts to the right, θ becomes larger, and the interplanar spacing becomes smaller. That is to say, the lattice is distorted, causing the lattice constant to become smaller, in the (111), (220), and (311) directions The diffraction intensity of the (200) direction is obviously increased while the diffraction intensity of the (200) direction is decreased, and the diffraction intensity is related to the grain content in this orientation, indicating that the (200) direction grains are twisted into grains that

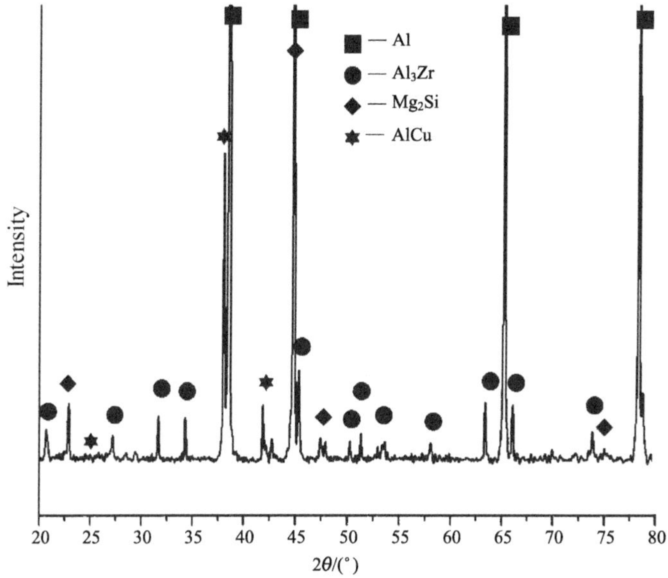

Fig. 8.14 X-ray diffraction patterns of 15 wt% Al_3Zr/6063Al wrought composite

are favorable for forging and extending the flow direction, especially the (111) and (220) directions.

8.2.3 *The Effect of Rolling on the Microstructure of Al–Zr–B Composites*

8.2.3.1 The Effect of Rolling on the Microstructure of ZrB_2/6063Al Composite

Figure 8.16 shows the corrosion of the optical structure of the ZrB_2/6063Al composite before and after rolling. After rolling deformation, the microstructure of the matrix material is fibrous, similar to the microstructure after forging, except that the width of the grain strips of the material after rolling is 3–5 μm, while the grain strips after forging. The width is between 5 and 15 μm. The reason is that the alloy at the edge of the rolling strip is not only subjected to the compressive stress of the roller, but also subjected to the extrusion stress from the center of the strip, and the matrix and reinforcing particles inside the material slip; The surface of the alloy is affected by the roller's strong plastic deformation occurs, and shear stress is generated due to friction. The ZrB_2 reinforcement is quickly broken into fine particles under the action of the surface and internal stress, which are evenly distributed along the rolling direction.

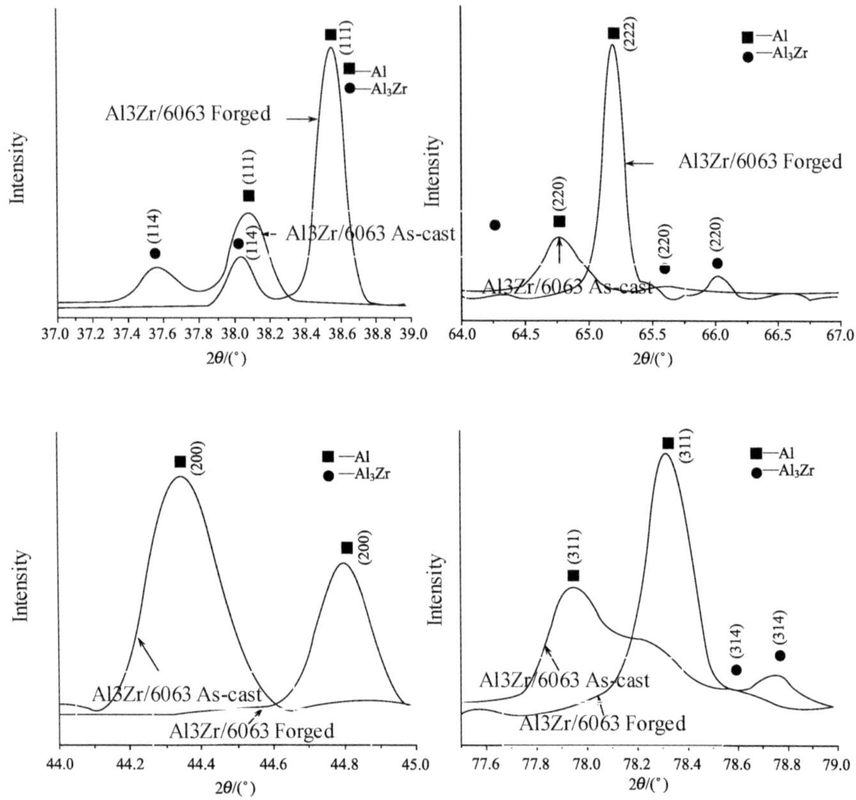

Fig. 8.15 Comparison of X-ray diffraction patterns of 15 wt% Al_3Zr/6063Al composites before and after forging: **a** (111); **b** (220); **c** (200); **d** (311)

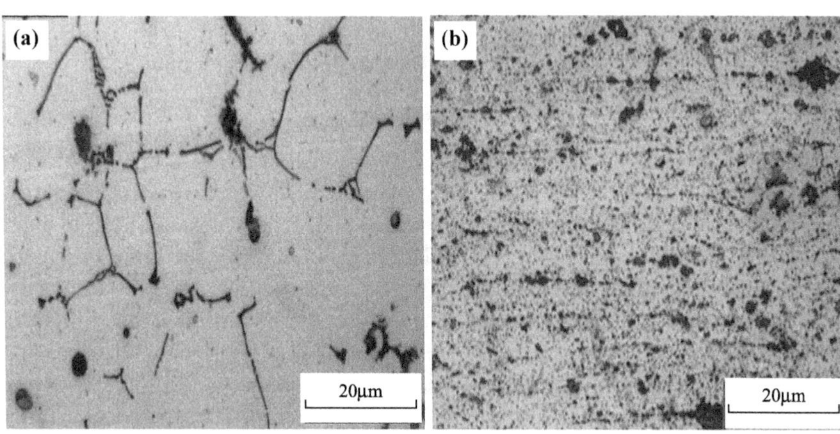

Fig. 8.16 Optical microstructure original specimen and rolled specimen: **a** original specimen; **b** rolling specimens

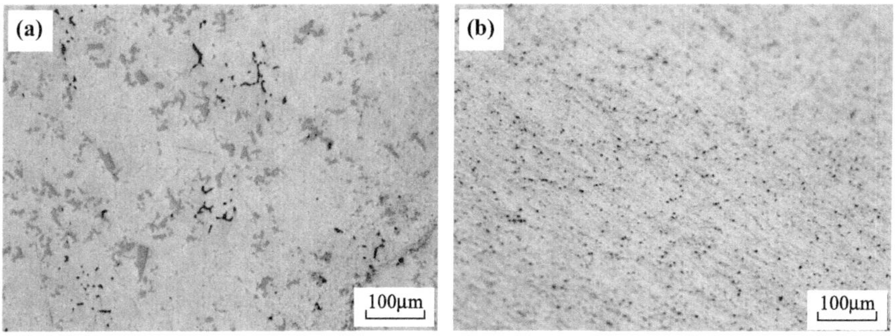

Fig. 8.17 Optical microstructure of Al_3Zr/6063Al composites: **a** original as-cast; **b** rolling state

8.2.3.2 The Effect of Rolling on the Microstructure of Al_3Zr/6063Al Composites

Figure 8.17 shows the comparison of the microstructure of the Al_3Zr/6063Al composite in the solidified state and the rolled state. From the comparison of the microstructure of the Al_3Zr/6063Al as-cast and the as-rolled in Fig. 8.17, it is found that the microstructure of the material after rolling has a certain change: the Al_3Zr particles are streamlined with the deformation of the grains, and the aggregated Al_3Zr particles are dispersed. Under the action of rolling, the original as-cast grains are 80–100 μm dendrites, while the rolled grains become strip-like grains with a width of 3–5 μm and a length of 90–100 μm. The direction is the same, and the lath and short rod Al_3Zr-reinforced particles in the original as-cast material are all broken and refined. After rolling, the size of the Al_3Zr-reinforced particles becomes a rectangular block with a width of about 0.5 μm and a length of 3 to 5 μm. The matrix of the composite material casting defects, porosity, and segregation is eliminated, and the matrix structure is improved.

8.2.3.3 XRD Phase and Grain Changes of Composite After Rolling Plastic Deformation

During the rolling process, each crystal grain slips, and the crystal grain orientation rotates regularly with respect to the external force. Due to the connection of grain boundaries, this kind of rotation is limited to a certain extent, but when the amount of deformation is large, the original grains of arbitrary orientation will gradually adjust, causing a certain degree of ordering of grain orientation and the lattice is distorted. The process in which this kind of polycrystalline crystal grains are arranged in disorderly orientation from the original orientation to the general orientation of each crystal grain tends to be the same is called "preferred orientation." Then the grains are rolled into a strip shape, and all show one direction. Figure 8.18 shows the X-ray diffraction pattern of rolled sample of Al_3Zr/6063Al composite.

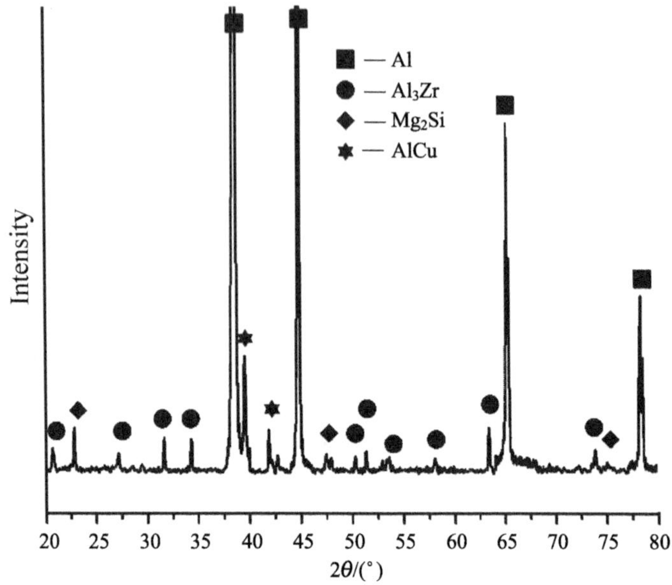

Fig. 8.18 X diffraction patterns of rolled Al_3Zr/6063Al composite

It can be seen from Fig. 8.18 that the phase composition of Al_3Zr/6063Al composite has not changed after rolling. However, careful analysis of it reveals that the crystal structure of the matrix has undergone a certain change.

Figure 8.19 shows a comparison of the diffraction peaks for Al_3Zr/6063Al composites after rolling. In the figure, comparing the X-ray diffraction curves of Al_3Zr/6063Al composites as-cast and rolled samples, it is found that after Al_3Zr/6063Al composites are forged, the matrix grains are twisted and the α-Al phase crystal lattice is distorted. It can be seen from the figure that the diffraction peaks in the directions (111), (220), (200), and (311) have shifted to the right. Among them, the diffraction intensity in the (111) direction increased most obviously, followed by the diffraction intensity in the (220) and (311) directions; the diffraction intensity in the (200) direction remained basically unchanged. According to the Bragg equation $2d\sin\theta = n\lambda$, the diffraction peak shifts to the right, θ becomes larger, and the interplanar spacing becomes smaller, that is, the lattice is distorted, causing the lattice constant to become smaller, in the (111), (220), and (311) directions the diffraction intensity of (200) direction is significantly increased, while the diffraction intensity of the (200) direction remains unchanged, and the diffraction intensity is related to the grain content of this orientation, indicating that the (200) direction grains are twisted to the (111) orientation grains demonstrating that the process of "preferred orientation" has occurred.

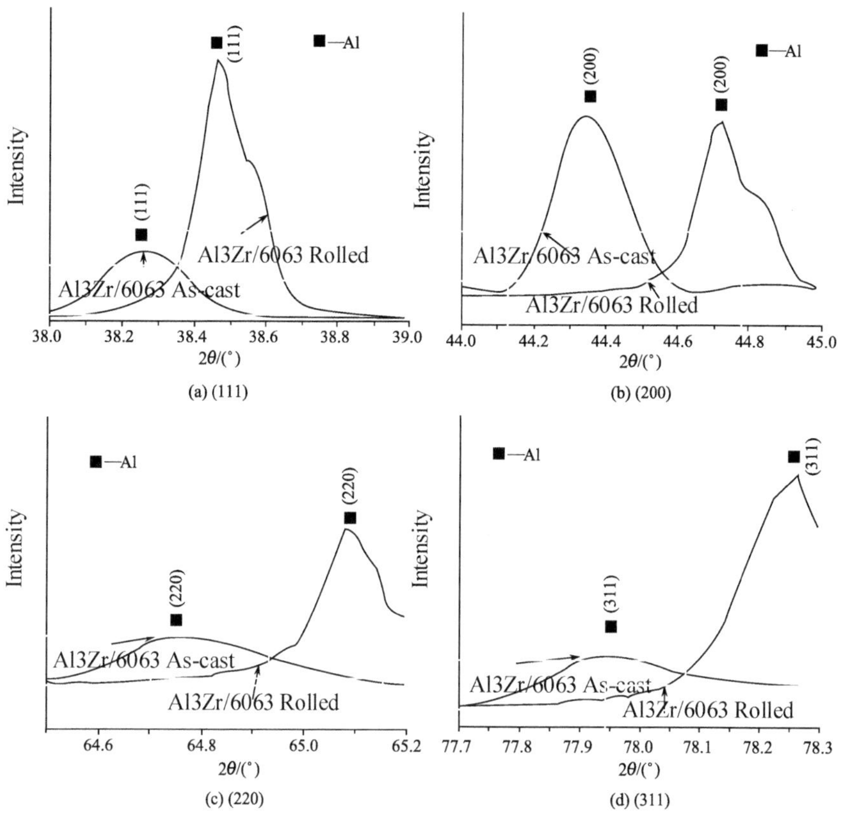

Fig. 8.19 Comparison of X-ray diffraction patterns of Al_3Zr/6063Al composites before and after rolling: **a** (111); **b** (200); **c** (220); **d** (311)

8.2.4 The Influence of Forging and Rolling on the Structure of Al–Ti–B Composites

8.2.4.1 Analysis of the Influence of Forging on Composite Reinforcement Particles

Figure 8.20a–d shows the microstructure of core and edge regions of the composite material with a particle content of 5 wt.% and 10 wt.%, respectively. After forging, the size of the reinforcing particles in the core is about 10–20 μm, and the size of the reinforcing particles at the edge is about 5–10 μm, and the particles tend to agglomerate. After forging, most of the reinforcing phases in the composite material have obvious fracture marks, that is, the long strip particles in the as-cast state are broken and fractured under the force during the forging process, and the particles are refined, but the existing small particles in the as-cast state do not have obvious fractures. That is, forging has a poor effect on the refinement of small particles. After

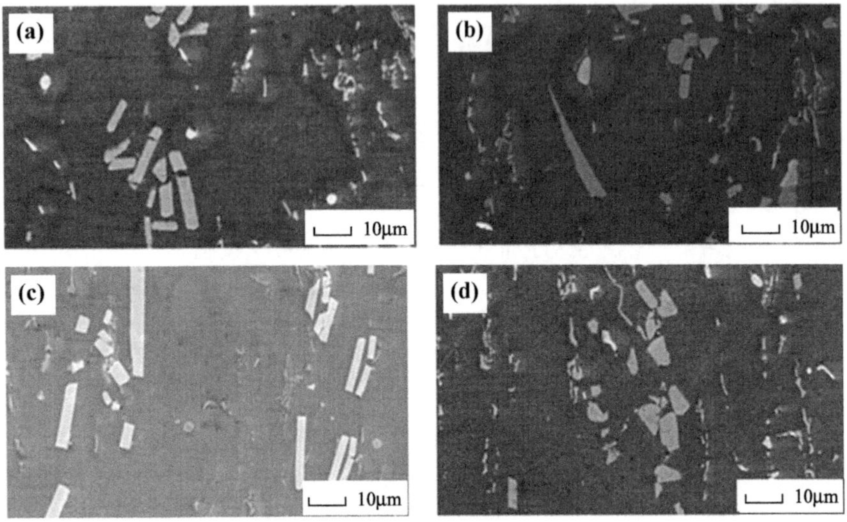

Fig. 8.20 SEM images of different parts of Al_3Ti/2024Al composite after forging: **a** 5 wt.% Al_3Ti/2024Al core; **b** 5 wt.% Al_3Ti/2024Al edge; **c** 10 wt.% Al_3Ti/2024Al core; **d** 10 wt.% Al_3Ti/2024Al edge

fracture, the reinforcing phase is oriented and dispersed in the matrix metal within a certain range and has a tendency to agglomerate. This is because the fractured reinforcing particles move and twist with the matrix during the forging process, and part of the reinforcing particles move together, thus showing the tendency of agglomeration. The small particles broken from a long rod-shaped particle will be twisted with the forging force, and a small range of deviation will occur in the original direction, showing a macroscopic directional distribution.

During the forging process, the internal particles of the material follow the law of least resistance, but due to factors such as friction, temperature unevenness and uneven force on the contact surface, uneven deformation occurs during forging. Figure 8.21 is a schematic diagram of forging, which can be divided into three regions according to the different stress and deformation degree. However, since the forging stage is multi-directional forging, the deformation degree of the material edge region, namely zone 1 and zone 3, which is similar, while the deformation degree of zone 2 is weaker, the particle fracture effect is weak, the particle size after fracture is also larger, the overall distribution and size of the particles are not uniform [12].

In summary, the reinforcement particles are broken and fractured during forging, which refines the reinforcement phase of the in-situ Al_3Ti/2024Al composite but has a poor refinement effect on small particles; particle movement may cause agglomeration during forging, and the particles may twist and stay within a certain range internally oriented distribution; due to the uneven deformation, the overall distribution and size of the reinforcing particles after forging are also uneven. In addition, after forging, a large number of second phases precipitated in the composite material

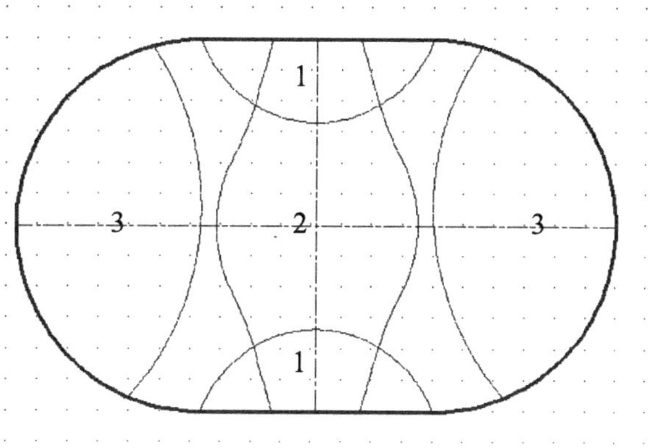

Fig. 8.21 Plastic deformation diagram of forging

are linearly arranged along the forging elongation direction in the matrix, and there is a phenomenon of aggregation.

8.2.4.2 Influence of Forging and Rolling on the Matrix Grains of Composite

Figure 8.22 is an ultra-depth of field image of the in-situ Al_3Ti/2024Al composite after corrosion. Figure 8.22a–c respectively shows microstructure of as-cast 2024Al alloy, as-cast Al_3Ti/2024Al composite, and the composite after forging. It can be seen from the figure that the grain size of 2024Al alloy is 50–150 μm, and the grain size of the as-cast Al_3Ti/2024Al composite material is 50–100 μm. After forging, the grains are deformed into fibrous shape with a length of about 150 μm and a width of 30–50 μm. The aluminum alloy and composite materials are dendrites in the as-cast state, and the crystal grains are fibrous after forging.

The grain size of the Al_3Ti/2024Al composite material in the as-cast state is slightly smaller than that of the Al alloy. This is because the lattice constant mismatch between Al_3Ti and α-Al is 5.2%, and Al_3Ti is easy to become α-Al. The core of the nucleation, that is, the reinforcement phase Al_3Ti, has a certain effect of crystal refinement. During forging, the grains undergo plastic deformation, which is mainly divided into intragranular deformation and grain boundary deformation. The intragranular deformation is dominated by slip. During forging, a part of the crystal slips along the crystal plane and crystal direction with the highest atomic density, which makes the crystal grain elongate along its deformation direction, and the lattice constant decreases in the direction perpendicular to it. The increase in deformation will eventually form a fibrous structure. Grain boundary deformation is mainly the mutual sliding and rotation of crystal grains, which gradually adjusts the phase of crystal grains and reduces the distance between crystal planes. When the stress at the front end of the plugging dislocation reaches a certain level, coupled with the rotation of adjacent crystal grains, the slip is transferred from one batch of crystal grains to another batch of crystal grains, and the orientation of all crystal grains tends

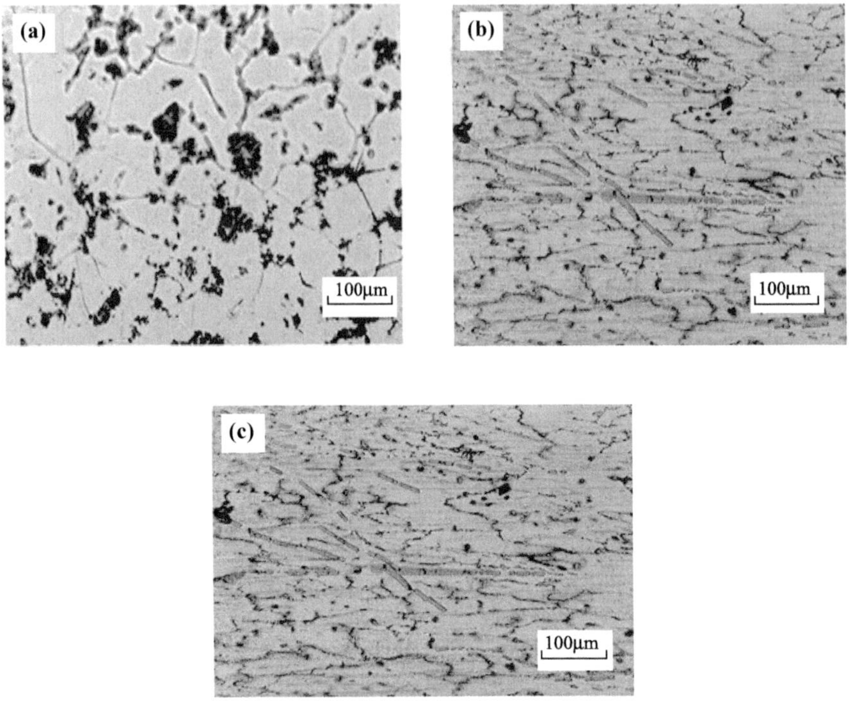

Fig. 8.22 Grain depth of in-situ Al_3Ti/2024Al composite: **a** 2024 Al alloy; **b** as-cast; **c** after forging

to be the same, showing obvious plasticity Deformed. Due to the uneven degree of deformation in each region during forging, there will be some differences in grain slip and rotation amplitude in different parts, which is shown in the figure as the size and direction of the grains are slightly different.

Figure 8.23a, b shows the superdepth of field diagrams of the composite material grain changes before and after rolling. The comparison shows that the composite material matrix α-Al grains elongate along the plastic deformation direction after rolling, showing a fibrous stripe structure. The texture of rolling deformation

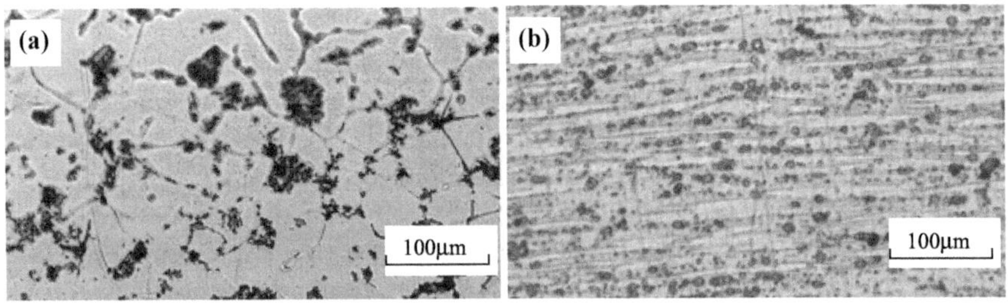

Fig. 8.23 Ultra-depth view of grain variation of composite matrix: **a** before rolling; **b** after rolling

appears. Reinforced particles are crushed into chains along the deformation direction, uniformly distributed, with high degree of dispersion. After rolling, the structure is fibrous slender grains and overlapped and interlaced with each other. The grains are arranged compactly, the grain boundary line is relatively straight, and the casting defect holes are fewer. It can be seen from Fig. 8.23a that the matrix grains of the composite material before rolling are dendrites with a size of 100–120 μm, and the grains become strips after rolling. The grain size is measured by the intercept method, and the strips are measured, the grain width is 3–5 μm, and the length is about 80–100 μm.

8.2.4.3 Influence of Forging Rolling on the Dislocation of Composite Materials

Figure 8.24 shows TEM microstructure observation results of 5 wt.%Al_3Ti/2024Al composite in as-cast state and after forging. In the as-cast state (Fig. 8.24a), the composite has coarse grains, and the reinforcing grains are elongated and larger in size, with fewer dislocations in the grains and bending of grain boundaries. After forging processing (Fig. 8.24b), a large number of dislocations appeared in the crystal grains, and they were entangled and intertwined into a network, resulting in blurred grain boundaries.

As shown in Fig. 8.25, there are some deformed substructures in the matrix grains of the rolled Al_3Ti/2024Al composite material. The boundary of the deformed substructures is the lattice distortion zone, and a large number of dislocations are accumulated. The internal lattice of substructured grains is relatively complete, and the deformation substructure of Al_3Ti/2024Al is formed during the rolling process. Under the action of the shear stress of the rolling plastic deformation processing, the dislocation source produces a large number of dislocations. When it encounters various obstacles to the grain boundary, Al_3Ti particles obstacles of equal dislocation movement produce a large number of dislocation entanglements.

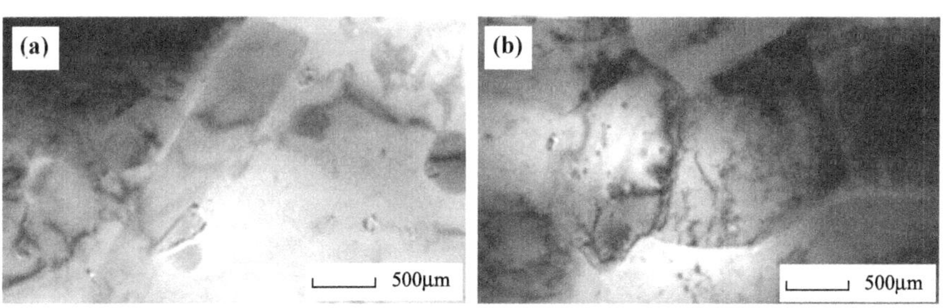

Fig. 8.24 TEM microstructure images of composite under different conditions: **a** as-cast; **b** after forging

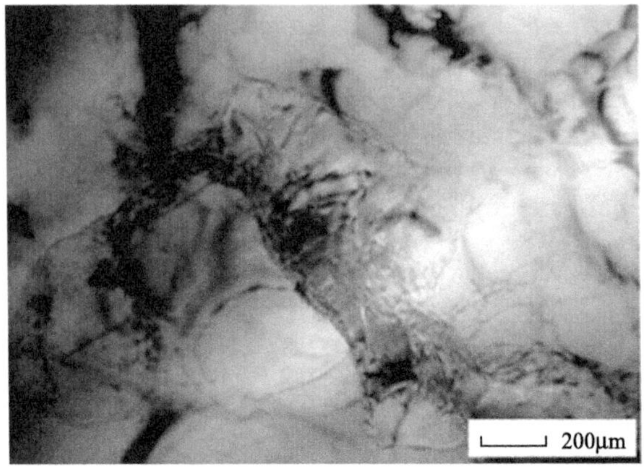

Fig. 8.25 TEM image of rolled Al_3Ti/2024Al composite

8.2.5 The Effect of Forging on ZrB_2/2024Al Composite

Microstructure of ZrB_2/2024Al composite before and after forging is shown in Fig. 8.26. It can be found that the as-cast composite material grains are dendritic crystals with irregular morphology, and the size range is approximately 80–100 μm. After the multi-directional forging pretreatment, the matrix grains are obviously

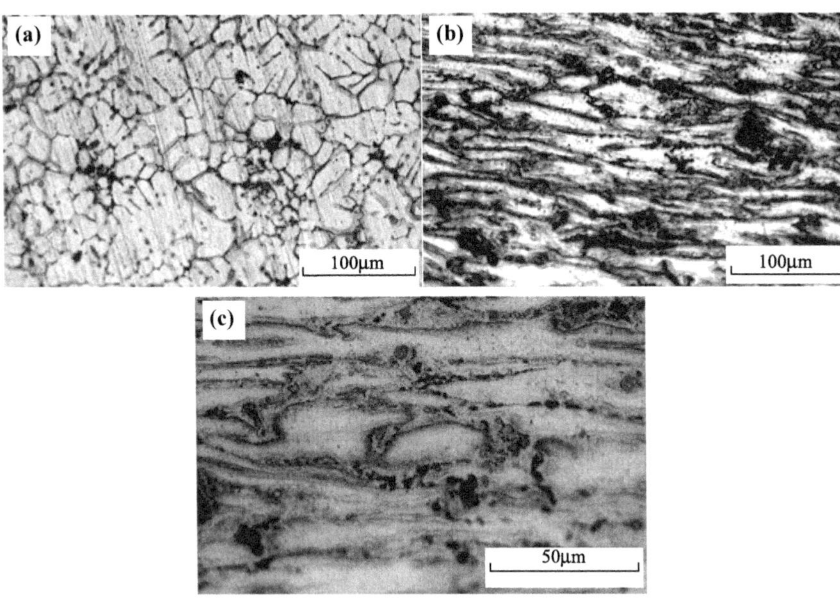

Fig. 8.26 Microstructure morphology of composites before and after multi-direction forging pretreatment: **a** as-cast; **b** forged; **c** forged

elongated, with sharp corners, and the size presents a fiber structure with a width of 5–15 μm and a lateral length of 90–120 μm; the direction of grain extension is perpendicular to the forging compression direction. Most of the reinforcing particles in the material are located near the grain boundaries, and the number of particles inside the grains is reduced, as shown in Fig. 8.26c. The distribution direction of the particles near the grain boundary is also the same as the extension direction, so there will be a phase difference in each crystal grain after forging, and the material will appear anisotropy. After the forging process, the void defects formed during the casting process are closed, and the inclusions in the structure will also be broken and dispersed into the matrix under the influence of the huge deformation force. In summary, after multi-directional forging, the grain size of the matrix grains is reduced to a certain extent in the forging force direction, and an anisotropic structure will be formed at the same time.

The changes of the main characteristic peaks of composite materials before and after the multi-directional forging pretreatment are shown in Fig. 8.27. After forging

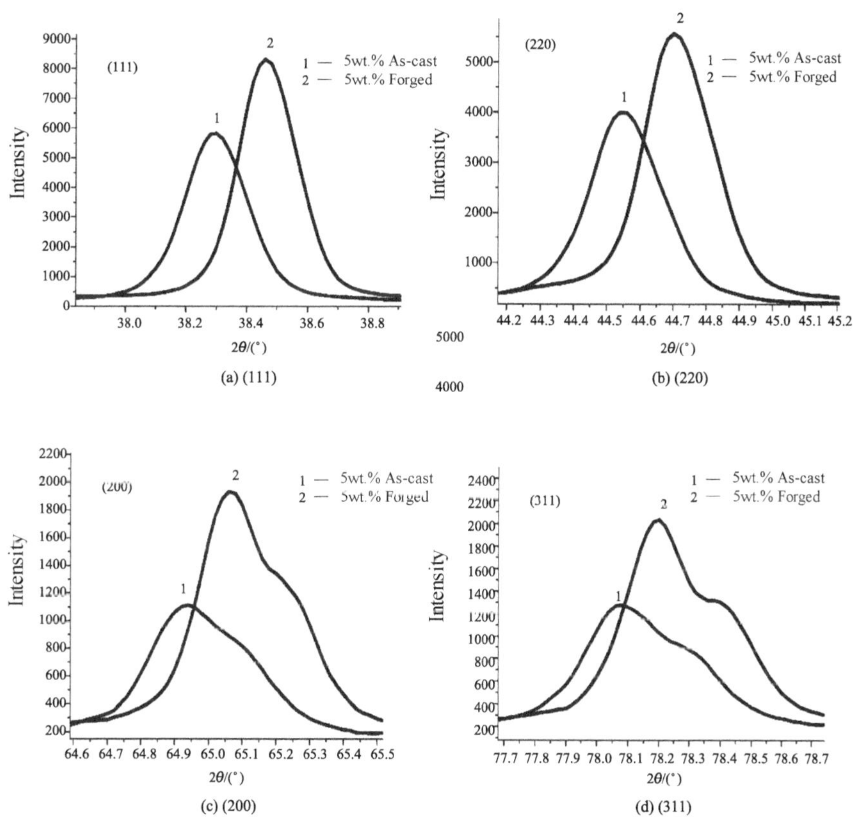

Fig. 8.27 XRD peaks shifting of ZrB_2/2024Al composite before and after forging: **a** (111); **b** (220); **c** (200); **d** (311)

Table 8.1 XRD analysis of composites before pretreatment

Peak corresponds to crystal plane index	(1 1 1)	(2 0 0)	(2 2 0)	(3 1 1)
2θ (°)	38.299	44.540	65.921	78.080
FWHM(peak half-height width)	0.236	0.248	0.320	0.374
Grain size (A)	394	378	310	284
Crystallinity/%	65.1	74.06	91.37	87.96
Al lattice constants		4.05943	4.05943	4.05943

Table 8.2 XRD analysis of composites after multi-direction forging

Peak corresponds to crystal plane index	(1 1 1)	(2 0 0)	(2 2 0)	(3 1 1)
2θ (°)	38.460	44.700	65.068	78.200
FWHM(peak half-height width)	0.216	0.242	0.285	0.338
Grain size (A)	440	389	353	317
Crystallinity/%	63.55	35.90	25.24	12.15
Al lattice constants		4.05186	4.05186	4.05186

pretreatment, the peak displacement of the Al matrix moves in a large angle direction. According to the formula, it can be seen that when θ rises, d will decrease. That is, the spacing between the crystal planes of the matrix decreases. At the same time, after the multi-directional forging process, the peak intensity of the Al matrix diffraction peak increases significantly, while the half-height width decreases, which is a typical compression distortion characteristic.

It can be seen from Fig. 8.27 that the diffraction intensity of each peak has increased. The peak intensity of the diffraction peak is related to the lattice structure, size, and crystallinity of the material. Use Jade software to fit the XRD curve and perform calculations. The data are shown in Tables 8.1 and 8.2. From the results of the fitting calculation, it can be seen that after the forging process, the crystal lattice of the matrix undergoes compression distortion, and the grain size grows in some directions. At the same time, in some phases, the crystal grains will be squashed, the complete crystal grains in the as-cast state are destroyed, and the crystallinity is greatly reduced. After forging, the lattice constant of the aluminum matrix also decreases, and the overall lattice collapses. From the perspective of the XRD curve analysis, it can be seen that the matrix grains are compressed and deformed, resulting in a flattened and pointed structure, and the grains are elongated in some directions and the size becomes larger.

8.2.6 Mechanism and Plastic Deformation Model of Forging on In-Situ Composites

As shown in Fig. 8.28, the macroscopic deformation process of the matrix grains is essentially carried out through the movement of microscopic dislocations. When the dislocation cuts across the slip surface, the crystal will produce a macroscopic deformation. The presence of reinforcing particles in composite materials will hinder the movement of dislocations. When the reinforcement particles are encountered during the movement of the dislocation line, the reinforcement particles will hinder the advancement of the dislocations, causing the dislocations to accumulate here and form a high stress concentration. If the stress concentration generated is greater than the strength of the reinforcing particles, the Al_3Zr particles will be broken and the dislocation will continue to move forward. In addition, if the maximum stress generated is also less than the strength of the reinforcing particles, the stress at this time may push the reinforcing particles to rotate to move forward in the direction of least resistance. After the particles such as ZrB_2 or Al_3Zr in the aluminum matrix composite material become fine due to crushing, it is easy to rotate and flow with the matrix grains under the action of compressive stress, and some places where the reinforcing particles agglomerate may be dispersed by rotation and flow. Therefore, forging can refine the reinforcing particles and improve the agglomeration of the reinforcing particles in the composite material.

The mechanism of forging plastic deformation on the grain refinement of the composite matrix is completed by the mechanical crushing refinement mechanism, the deformation induced dynamic recrystallization refinement mechanism and the thermally activated grain growth induced dynamic recrystallization refinement mechanism. Under hot forging conditions, the deformation and movement of the matrix grains and the reinforcement particles are subject to a certain torque, and the rotation occurs under the action of the torque. At the same time, the matrix is accompanied by dynamic recrystallization. The nucleation mechanism is subcrystalline rotation, aggregation nucleation, and the interaction between the dispersed reinforcing particles and the high-density dislocations slows down the growth rate of the matrix grains by applying resistance to the migration of the grain boundaries or subgrain boundaries, so that smaller matrix grains are obtained to realize the matrix grains a

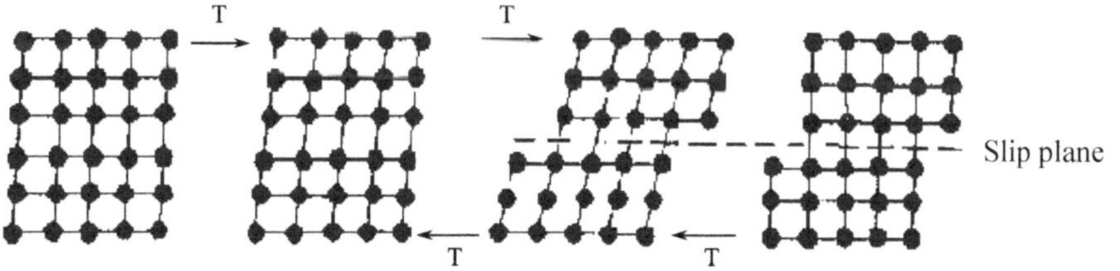

Fig. 8.28 Schematic illustration of dislocation movement in matrix lattice of composites

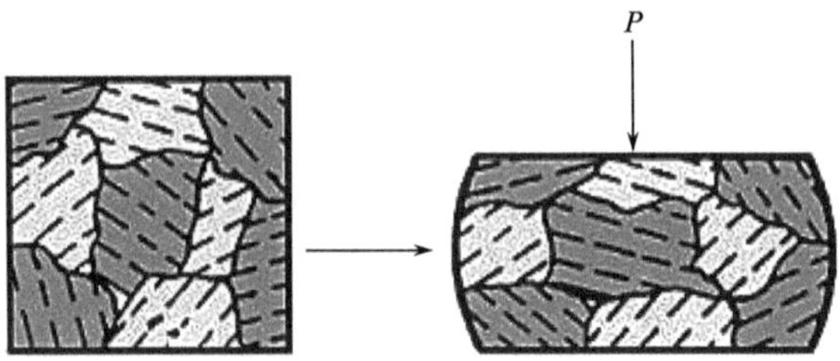

Fig. 8.29 Schematic diagram showing matrix grain changes during forging plastic deformation of in-situ particle-reinforced aluminum matrix composite

certain degree of refinement. At the same time, the ZrB_2 or Al_3Zr aggregated reinforcement particles are distributed in an orderly directional dispersion perpendicular to the forging direction.

The forging plastic deformation process of aluminum matrix composites is the comprehensive result of the deformation of all the crystal grains of the matrix, that is, a single crystal grain not only undergoes intragranular slip but also intergranular rotation during the deformation process. The process model is shown in Fig. 8.29.

The main way of rotation between the grains is the mutual sliding and rotation of the grains. When the composite material is deformed by the forging pressure *P*, a shear stress is generated along the grain boundary. When the shear stress is sufficient to overcome the resistance of the grains to sliding relative to each other, it will relative sliding occurred. In addition, due to the different phases of each crystal grain, its deformation and degree of difficulty are also different. In this way, force interaction is inevitably caused between adjacent crystal grains, which may produce a pair of force couples, causing mutual rotation between crystal grains. Therefore, the influence model of forging plastic deformation of Al–Zr–B in-situ particle-reinforced aluminum matrix composites on the microstructure is shown in Fig. 8.30.

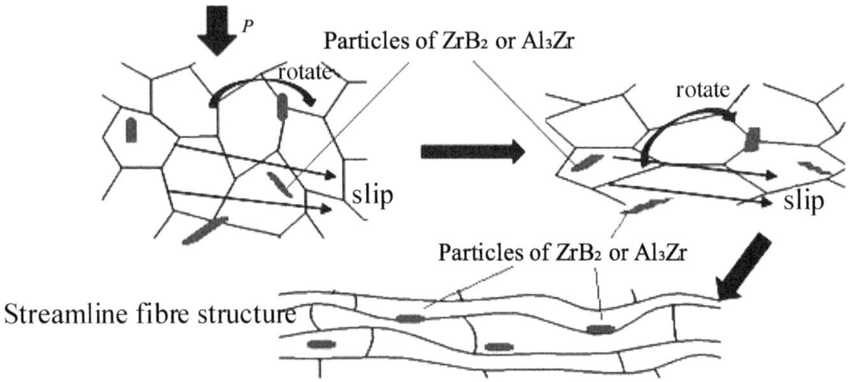

Fig. 8.30 Effect of forging plastic deformation on microstructure of Al–Zr–B series in-situ particle-reinforced aluminum composites

It can be seen from Fig. 8.30 that the forging plastic deformation process of aluminum matrix composites, that is, the matrix grains under the forging pressure *P*, not only occur intragranular slip, but also coordinate other grains to rotate intergranularly. The matrix grains slip along the easy-slip surface under the action of forging pressure. At the same time, the grains whose slip system direction does not coincide with the deformation flow direction rotate. The result of the rotation makes the easy-slip direction parallel to the deformation flow of the matrix direction.

8.3 Friction Stir Processing of In-Situ Aluminum Matrix Composites

Friction stir processing is a processing method evolved from friction stir welding, also known as friction stir processing [13–16]. The basic principle is to make the processed material undergo violent plastic deformation, mixing, and crushing through the strong stirring action of the stirring head, so as to realize the densification, homogenization and refinement of the microstructure. Friction stir processing can eliminate defects such as shrinkage porosity and shrinkage cavities in cast products, and can also refine grains, thereby improving material properties [17, 18].

8.3.1 *The Influence of Friction Stir Processing on ZrB_2/2024Al*

The composite sheet after friction stir treatment was cut perpendicular to the direction of the stirring process. After corrosion of the metallographic sample, the cross-sectional morphology as shown in Fig. 8.31 was observed. The morphology of the area near the stirring needle is generally "pot shape." The area where the material inside the pot is affected by friction and mechanical stirring shear stress is the friction stir processing zone [19, 20].

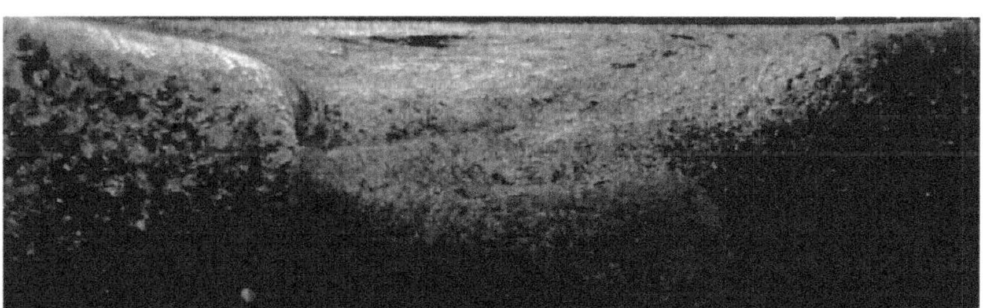

Fig. 8.31 Macroscopic morphology of composite cross section after friction stir treatment

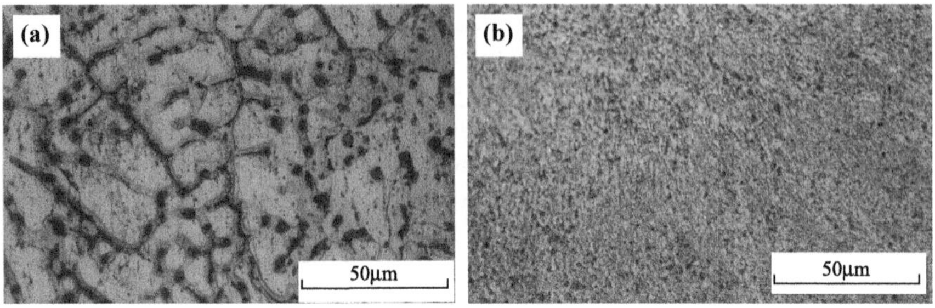

Fig. 8.32 Changes in reinforcement particles morphology: **a** before friction stir treatment; **b** after friction stir treatment

The material in the friction stir processing zone will experience strong plastic deformation under the direct shear stress of the stirring head. The heat released by the friction between the metal and the stirring head directly heats the metal under the shoulder of the stirring head to a temperature that produces plastic deformation or plastic flow. The mixing head will directly squeeze the material in front of the mixing head during the forward movement. At the same time, the composite material will be subject to the huge shear stress caused by the rotation of the mixing head. Therefore, the actual temperature in this area will exceed the liquid phase of the composite material, but lower than its complete melting temperature, at this time the composite material is in a semi-solid state. Under the action of stress and frictional heat, large grains will be broken into fine equiaxed grains. At the same time, some of the distortion energy will be retained after the grains are broken and become the driving force for recrystallization. The fine grains formed after crushing will undergo a recrystallization process and finally form a very fine, uniformly distributed equiaxed crystal structure. The matrix structure before and after friction stir pretreatment is shown in Fig. 8.32.

The reinforced particles in the composite material will also be subjected to huge external force and high temperature during the stirring process, and will deform and redistribute in the semi-solid aluminum melt: larger particles and part of the lath-like precipitated phase at the grain boundary will be affected The large plastic deformation force will break, and the agglomerated particles will redistribute and become dispersed with the violent flow of the semi-solid aluminum liquid, as shown in Fig. 8.33.

Figure 8.34 shows the SEM image of composite before and after friction stir treatment. SEM images in Fig. 8.34b, c after friction stir treatment show that the reinforcement particles are obviously dispersed, the agglomeration phenomenon is relieved or even disappeared, the lath structure at the grain boundary is broken, and the ZrB_2 reinforcement particles are obviously fine. What is more, the final composite material reinforcement distribution is more uniform.

Figure 8.35 is a high-resolution transmission image of the interface after friction stir processing. The friction stir processing not only improves the distribution of the reinforcement phase particles but also does not pollute the interface bonding

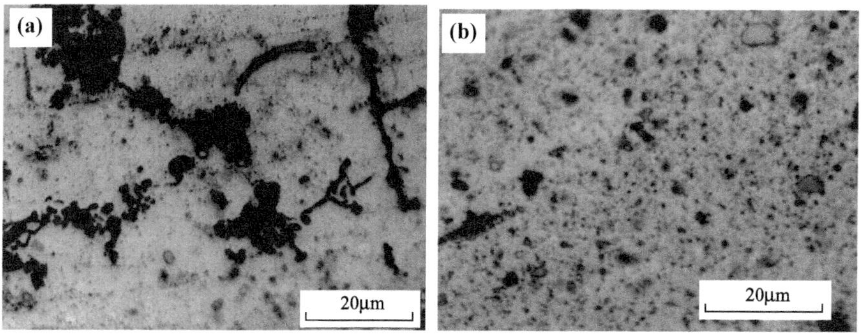

Fig. 8.33 Granular changes in reinforcement: **a** before friction stir treatment; **b** after friction stir treatment

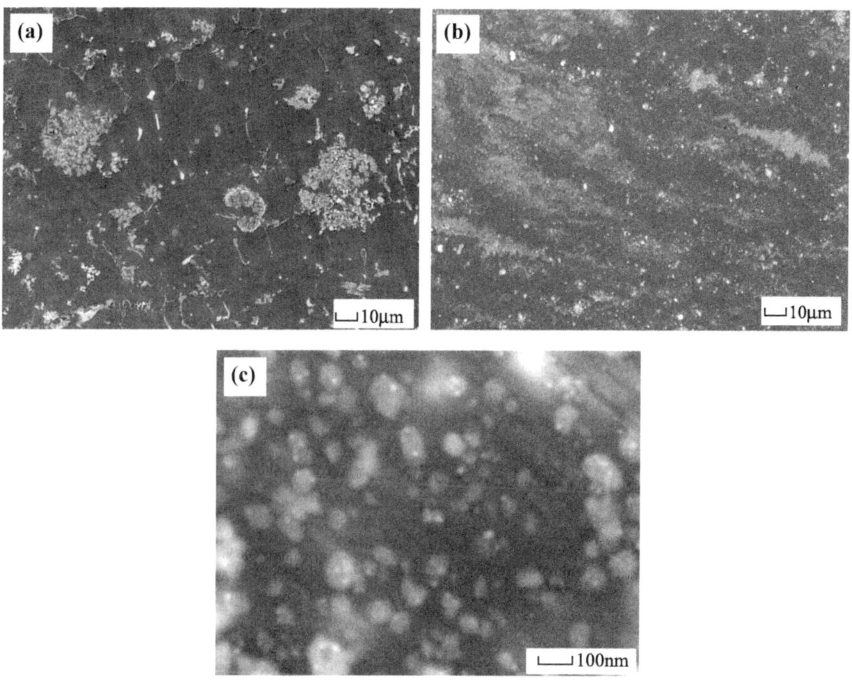

Fig. 8.34 SEM images showing changes in reinforcement phase particles: **a** before treatment; **b**, **c** after treatment

advantages of the in-situ composite for the interface is well bonded and there is no inclusion phase.

It can be found from Fig. 8.36 that after friction stir treatment, the peaks move to a large angle, that is, the distance between the crystal planes of the matrix decreases, but the peak intensity of each peak decreases, and the peak width increases significantly.

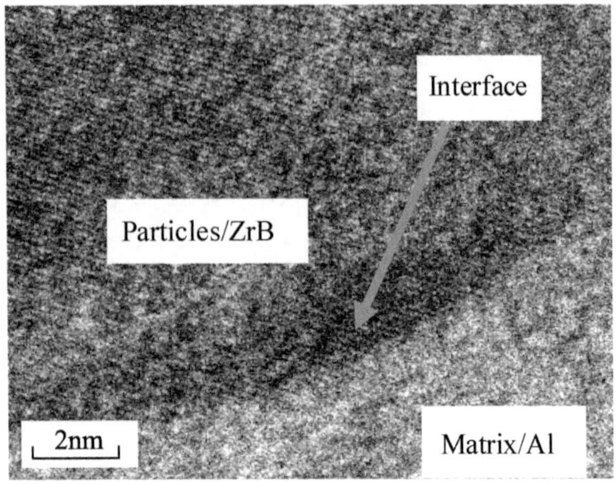

Fig. 8.35 High-resolution TEM image of matrix-reinforcement particle interface

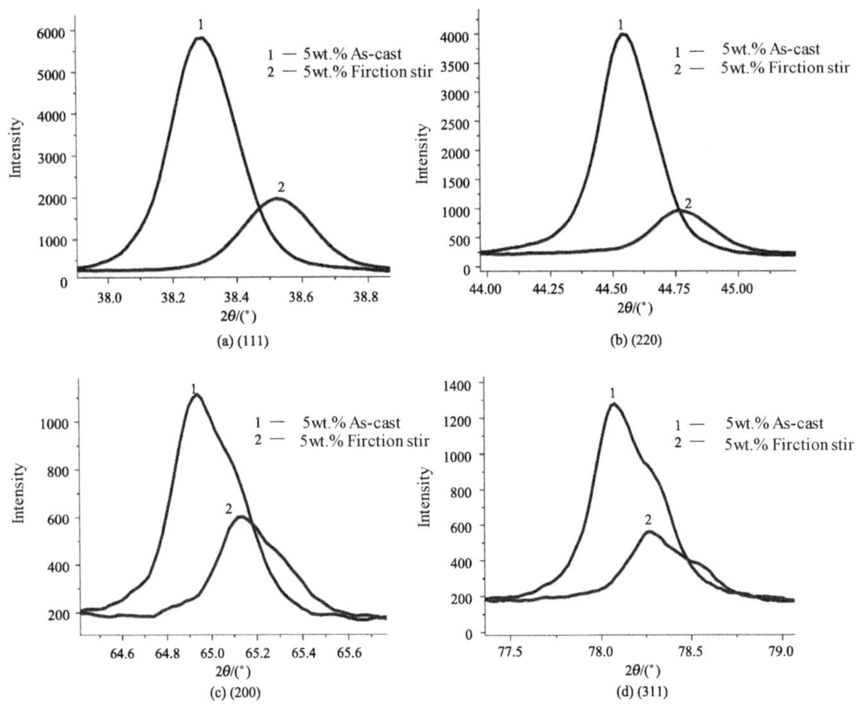

Fig. 8.36 XRD peak shifting before and after friction stir treatment: **a** (111); **b** (220); **c** (200); **d** (311)

Use Jade software to fit and calculate the XRD curve, the results are shown in Table 8.3.

It can be seen from the fitting calculation results that the lattice size of the matrix has been reduced after the friction stir treatment. The high degree of crystallinity among the various crystal planes may be due to the fact that the grain refinement of the friction stir treatment is the crushing of large crystal grains through physical methods at high temperatures and the mechanical action of the stirring head, accompanied by a certain degree of recrystallization, so the crystallinity of each crystal face family is higher. After friction stir treatment, the lattice constant of the aluminum matrix will also decrease, and the overall lattice will collapse. From the perspective of XRD analysis, after friction stir pretreatment, the matrix grains are

Table 8.3 Analysis of XRD of friction stir treated composites

Peak corresponds to crystal plane index	(1 1 1)	(2 0 0)	(2 2 0)	(3 1 1)
2θ	38.460	44.700	65.068	78.200
FWHM (peak half-height width)	0.479	0.245	0.237	0.314
Grain size (A)	173	168	164	145
Crystallinity/%	83.83	79.64	76.28	71.59
abc Al lattice constants		4.03796	4.03796	4.03796

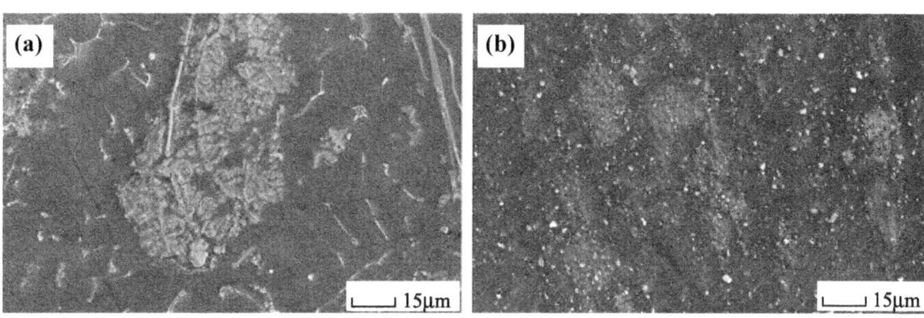

Fig. 8.37 SEM micrographs of ZrB_2/6063 composites: **a** before friction stir; **b** after friction stir

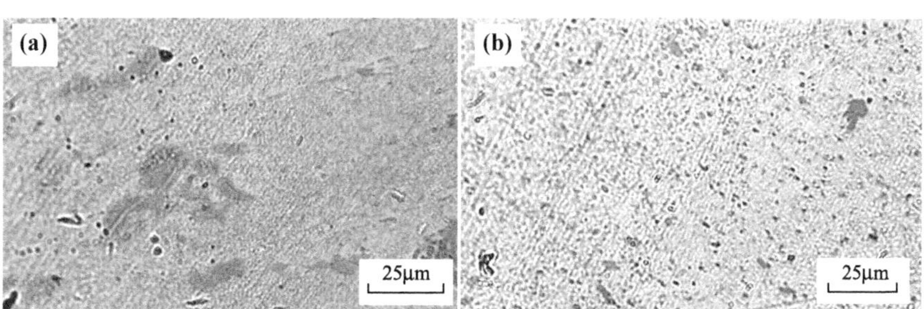

Fig. 8.38 ZrB_2/6063Al ultra-depth view of composite: **a** forging; **b** friction stir

deformed, and the relative as-cast grains become finer, and the degree of deformation is much greater than the degree of refinement after forging treatment, and the effect of recrystallization is small.

8.3.2 The Influence of Friction Stir Processing on ZrB_2/6063 Aluminum Matrix Composites

8.3.2.1 The Effect of Friction Stir on the Particle Morphology of ZrB_2/6063Al Composite

Figure 37a, b shows SEM images of the microstructure of the material before and after friction stir processing. It can be clearly seen from the figure that under the action of friction stir, the ZrB_2-reinforced particles are obviously dispersed and the agglomeration phenomenon disappears completely. The size of ZrB_2 particles is further reduced and the refinement effect is obvious. For particle-reinforced metal matrix composites, fine and uniform nanoreinforcement plays an important role in improving the performance of the material. The matrix grains of the ZrB_2/6063Al composite material are obviously refined after friction stir, and the matrix grain size is at least 5 μm or less, and a few grains may reach the nanometer level. At the same time, comparing the microstructure of the forged state, it is found that there is no obvious processing streamline, and the dislocation aggregation and stress concentration of the original sample are significantly improved, indicating that the friction stir processing can refine the microstructure and improve the performance of composite matrix.

Traditional plastic processing methods (forging, extruding, rolling) have a certain effect on the particle distribution of the reinforcement phase of the composite material, but compared with large plastic deformation (FSP), it is quite limited and can only extend and disperse the agglomeration along the plastic deformation direction to a certain extent, such as forging deformation. On the other hand, since the agglomeration size of the in-situ composite reinforcement phase is greatly reduced compared with the external particle method, it is difficult for the traditional plastic deformation method to further improve its particle distribution. Because friction stir processing will produce violent material flow in the stirring zone, which has a greater impact on the modification of the material's microstructure, and it significantly improves the distribution of the reinforcing phase particles of the in-situ aluminum matrix composites. As shown in Fig. 8.38, the comparison shows that the deformation texture of the forged material has completely disappeared in the friction stir state, and the particle clusters elongated along the texture direction have been dispersed evenly.

In order to further determine the changes of the reinforcing particles after friction stir, the scanning electron microscope was used to observe. Figure 8.39 shows the SEM comparison of the ZrB_2/6063Al composites in the as-cast state and after friction stir. Figure 8.39a shows the forged state and Fig. 8.39b shows the friction

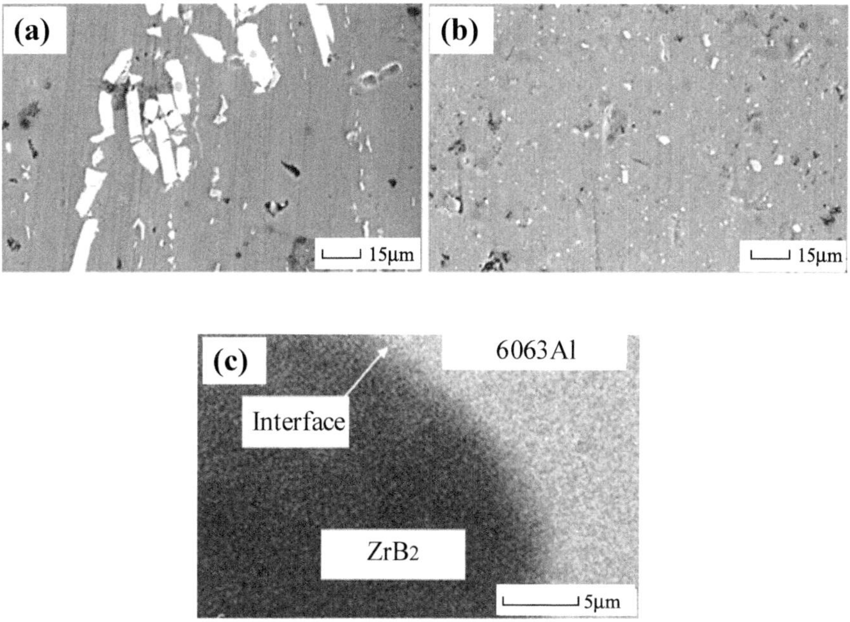

Fig. 8.39 SEM and TEM images of ZrB_2/6063Al composite after plastic deformation: **a** forged; **b** friction stir treated

stir state. Compared with forged composites, the agglomeration of reinforcements in the composites after friction stir has been significantly improved. After friction stir processing, the reinforcement particles are evenly dispersed in the matrix, and no agglomeration of the reinforcement particles can be seen. Some of the larger slab-shaped ZrB_2 have been completely crushed, but the edges and corners are still clear, indicating that friction stir processing improves the dispersion of reinforcing phase particles without compromising the unique interface advantages of the in-situ composite preparation method, such as shown in Fig. 8.39c, the sharp edges and corners of the particles become rounded and blunt, the interface is pure, and there is no brittle phase.

8.3.2.2 The Effect of Friction Stir on the Matrix Grains of ZrB_2/6063Al Composite

Figure 8.40 shows deep etched optical micrographs of the microstructure of ZrB_2/6063Al composite before and after friction stir. Figure 8.40b shows that the ZrB_2 particles in the original as-cast state are distributed along the grain boundaries of the matrix grains, and after friction stir, the ZrB_2 particles are agglomerated and dispersed, the streamline morphology of the matrix grains disappears, and the

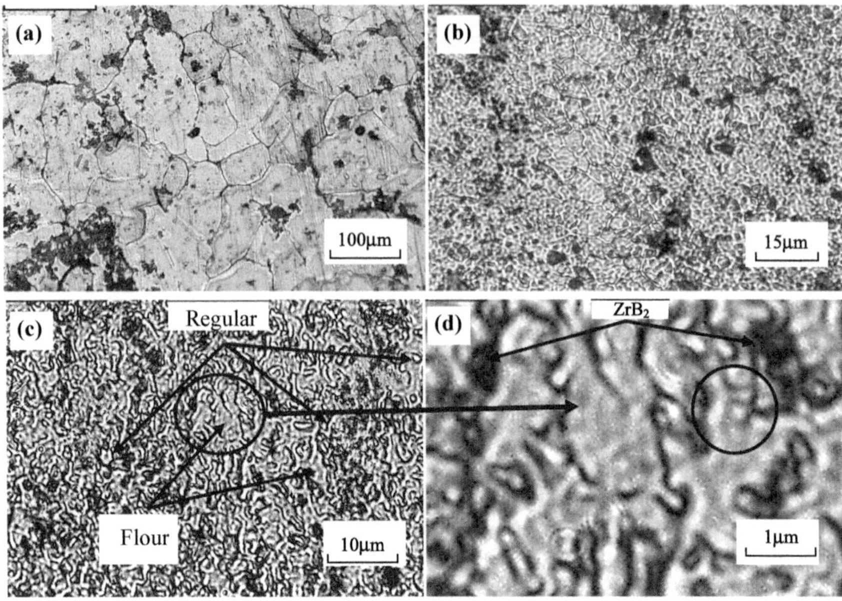

Fig. 8.40 Deep etched optical micrographs of ZrB_2/6063Alcomposite: **a** before friction stir, **b–d** after friction stir

original as-cast state is large. The crystal grains are broken during the violent deformation process, forming irregular very fine flocculent crystal grains with a size of 1–5 μm, even reaching nanometer level. From Fig. 8.40, it is impossible to accurately measure the grain size. It is judged that the ZrB_2/6063Al composite material has low matrix recrystallization during the friction stir process, and the formation of fine and regular flocculent equiaxed small grains. After friction stir treatment of ZrB_2/6063Al composite, the as-cast defects of the grains have been largely eliminated.

Analysis of the reason for the formation of flocculent crystals lies in one aspect: according to the research in the literature, the temperature of the matrix near the stirring head can reach 500 °C due to the action of friction during the friction stir process, which is much higher than the recrystallization temperature of 6063 aluminum alloy. In the process of friction stir, the matrix material deforms violently, which can produce a great stress concentration, that is, the strain energy in the matrix grains is very high, which is very conducive to the occurrence of recrystallization. Considering the advancing speed of the stirring head is 60 mm/min, the period of time during which the stirring head passes and is cooled to below the recrystallization temperature should be sufficient for a large amount of recrystallization. According to enlarged image in Fig. 8.40d, there are very few regular fine crystal grains around the flocculent crystal grains, and most of the regular crystal grains mostly exist alone. Observing the crystal grains in the circle, it can be found that regular equiaxed crystal grains are surrounded by flocculent crystal grains, and the grain boundary between the two is very fuzzy. It may be that small equiaxed crystals and flocculent crystal grains have fused together, but it may also be possible that the flocculent grains recrystallize to form equiaxed grains.

On the other hand: after the recrystallization is completed, the grain growth of the α-Al phase is uniform and continuous, but in some cases, a small number of energy-obtained grains suddenly and rapidly coarsen. The size difference between the crystal grains is significantly increased. The larger the difference in grain size, the more advantageous it is for large grains to swallow small grains. Therefore, as the initial growth rate of individual grains is faster and the difference in size from the surrounding small grains increases, the growth rate will increase. The faster, a structure with extremely uneven grain size or a structure composed of only coarse grains will eventually be formed. The main reason why a small number of crystal grains can grow up quickly in recrystallization is that there are a small number of factors that facilitate migration of crystal grain boundaries in the structure, such as the fine ZrB_2 particles in the composite material. The presence of ZrB_2 particles cause the grain boundaries around the particles to be pinned, but the grain boundaries that are not pinned by ZrB_2 can migrate, causing grain growth and merging the surrounding small grains. The grains will also be pinned when they encounter ZrB_2 grains in the process of growing up, and grain boundary migration will be blocked and bent here, and flocculent grains may eventually form.

8.3.2.3 Influence of Friction Stir Processing on the Lattice Structure of the Composite Matrix

Figure 8.41 shows the XRD patterns of as-cast, forged and friction stir ZrB_2/6063Al-based composites. Comparative analysis shows that the phase composition of the composite has not changed, but after friction stir processing and forging deformation, the matrix can be clearly observed. The diffraction peak of α-Al phase shifts to the right. It can be seen from the Bragg formula $2d\sin\theta = n\lambda$, where d is the interplanar spacing and θ is the diffraction half angle. After plastic deformation, the value of θ becomes larger, and the lattice constant d becomes smaller, that is, α-Al The phase lattice is distorted. The XRD diffraction peak shape of the ZrB_2/6063Al-based composite before and after the friction stir was analyzed by partial magnification. The result is shown in Fig. 8.42. It can be seen that the α-Al phase diffraction peak of the composite matrix after the friction stir has shifted, indicating that the friction stir caused the lattice distortion of the composite material matrix. At the same time, it was found by comparison that the offset of the α-Al phase XRD diffraction peak after friction stir was smaller than the forging offset, and the α-Al phase XRD diffraction peak after friction stir The FWHM is also wider than the original as-cast state, and the in-situ ZrB_2/6063Al composite material has a small amount of recovery and recrystallization during the friction stir process.

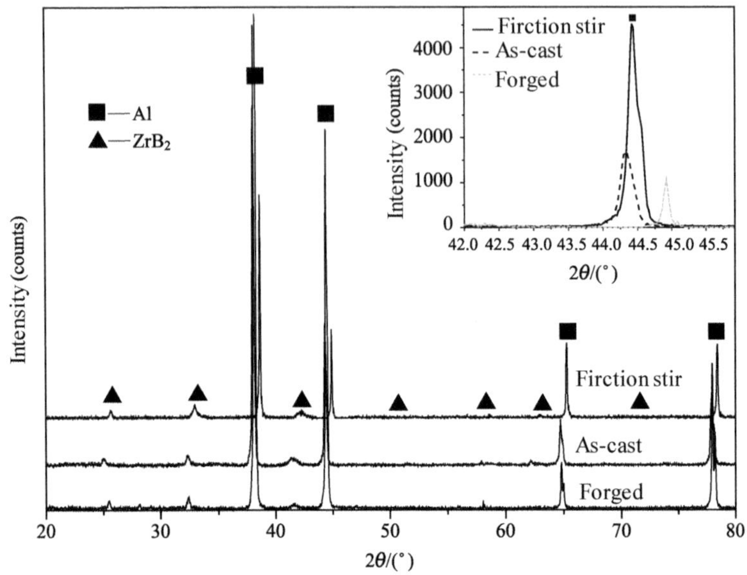

Fig. 8.41 XRD pattern of as-cast, forged and friction stirred composites

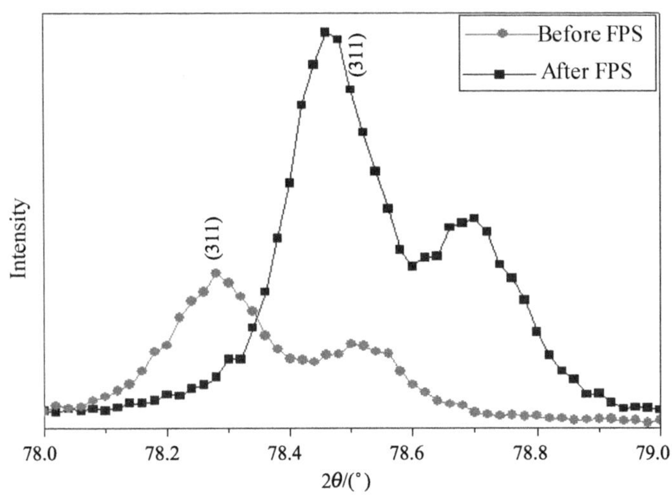

Fig. 8.42 XRD diffraction peaks after friction stir processing

8.3.3 The Effect of Friction Stir Processing on Al_3Zr/6063Al Composites

8.3.3.1 Effect of Friction Stir Processing on Particles Morphology in Al_3Zr/6063Al Composites

Figure 8.43 shows the optical micrographs of Al_3Zr/6063Al composite before and after FSP. Figure 8.43a, c, and e shows three optical micrographs taken from the upper middle area of Al_3Zr/6063Al ingot respectively (cross section perpendicular to the direction of advancement of the mixing head), and Fig. 8.43b, d and f shows optical

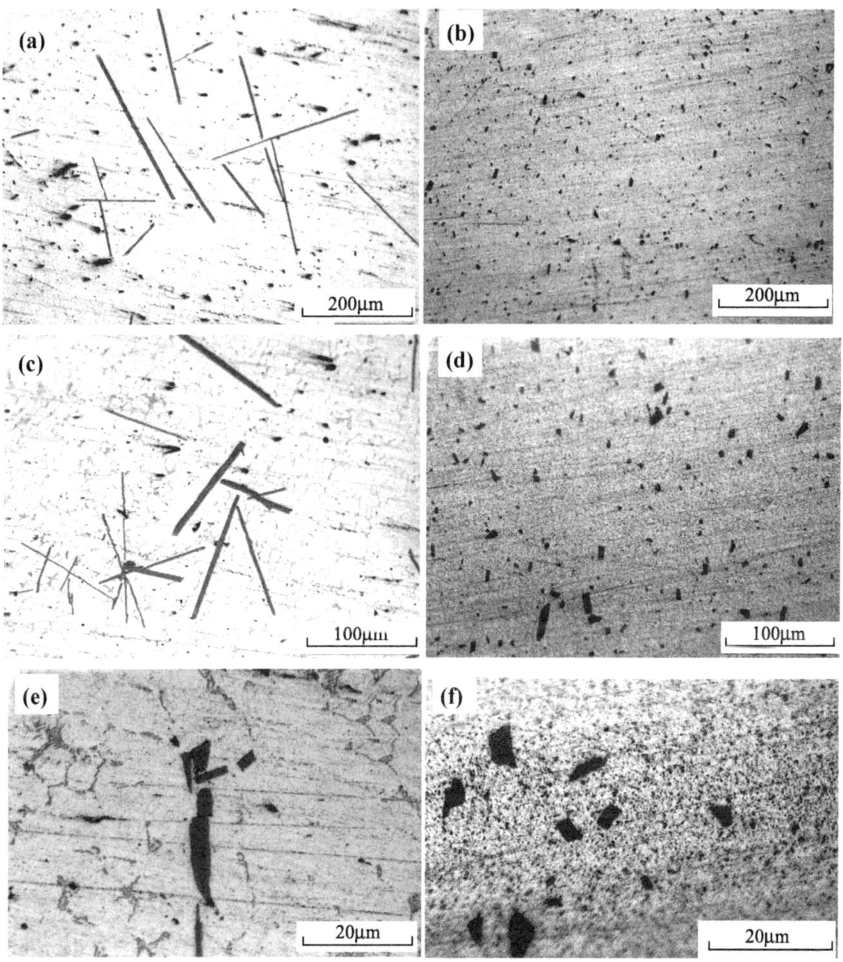

Fig. 8.43 Optical micrographs of Al_3Zr/6063Al composites: **a**, **c** and **e** before friction stir; **b**, **d** and **f** after friction stir

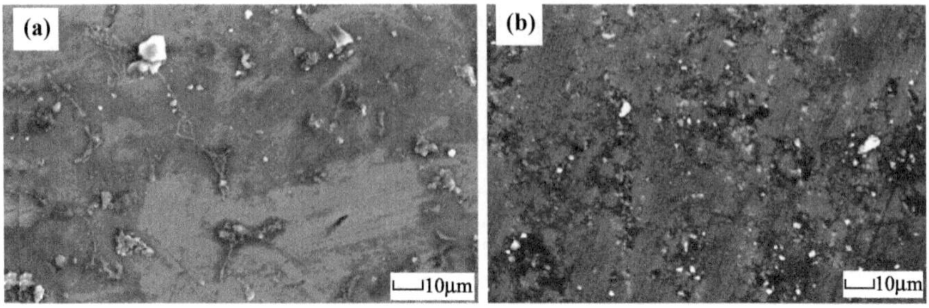

Fig. 8.44 SEM image of Al_3Zr/6063Al composites: **a** before friction stir; **b** after friction stir

micrographs of cross section of the friction stir processed area. As shown in Fig. 8.43, before FSP the reinforcement Al_3Zr has a long strip shape with a size range of 10–350 μm. After friction stir plastic deformation, the size of the reinforcing phase is obviously reduced, the particles are broken, about 1–7 μm, and are evenly distributed in the matrix. The heat exposure and strong plastic deformation produced by FSP cause the thick and long reinforcing phase to break. At the same time, because the material flows at higher temperatures, FSP makes the broken reinforcements more uniformly distributed in the matrix.

Through further scanning electron microscope observation and analysis, it can be seen that the prepared composite material has locally distributed Al_3Zr-reinforced particles with uniform distribution and fine size, and its morphology is long and the size is about 100 μm or more. After friction stir processing, the uniformly distributed reinforcement particles have a size of 1–2 μm, as shown in Fig. 8.44.

8.3.3.2 The Effect of Friction Stir on Al_3Zr/6063Al Matrix Grains

Figure 8.45 shows etched optical micrographs of Al_3Zr/6063Al composite before and after friction stir deformation. As shown, the matrix grain size before deformation is at least 30 μm, the grains are regular and the grain boundaries are obvious. It can be seen from the figure that the matrix crystal grains of the composite material after FSP treatment are very small, which are composed of equiaxed crystals or flocculent crystals, with uniform structure, and an average size of 5 μm or less, without obvious directionality. During the friction stir processing, a large amount of heat is generated between the stir needle and the workpiece and between the shaft shoulder and the workpiece, which plasticizes the surrounding materials and flows fully. In addition, the temperature of the processed area rises to the recrystallization temperature. Under the action of the stirring force, the dislocation density continues to increase, and when the dislocation energy is increased enough to recrystallize, crystal nuclei will continue to form inside the material. This kind of crystallization process is a dynamic recrystallization process that occurs under the action of the stirring heat cycle. The formed crystal grains will be broken under the action of the stirring needle before they can grow up to form equiaxed and fine crystal grains.

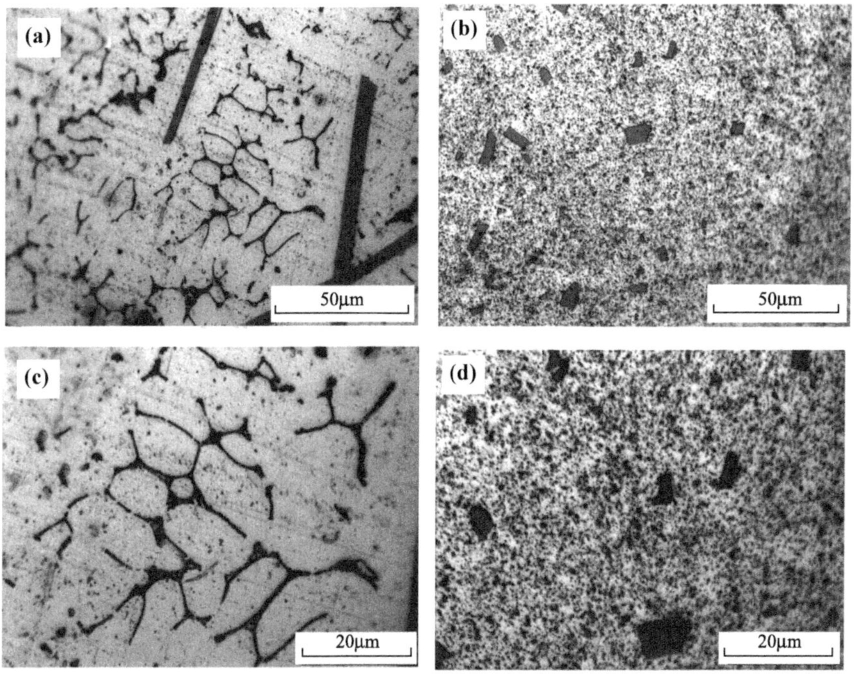

Fig. 8.45 Etched optical micrographs of Al_3Zr/6063Al composite: **a**, **b** before friction stir; **c**, **d** after friction stir

8.3.3.3 Influence of Friction Stir Processing on Dislocations in Al_3Zr/6063Al Composites

Figure 8.46 shows the distribution of dislocations inside the crystals of the in-situ endogenous Al_3Zr/6063Al composite after different plastic processing treatments, (a) is the as-cast state, (b) is the forged state, (c) is the friction stir state, (d) is the rolled state. It can be seen from the comparison of the figures that most of the dislocations in the as-cast state are packed near the interface between the Al_3Zr-reinforced particles and the aluminum matrix. The number of dislocations in the forged composite crystals increases, showing a regular, unidirectional arrangement.

In the friction stir sample, the dislocations are staggered and disordered, and there are obvious dislocation lines and slip bands in the grains of the rolled composite material (as shown by the arrow in the Fig. 8.46), which indicates that the matrix crystals during the rolling process. A large amount of slip deformation occurred in the particles. The comparison shows that friction stir plastic processing makes the dislocation density multiplication in the material the largest, the dislocation relationship is more complicated, the rolling dislocation density is the second, and the number of dislocations in the as-cast state is the least.

Figure 8.47 shows the TEM image of the composite material after friction stir processing. It can be seen from the TEM image that there are obvious dislocation

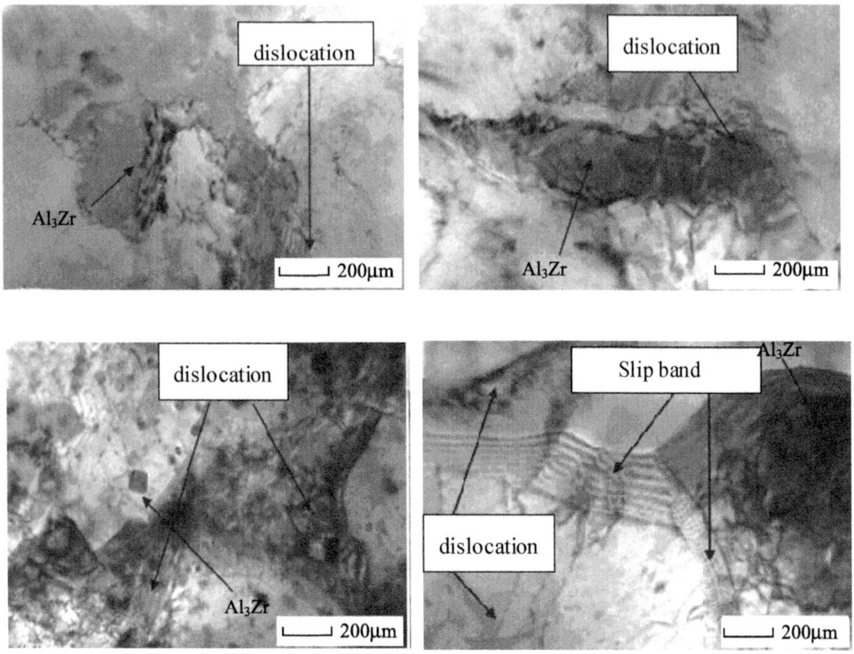

Fig. 8.46 TEM images showing dislocations in Al_3Zr/6063Al composite: **a** as-cast state, **b**) forged state, **c** friction stir state, **d** rolled state

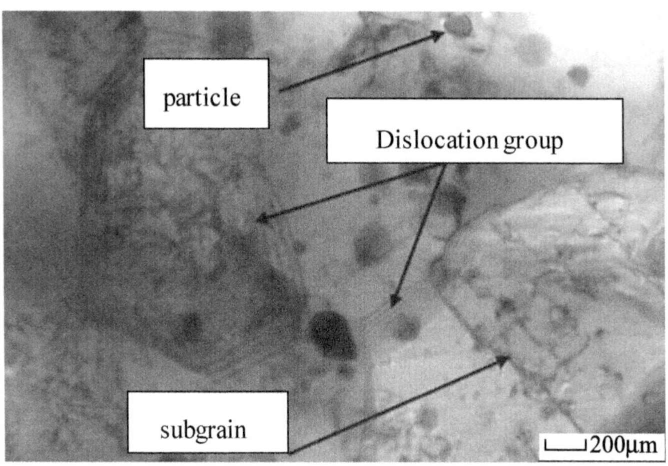

Fig. 8.47 TEM image of Al_3Zr/6063Al composite after friction stir processing

groups at the grain boundary (pointed by the arrow), indicating that the degree of recrystallization is very low here. There are subcrystals inside the crystal grains, indicating that the material is still in the recovery stage. It can be inferred from this that the degree of recrystallization of the composite material matrix is not high during the friction stir process, and there are fewer fine and regular equiaxed grains formed, and most of the grains have not recrystallized, but the violent plastic deformation of the friction stir. In the process, it was fully broken by the friction stir head, forming irregular flocculent crystal grains.

8.3.4 The Effect of Friction Stir Processing on Al_3Ti/2024Al Composites

Figure 8.48 shows the X-ray diffraction pattern of as-cast 5 wt.% Al_3Ti/2024Al composite after forging, and after friction stir. It can be seen from the figure that compared with the as-cast material, the shape of the α-Al phase XRD diffraction peak changes significantly after forging and friction stir, and the diffraction peaks are shifted to the right. Compared with the friction stir forging state, α-Al phase diffraction peaks shift is smaller. According to Bragg's law $2d\sin\theta = n\lambda$, after forging, the diffraction peak of the material is shifted to the right, and θ becomes larger, so the interplanar spacing d in the unit cell parameters becomes smaller, that is, lattice distortion also occurs. Therefore, it can be preliminarily determined that forging

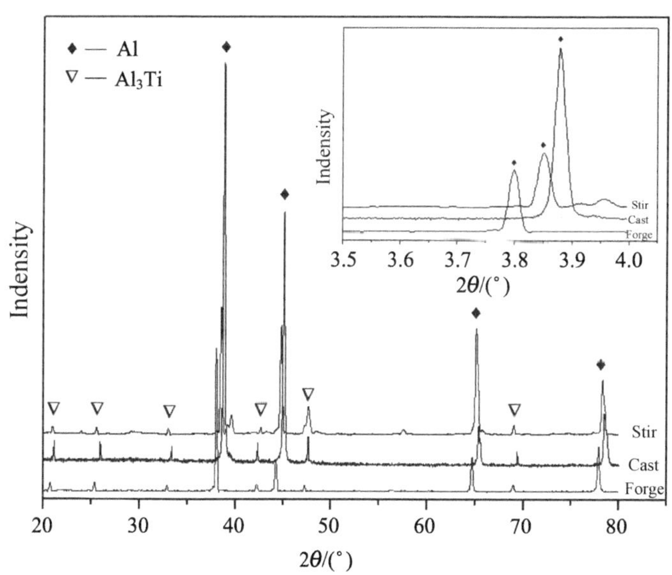

Fig. 8.48 XRD analysis of in-situ Al_3Ti/2024Al composites in as-cast, forged and friction stirred states

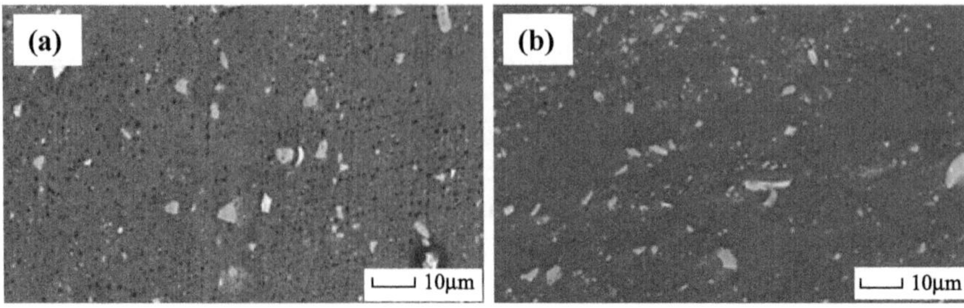

Fig. 8.49 SEM images of Al_3Ti/2024Al composites with different particle content after friction stir: **a** 5 wt.%; **b** 10 wt.%

and friction stir cause lattice distortion of the Al_3Ti/2024Al composite. In addition, according to the obvious change in the peak height of α-Al after plastic deformation processing, the half-height width of the diffraction peak of α-Al phase becomes larger. It is inferred that the α-Al phase of the Al_3Ti/2024Al composite matrix undergoes dynamic renewal during the plastic deformation process.

Figure 8.49 shows SEM images of Al_3Ti/2024Al composites with different particle content after friction stir plastic deformation. It can be seen from the figure that after friction stir plastic deformation, most of the Al_3Ti-reinforced particles have a size of about 1–5 μm, and a small part of the particles have a size of 0.1–1 μm. The reinforced particles are roughly oriented and distributed in the matrix, and the distribution is relatively uniform. The effect of friction on the refinement of the reinforcement phase is obvious. During friction stirring, the large Al_3Ti particles are broken under the stirring and extrusion action of the high-speed rotating stirring head, and the agglomerated particles are also broken up. Because friction stir is linear processing, and the friction heat during processing makes the surrounding metal matrix form a plastic softening layer, the fractured reinforcing particles are squeezed and distributed along the path of friction stir.

Figure 8.50 is deep etched optical micrographs of the in-situ Al_3Ti/2024Al composite after corrosion. Figure 8.50a–d shows respectively the as-cast 2024Al alloy, the as-cast Al_3Ti/2024Al composite, the composite material after forging, and the composite material after friction stir. Figure 8.50e shows the morphology of Al_3Ti/2024Al composite after friction stir at 5000 times ultra-depth of field. It can be seen from the figure that the grain size of 2024Al alloy is 50–150 μm, and the grain size of the as-cast Al_3Ti/2024Al composite is 50–100 μm. After forging, the grain is deformed into fibrous shape with a length of about 150 μm and a width of 30–50 μm, the grains are fine after friction stir, the size is 5–10 μm. In the as-cast state, both the aluminum alloy and the composite material are dendrites, the crystal grains are fibrous after forging, and the matrix grains of the composite material after friction stir are fine equiaxed crystals.

In Fig. 8.50d, the matrix grains after friction stir have obvious streamline marks, which are the same as the direction of friction stir, that is, the matrix grains undergo plastic flow along the direction of the stirring head under the action of stirring and

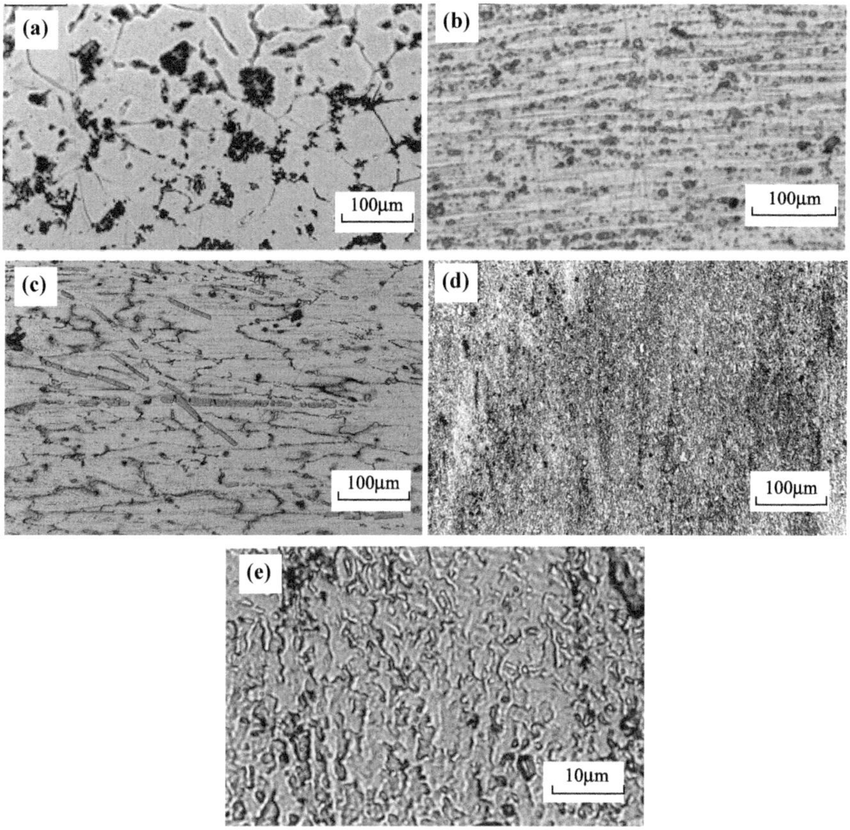

Fig. 8.50 Grain structure of in-situ Al_3Ti/2024Al composite: **a** 2024Al alloy; **b** as-cast; **c** after forging; **d** after friction stir; **e** after friction stir

extrusion during friction stir. Frictional heat is generated during vigorous stirring, and the temperature in the stirring zone reaches the recrystallization requirement; at the same time, the dislocation density increases under the action of stirring force, and recrystallization occurs when the energy storage reaches a certain level. This kind of recrystallization is a dynamic recrystallization process of plastic materials under the action of welding heat cycle. The crystal grains formed by recrystallization are quickly broken by the stirring needle, and the cycle advances, finally forming equiaxed and fine crystal grains.

Figure 8.51 shows TEM observation results of 5 wt.% Al_3Ti/2024Al composite in as-cast state, forging and friction stir state. In the as-cast state (Fig. 8.51a), the composite has coarse grains, and the reinforcing grains are elongated and larger in size, with fewer dislocations in the grains and curved grain boundaries. After forging processing (Fig. 8.51b), a large number of dislocations appeared in the grains, and they were entangled and intertwined into a network, resulting in blurred grain boundaries. After friction stir, the composite material has fine grains, some of the grain

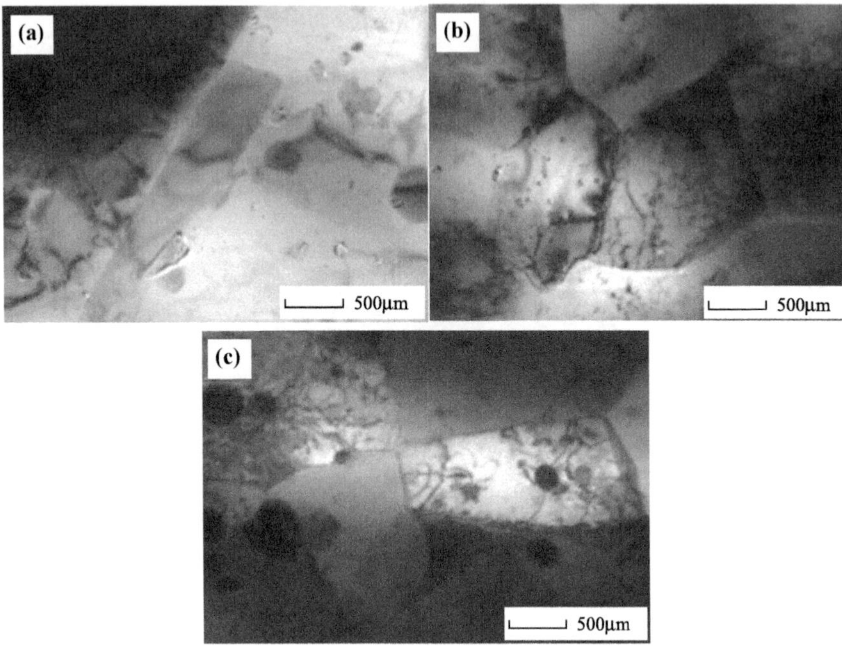

Fig. 8.51 TEM images of composites under different conditions: **a** as-cast; **b** after forging; **c** after friction stir

size is 0.5–1 μm, the grain shape is relatively regular, and the grain boundaries are straight, indicating that the material has undergone relatively complete dynamics at the dislocation plugging during the friction stir process Recrystallize, and partially recrystallized grains have a tendency to grow. In Fig. 8.49c, some small, nearly spherical Al_3Ti reinforcement particles can be found in the lower grains, indicating that the reinforcement phase in the composite material is significantly broken during the friction stir process, and is more evenly distributed in the grains. After recrystallization, the dislocation density in some grains is reduced, but there are still a large number of dislocations in the grain boundaries and places where there are more grains.

References

1. Evans A, San Marchi C, Mortensen A. Metal matrix composites in industry: an introduction and a survey. Springer Science & Business Media, 2013; 242.
2. Jiao L, Zhao YT, Wang XX. The research progress of the in-situ metal matrix composites. Key Eng Mater. 2014;575:137–41.
3. Fayun Z, Hong Y, Tianrui Z, et al. Research progress of metal matrix composites preparation technology. Forging Technol. 2007;31(6):100–5 (in Chinese).
4. Guirong L, Yutao Z, Qixun D, Xiaonong C, Hongming W, Zhongde L, Gang C. Solidification structure of squeeze casting particle reinforced aluminum matrix composites. Special Casting Nonferrous Alloys 2006;10:672–3, 607–8 (in Chinese)

5. Yongliang G, Xingqiu L, Jiangli N, et al. The effect of hot extrusion on the tensile properties and microstructure of Al_2O_{3p}/6061 aluminum matrix composites. Mater Mech Eng. 2007;31(3):65–7 (in Chinese).
6. Włodarczyk-Fligier A, Dobrzański L A, Kremzer M, et al. Manufacturing of aluminium matrix composite materials reinforced by Al_2O_3 particles. J Achievements Mater Manuf Eng 2008, 27(1).
7. Meng Z, Yutao Z, Dengbin C, Zhao Z, Hongbiao D. Preparation and microstructure of in-situ nano-Al_2O_3 particle reinforced aluminum matrix composites. Functional Mater 2011;S4:748–750, 754 (in Chinese)
8. Yida Z, Yutao Z, Lei J, Kangle T. The effect of extrusion and heat treatment on the mechanical properties of in-situ synthesized $(Al_3Zr+ZrB_2)_p$/6063Al composites. Funct Mater. 2014;19:19128–33 (in Chinese).
9. Guirong L, Yutao Z, Qixun D, et al. Solidification structure of squeeze casting particle reinforced aluminum matrix composites. Special Casting Nonferrous Alloys, 2006;26(10):672–3 (in Chinese)
10. Ceschini L, Minak G, Morri A. Forging of the AA2618/20 vol.% Al_2O_{3p} composite: effects on microstructure and tensile properties. Compos Sci Technol 2009;69(11):1783–9.
11. Alizadeh M, Paydar MH. Fabrication of Al/SiC P composite strips by repeated roll-bonding (RRB) process. J Alloy Compd. 2009;477(1):811–6.
12. Minhua Z, Gang C, Pengfei J, et al. Effects of heat treatment and forging on the microstructure and properties of in-situ Mg_2Si/Al composites. Hot Work Technol 2014;43(24):125–8 (in Chinese)
13. Lei J. Research on plastic deformation behavior and properties of in-situ particle reinforced aluminum matrix composites. Jiangsu University 2014 (in Chinese)
14. Dixit M, Newkirk JW, Mishra RS. Properties of friction stir-processed Al1100–NiTi composite. Scriptamaterialia. 2007;56(6):541–4.
15. Shafiei-Zarghani A, Kashani-Bozorg SF, Zarei-Hanzaki A. Microstructures and mechanical properties of Al/Al_2O_3 surface nano-composite layer produced by friction stir processing. Mater Sci Eng, A. 2009;500(1):84–91.
16. Lei J, Yutao Z, Xiaolu W, et al. Microstructure and properties of friction stir treated Al_3Zr/6063 in-situ composites. Rare Metal Mater Eng 2014;7:046 (in Chinese)
17. Ma ZY. Friction stir processing technology: a review. Metall Mater Trans A. 2008;39(3):642–58.
18. Sharifitabar M, Sarani A, Khorshahian S, et al. Fabrication of 5052Al/Al_2O_3 nanoceramic particle reinforced composite via friction stir processing route. Mater Des. 2011;32(8):4164–72.
19. Bauri R, Yadav D, Suhas G. Effect of friction stir processing (FSP) on microstructure and properties of Al–TiC in situ composite. Mater Sci Eng, A. 2011;528(13):4732–9.
20. Yang M, Xu C, Wu C, et al. Fabrication of AA6061/Al_2O_3 nano ceramic particle reinforced composite coating by using friction stir processing. J Mater Sci. 2010;45(16):4431–8.

Chapter 9
Wear Properties of In-Situ Aluminum Matrix Composites

As a new type of wear-resistant material, particle-reinforced aluminum matrix composites have been used in many fields requiring wear resistance, such as engine pistons, cylinder liners, rotary bearings, automobile brake systems, and oil separation caps. Wear performance is important performance for particle-reinforced aluminum matrix composites, which is affected by many factors: ① external factors, such as the load, the relative sliding speed, the sliding distance, the temperature, the surface state, the grinding parts, and the wear mode; ② internal factors, such as the type of reinforced particles, the particle size, the volume fraction, and the matrix microstructure. These factors often interact with each other, which make it difficult to study the wear behavior of particle-reinforced aluminum matrix composites.

9.1 Wear Performance of In-Situ Aluminum Matrix Composites at Room Temperature

9.1.1 Wear Performance of Hyper-Eutectic Al-Si Alloy Matrix Composites

The wear resistance of hyper-eutectic Al-Si alloy matrix in-situ composites was studied systematically. The wear resistance test was carried out on a domestic MM-2000 pin-plate testing machine, the working principle of which is shown in Fig. 9.1. Before the experiment, the sample was wire-cut with a size of 19.5 mm × 10 mm × 8 mm, and the grinding surface was 19.5 mm × 10 mm. The grinding material is 40Cr steel (HRC40) with the size of ϕ 50 × 10 mm. The wear resistance was measured by the weight difference between the samples before and after the test. MA110 electronic analytical balance was used for weighing with an accuracy of 0.0001 g. The friction type is dry sliding friction.

Y. Zhao, *In-Situ Synthesis of Aluminum Matrix Composites*,
https://doi.org/10.1007/978-981-16-9120-1_9

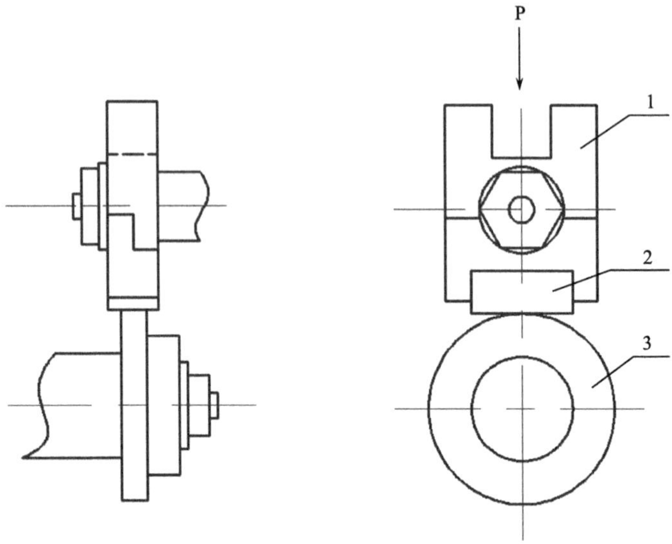

Fig. 9.1 Schematic diagram of MM-2000 friction and wear testing machine 1—fixture; 2—sample; 3—grinding parts; *p*—load

In the experiment, the rotating speed of the upper sample was 180r/min, and that of the lower sample was 200r/min. Read the torque value from the torque scale, and then, calculate the change of friction coefficient according to Eq. (9.1):

$$\mu = T/(R \times P) \quad (9.1)$$

where μ is the friction coefficient; T is the moment, unit: N·m; R is the radius of the lower sample, unit: m; and P is the load, unit: N.

Before each test, the sample and the grinding part should be cleaned with acetone. After each test, the grinding part should be grounded to 1200ppi before the next test to ensure that its surface roughness for each test is constant. Both the Al-22Si matrix alloy and the corresponding composites were annealed, while both the alloyed matrix alloy and the corresponding composites were T6 heat treated.

9.1.1.1 Wear Curve of (Al_2O_3 + Al_3Zr) $_p$/Al-22Si Composites

Figure 9.2 shows the weight loss curves of (Al_2O_3 $_+$ Al_3Zr)$_p$/Al-22Si composite and Al-22Si hyper-eutectic Al-Si alloy at different wear times when the sliding velocity is 0.523 m/s and the load is 60 N. It can be seen from Fig. 9.2 that the wear weight loss of both the materials increases and there is nearly a linear relationship between the wear loss and the wear time. The slope of the wear loss curve of hyper-eutectic Al-Si alloy is larger than that of the composite, which indicates that the increased amplitude of the wear of hyper-eutectic Al-Si alloy is larger than that

Fig. 9.2 Relationship between the wear loss and the wear time of Al-22Si matrix alloy and composite

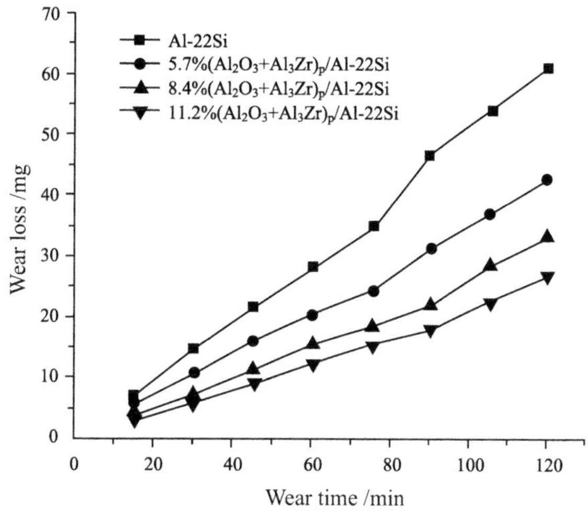

of the composite. After 75 min, the wear curve of the hyper-eutectic Al-Si alloy shows an obvious process of intensified wear with increasing slope, which is restored later. The composite also shows the same phenomenon at 75 min but with a small amplitude. This phenomenon indicates that the wear mode has changed and severe wear has begun to occur. However, the change in 8.4 vol.% $(Al_2O_3 + Al_3Zr)_p$/Al-22Si and 11.2 vol.% $(Al_2O_3 + Al_3Zr)_p$/Al-22Si composites did not occur until 90 min, and the change was very small. After 120 min, the wear amount of 11.2 vol.% $(Al_2O_3 + Al_3Zr)_p$/Al-22Si composite was 26.84 mg, and that of hyper-eutectic Al-Si alloy was 60.79 mg, about 2.3 times that of the corresponding composite. This was caused by the high silicon aluminum alloy being torn off in large pieces at the later stage of wear.

9.1.1.2 Influence of Load on Wear Performance of Composites and Al-22Si Alloys

Figure 9.3 shows the relationship between the applied load and the wear amounts of $(Al_2O_3 + Al_3Zr)_p$/Al-22Si composite and hyper-eutectic Al-Si alloy when the sliding velocity is 0.523 m/s and the sliding time is 60 min. It can be seen from Fig. 9.3 that with the increase of the applied load, the wear amount of both the hyper-eutectic Al-Si alloy and the composite increases with the increase of load, but the wear amount of the hyper-eutectic Al-Si alloy with the increase of load is larger than that of the composite. After the same sliding time, the wear amount of the composite is much less than that of the hyper-eutectic Al-Si alloy. Accordingly, under the same sliding condition, the wear amount of composite decreases with the increase of volume fraction. When the load is less than 80 N, the wear amount variation of hyper-eutectic Al-Si alloy and composite material is basically linear. However, when the

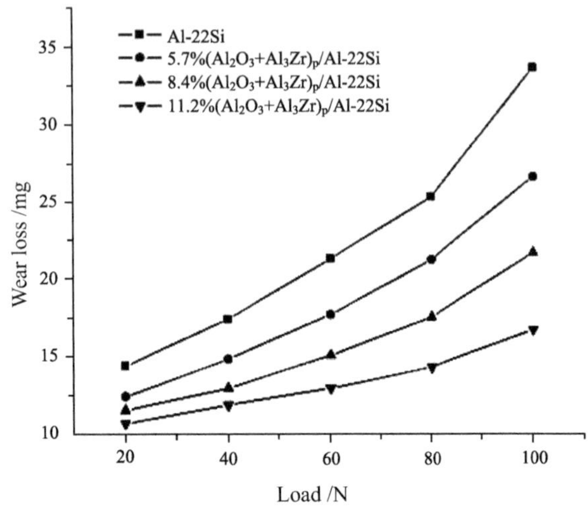

Fig. 9.3 Relationship between the wear loss and the applied load for Al-22Si matrix alloy and composite

load is greater than 80 N, the wear amount of hyper-eutectic Al-Si alloy increases sharply. Such a critical transition also occurs in 5.7 vol.% $(Al_2O_3 + Al_3Zr)_p$/Al-22Si, 8.4 vol.% $(Al_2O_3 + Al_3Zr)_p$/Al-22Si, and 11.2 vol.% $(Al_2O_3 + Al_3Zr)_p$/Al-22Si, but the amplitude is smaller than that of high Si-Al alloy. The larger the particle volume fraction is the less obvious the trend is. This phenomenon may be caused by the following two reasons: First, the formation and dispersion of fine in-situ particles improves the hardness of the material, prevents the plastic flow of the subsurface matrix alloy under the action of external forces, reduces the adhesion between the material and the grinding block, and leads to obvious hysteresis of the wear revulsion. The second, the reinforcement phase particles generated by the reaction are small and closely bonded to the matrix, which inhibits the tendency of particles to peel off from the matrix and speeds up the wear of abrasive particles, thus enhancing the resistance to adhesion and deformation of the composite materials and improving the wear resistance of composite [1–4]. When the load increases to a certain extent (>80 N), the large primary silicon particles on the grinding surface of hyper-eutectic Al-Si alloy will break and fall off under high surface shear stress, which will not play a good role in bearing the load. As a result, the friction surface will directly contact with the aluminum matrix and peel off under high shear stress, causing severe wear. Although the primary silicon particles in the composite are broken and detached, many fine in-situ particles play the role of bearing load, which increases the load needed for wear drastic changes. It can be seen that the composite exhibits a better wear performance than the hyper-eutectic Al-Si alloy.

9.1.1.3 Influence of Load on the Friction Coefficient of Composites and the Al-22Si Alloys

Figure 9.4 shows the changes in the friction coefficient of $(Al_2O_3 + Al_3Zr)_p$/Al-22Si composite and Al-22Si hyper-eutectic Al-Si alloy with the applied load as the sliding time is 60 $_{min}$. It can be seen from the figure that the friction coefficient of $(Al_2O_3 + Al_3Zr)_p$/Al-22Si composite material and Al-22Si matrix decreases with the increase of load. Under the same conditions, the friction coefficient of $(Al_2O_3 + Al_3Zr)_p$/Al-22Si composite is lower than that of Al-22Si hyper-eutectic Al-Si alloy. With the increase of the volume fraction, the friction coefficient of $(Al_2O_3 + Al_3Zr)_p$/Al-22Si composite decreases, but the variation amplitude of the friction coefficient of the composite is smaller than that of the matrix, showing a more stable friction behavior than the matrix.

With the increase of applied load, α-Al microconvex bodies on the contact surface are deformed plastically, and the primary silicon and in-situ reinforced particles begin to contact with the grinding block, where the adhesion coefficient between the reinforcement particles and the grinding block is lower than that between the α-Al microconvex body and the grinding block. When the load increases further, the Al microconvex body falls off, and the contact surface is mainly composed of reinforcement particles, which makes the actual contact surface decrease. Therefore, the friction coefficient of both the matrix and the composite material decreases with the increase of load. In addition, with the increase of load, Fe_2O_3 is formed after the oxidation of iron produced by the grinding of steel by in-situ hard particles, which plays a lubricating role. Figure 9.5a shows the SEM image of grinding particles, and Fig. 9.5b shows the EDS spectrum of point 1 in Fig. 9.5a. After the introduction of in-situ wear-resistant reinforcement particles, the contact area between the Al matrix and the grinding block is further reduced, and the adhesion between the matrix material and the grinding block is reduced, which is conducive to the reduction of friction

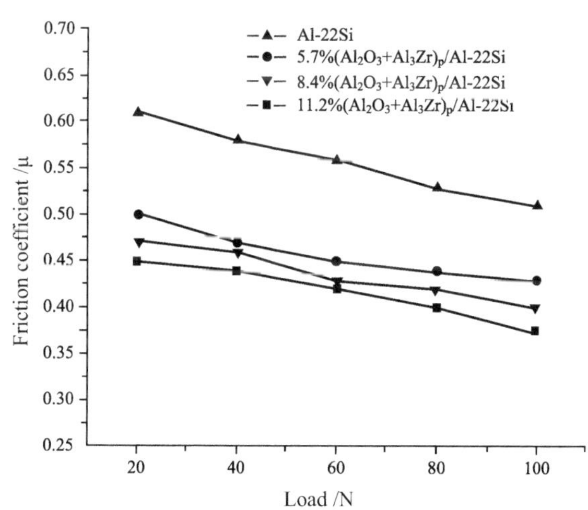

Fig. 9.4 Influence of applied load on the friction coefficients of Al-22Si matrix alloy and composite

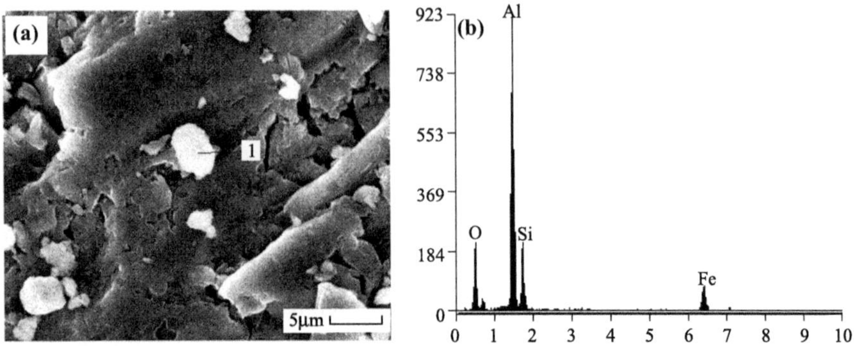

Fig. 9.5 SEM image and EDS analysis of abrasive particles in composites

coefficient. At the same time, due to the introduction of in-situ particles, the size of primary silicon particles in the in-situ composite material also decreases, so that the particles have a less scraping effect on the grinding block and the effect of abrasive wear on the friction force will be reduced, which reduces the friction coefficient. In addition, from the microscopic point of view, the increase of oxide film caused by the increase of load and temperature will also reduce the friction coefficient.

9.1.1.4 Influence of Alloying on Friction and Wear Performance of Composites

Figure 9.6 shows the influence of alloying on the wear performance of composites. In Fig. 9.6, M1 is Al-22Si matrix alloy, M2 is 11.2 vol.%$(Al_2O_3 + Al_3Zr)_p$/Al-22Si composite, M3 is Al-22Si alloy after alloying, and M4 is 11.2

Fig. 9.6 Influence of alloying on wear amount of matrix alloy and its composites

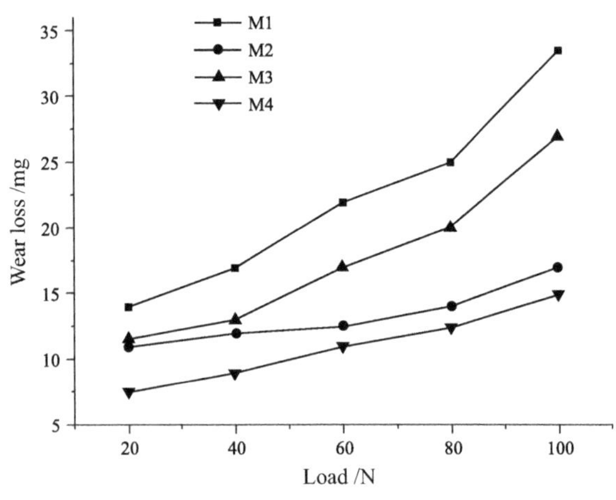

vol.%$(Al_2O_3 + Al_3Zr)_p$/Al-22Si composite after alloying. It can be seen from the figure that the wear amount after alloying is greatly reduced in both the matrix alloy and its composite. Under low load, the wear amount of alloyed composite changes greatly compared with 11.2 vol.%$(Al_2O_3 + Al_3Zr)_p$/Al-22Si composite, but the difference between the two kinds of composite is small when the load is higher than 80 N. This is because that the influence of matrix alloy on wear amount plays a major role under low load. After alloying, the hardness and wear resistance of matrix alloy are improved, so the wear amount of composite is greatly reduced. However, under the condition of high load, the in-situ particles are exposed to play a bearing role with the aggravation of wear, and the particles play a major role in the material wear. At this time, the difference of wear amount becomes smaller when the volume fractions of in-situ particles are same. By comparing the wear curves of M2 and M3, it can be seen that there is little difference in the wear between the alloyed matrix alloy and 11.2 vol.%$(Al_2O_3 + Al_3Zr)_p$/Al-22Si composite material under low load, which is due to the superior microstructure of alloyed matrix alloy. However, the difference of wear between these two increases with the increase of the load because of the effect of fine in-situ reinforcement particles.

Figure 9.7 shows the influence of alloying on the friction coefficient of matrix alloy and its composite. It can be seen from Fig. 9.7 that the friction coefficient of both the matrix alloy and its composite after alloying still decreases with increasing load. After alloying, the friction coefficient of matrix alloy is even lower than that of 11.2 vol.%$(Al_2O_3 + Al_3Zr)_p$/Al-22Si composite before alloying. The dispersion strengthening of alloying elements may be the reason why the wear resistance of Al-22Si alloy after alloying is higher than that of Al-22Si binary alloy. The hardness of the multi-component alloy is higher than that of the binary alloy, which will also reduce the impact of breakage and the deformation caused by severe surface wear, resulting in the reduced actual contact area. Therefore, the friction coefficient of the multi-component alloys is lower than that of the binary alloys under similar sliding

Fig. 9.7 Influence of alloying on friction coefficient of matrix alloy and its composite

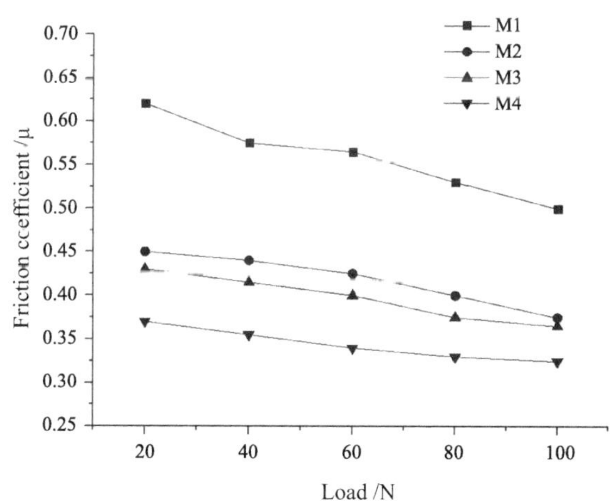

conditions. There are two main reasons for the friction rise: ① friction necessary to shear the microsolder joints of the contact points between the two friction surfaces, ② friction necessary to remove the soft surface on the hard friction surface. Therefore, the friction of binary Al-Si alloy is higher than that of multi-component alloy during the sliding process, which is attributed to the low hardness and strength caused by the lack of alloying elements in binary alloy and the increasing actual contact surface due to the low hardness and strength [5].

9.1.1.5 Stability of the Friction Coefficient of Composites

As a wear-resistant material, the stable friction coefficient is an important aspect of the safety of composite applications (such as those on brake discs). The stability of the friction coefficient can be expressed by the stability coefficient, which is the ratio of average friction coefficient to the maximum friction coefficient. This concept uses M1 (Al-22Si matrix alloy), M2 (11.2 vol.%$(Al_2O_3 + Al_3Zr)_p$/Al-22Si composite), M3 (alloyed Al-22Si alloy), and M4 (alloyed 11.2 vol.%$(Al_2O_3 + Al_3Zr)_p$/Al-22Si composite) for comparison. Table 9.1 shows the overall variation range of the friction coefficients of the four materials under loads of 20 ~ 100 N. As can be seen from the table, the friction stability coefficient of the four materials increases gradually. It is also found that the torque fluctuates within a certain range rather than being constant during the wear process.

Table 9.2 shows the variation range of the friction coefficient during a 60 min-long sliding at a sliding speed of 0.523 m/s under 80 N load. It can be seen from the table that, after sliding under the same load, the friction stability coefficient of the alloyed material is greater than that of the unalloyed binary alloy, while the stability coefficient of the alloyed composite is also higher than that of the unalloyed composite. The stability coefficient of the composite is greater than that of the matrix alloy. During the testing, the temperature of the matrix alloy can increase faster than that of the corresponding composite, and the severe plastic deformation and adhesion will appear. The greater the pressure, the more severe the adhesion, and there will be a large number of matrix adhesion to the grinding plate resulting in

Table 9.1 Variation of friction coefficient of four materials under different loads

Material	M1	M2	M3	M4
Friction coefficient range	0.50 ~ 0.62	0.37 ~ 0.45	0.365 ~ 0.430	0.32 ~ 0.37
Stability factor	90.3%	91.1%	92.4%	93.2%

Table 9.2 Variation range of friction coefficient of four materials under the same load

Material	M1	M2	M3	M4
Friction coefficient range	0.47 ~ 0.61	0.36 ~ 0.44	0.350 ~ 0.410	0.31 ~ 0.35
Stability factor	88.5%	90.9%	92.7%	94.3%

severe adhesive wear, and then the friction coefficient will be unstable. However, the adhesion extent of the composite is much less, and the friction coefficient is relatively stable. After alloying, especially the addition of Ni, the heat resistance of the material is improved, and the hardness of both the matrix alloy and its composites has been improved greatly. Meanwhile, less serious adhesion will appear due to the dispersion strengthening of the alloying element, and a smooth friction surface will be formed during the process, which makes the friction coefficient more stable.

9.1.2 Friction and Wear Performance of ZL101A Aluminum Matrix Composites

9.1.2.1 Wear Performance of $(Al_3Zr + Al_2O_3)_p$/ZL101A Composite

The wear properties of in-situ Al_3Zr and Al_2O_3 reinforced ZL101A aluminum matrix composites were tested to investigate the effects of the particle volume fraction, the load, and the matrix on the wear properties of composites, and the microstructures of worn surface and subsurface (SEM) were observed to understand the wear mechanism.

(1) Influence of particle volume fraction on wear performance of composites

Figure 9.8 shows the relationship between the sliding time and the composite materials with different particle volume fractions and the matrix ZL101A alloy at a sliding velocity of 0.42 m/s and a load of 58.8 N. It can be seen from the figure that the wear amount (wear weight loss) of the $(Al_3Zr_+ Al_2O_3)_p$/ZL101A composite is significantly less than that of the ZL101A alloy matrix at the same sliding time. Moreover, the wear amount of the composite decreases, and the wear resistance increases with the increase of particle volume fraction. In addition, the longer the sliding time is the

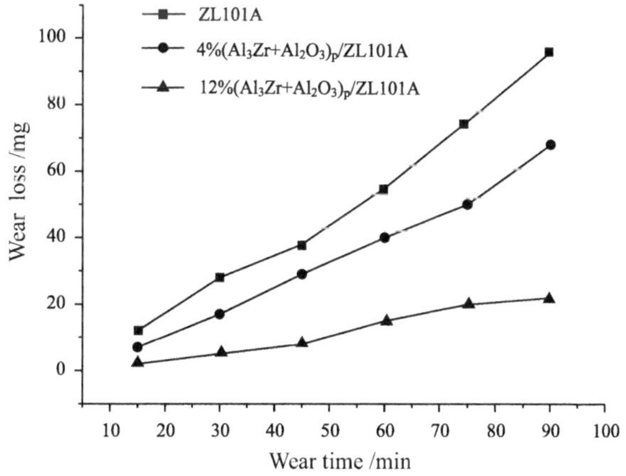

Fig. 9.8 Graphical relationship of wear amount of composite and matrix with different particle volume fractions and sliding time (sliding velocity 0.42 m/s, load 58.8 N)

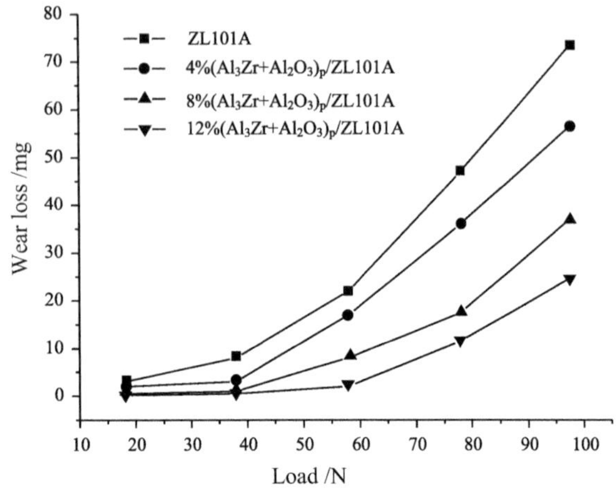

Fig. 9.9 Graphical relationship of wear amount of composite and matrix under different loads (sliding speed 0.42 m/s, sliding time 20 min)

more apparent the wear resistance of the composite becomes. When the sliding time is 90 $_{min}$, the wear loss of 12 vol.%$(Al_3Zr_+Al_2O_3)_p$/ZL101A composite is 26.7 mg, which is only 27.93% of that of the matrix alloy (98.6 mg).

(2) Influence of load on wear performance of composites

Figure 9.9 shows the relationship of wear amount of $(Al_3Zr_+Al_2O_3)_p$/ZL101A composite and ZL101A matrix alloy under different loads with a sliding velocity of 0.42 m/s and sliding movement time of 20 min. It can be seen from the figure that the wear amount of both the composite and the matrix increases rapidly with the increase of the load. However, the increase of wear amount of the composite is significantly lower than that of ZL101A alloy, and the increase of wear amount of the composite decreases with the increase of particle volume fraction. Especially, the composite has better wear resistance than the matrix when the load is greater than 58.8 N.

(3) Influence of matrix alloy on the wear performance of composites

In order to investigate the influence of different matrix alloys on the wear loss energy of composites, a comparative test of the wear performance of $(Al_3Zr_+Al_2O_3)_p$/Al and $(Al_3Zr_+Al_2O_3)_p$/ZL101A composites is conducted. The results are as follows:

1. The influence of sliding time on the wear performance of composites

Figure 9.10 shows the wear loss of $(Al_3Zr_+Al_2O_3)_p$/Al and $(Al_3Zr_+Al_2O_3)_p$/ZL101A composites with the sliding time with the sliding speed of 0.42 m/s and the load of 58.8 N. It can be seen from the figure that the wear amount of the composite with 12 vol.%$(Al_3Zr_+Al_2O_3)_p$/Al particles increases nearly linearly with the increase of sliding time. After the same sliding time, the wear amount of 12 vol.%$(Al_3Zr_+Al_2O_3)_p$/Al composite is significantly lower than that of ZL101A alloy, slightly lower than that of 4 vol.%$(Al_3Zr_+Al_2O_3)_p$/ZL101A composite, and significantly higher than that of 12 vol.%$(Al_3Zr_+Al_2O_3)_p$/ZL101A

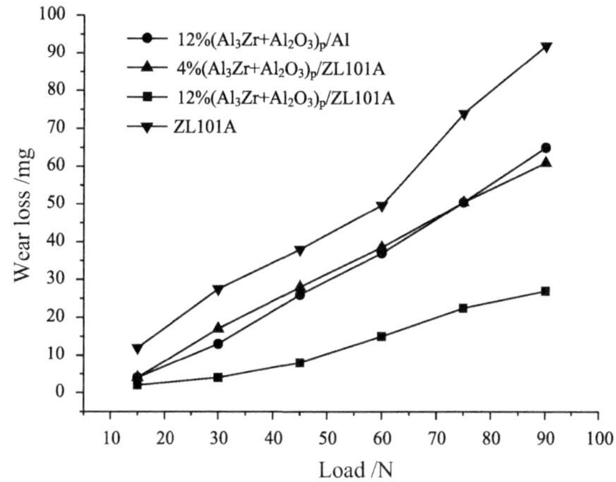

Fig. 9.10 Effect of different substrates and different sliding times on the wear performance of composite materials (sliding velocity 0.42 m/s, load 58.8 N)

composite with the same particle volume fraction and different matrix. It is shown that, under the constant sliding velocity and load, the wear resistance of 12 vol.%($Al_3Zr + Al_2O_3$)$_p$/Al composite is significantly better than ZL101A alloy, even better than 4 vol.%($Al_3Zr + Al_2O_3$)$_p$/ZL101A composite, but relatively worse than 12 vol.%($Al_3Zr + Al_2O_3$)$_p$/ZL101A composite.

2. The influence of load on the wear performance of composites

Figure 9.11 shows the curves of wear performance of ($Al_3Zr + Al_2O_3$)$_p$/Al and ($Al_3Zr + Al_2O_3$)$_p$/ZL101A composites under different loads. It can be seen from the figure that when the load is less than 39.2 N (4 kgf), the wear amount of 12 vol.%($Al_3Zr + Al_2O_3$)$_p$/Al composite is not only higher than that of 4 vol.%($Al_3Zr + Al_2O_3$)$_p$/ZL101A composite, but also slightly higher than that of ZL101A alloy. When the load is between 39.2 N and 58.8 N (i.e., 4 ~ 6 kgf), the wear of 12 vol.%($Al_3Zr + Al_2O_3$)$_p$/Al composite is lower than that of ZL101A alloy,

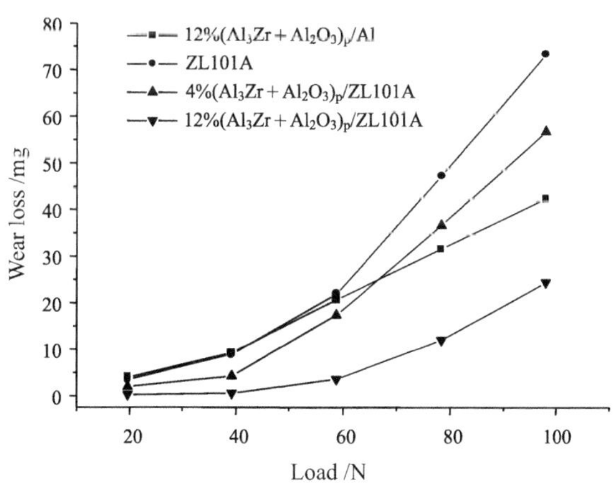

Fig. 9.11 Influence of different substrates and different loads on the wear performance of composite materials (sliding speed 0.42 m/s, sliding time under each load 20 min)

but still higher than that of 4 vol.%$(Al_3Zr + Al_2O_3)_p$/ZL101A composite. However, when the load $L \geq 78.4$ N (8 kgf), the wear amount of 12 vol.%$(Al_3Zr + Al_2O_3)_p$/Al composite is not only lower than that of 4 vol.%$(Al_3Zr + Al_2O_3)_p$/ZL101A composite but also lower than that of ZL101A alloy. As a result, it shows that under the low load, the wear resistance of $(Al_3Zr + Al_2O_3)_p$/Al composite is poor, even worse than ZL101A alloy, while the wear resistance is good under high load (≥78.4 N). The phenomenon is different from that of 12 vol.%$(Al_3Zr + Al_2O_3)_p$/ZL101A with the same particle volume fraction, which shows excellent wear resistance no matter under low or high load and the wear resistance becomes better as the load increases.

3. Morphology of worn surface and subsurface

Friction and wear is a dynamic process of the system, which is in continuous development and change, and the relationship is complex. The instantaneous change of wear is not yet traceable. However, as the wear of materials is a kind of surface failure phenomenon, the wear surface morphology is the final direct record of the whole process of friction and wear, especially the wear subsurface has become one of the important bases to understand the wear process. Therefore, the observation and analysis of wear surface and subsurface morphology is the direct basis for studying the wear mechanism.

Figure 9.12 shows the SEM images of ZL101A alloy worn under a sliding speed of 0.42 m/s and a load of 58.8 N for 90 min. It can be seen from Fig. 9.12a that the wear surface of ZL101A alloy has large pits accompanied by deformation and migration layer caused by extrusion. In addition, there are many pits left along the sliding direction, and the high-resolution image of pits is shown in Fig. 9.12b. It can be seen from Fig. 9.12c that a crack band appears along the sliding direction at the worn subsurface. The crack is mainly generated on the worn surface and extends to the subsurface layer, connecting with the crack generated in another part, so that the small pieces flake off to form pits. When the friction pair in contact is relatively sliding with each other, due to the low hardness of matrix ZL101A alloy, coupled with the friction heat, the wear surface is heated up, and the surface layer is softened, resulting in plastic flow and deformation on the wear surface of ZL101A, which is more likely to occur especially when the load is large and the sliding speed is fast. On the other hand, in the wear surface layer, due to the inconsistency of the Al phase and Si phase in the matrix during the deformation process, the interface is separated, and cracks are generated and extended to the subsurface layer, and then, shear tear or spalling occurs under the shear force, which is mainly characterized as adhesive wear.

Figures 9.13 and 9.14 show the SEM images of worn surfaces of $(Al_3Zr + Al_2O_3)_p$/ZL101A composites with particle volume fractions of 8 vol.% and 12 vol.%, respectively. It can be seen from Fig. 9.13 that the worn surface has obvious plastic deformation and ploughing characteristics. The proportion of plastic deformation is relatively large, while the ploughing characteristics are relatively weak. However, compared with ZL101A matrix (Fig. 9.12a), the degree of plastic deformation is significantly reduced. With the increase of particle volume fraction, the characteristics of ploughing or cutting become more and more obvious. There are

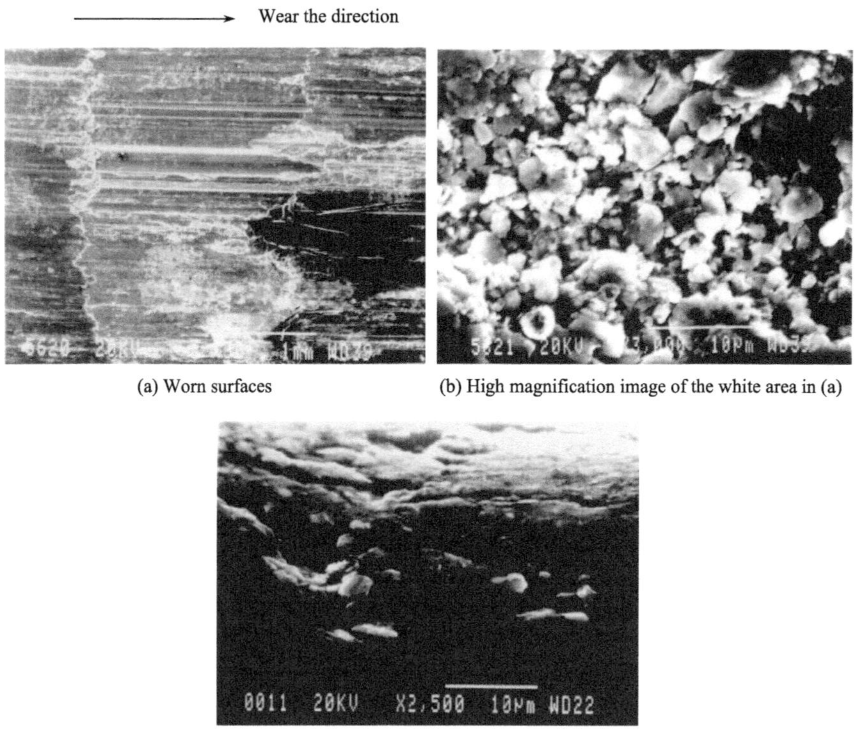

(a) Worn surfaces (b) High magnification image of the white area in (a)

(c) Cracks and pits in the worn subsurface

Fig. 9.12 SEM images of worn surface and subsurface of ZL101A matrix alloy ($V = 0.42$ m/s, $L = 58.8$ N, $t = 90$ min)

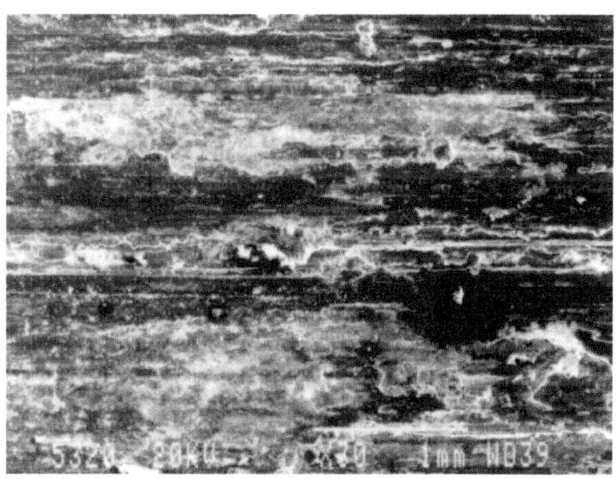

Fig. 9.13 SEM image of worn surface of 8 vol.% $(Al_3Zr + Al_2O_3)_p$/ZL101A composite ($V = 0.42$ m/s, $L = 58.8$ N, $t = 90$ min)

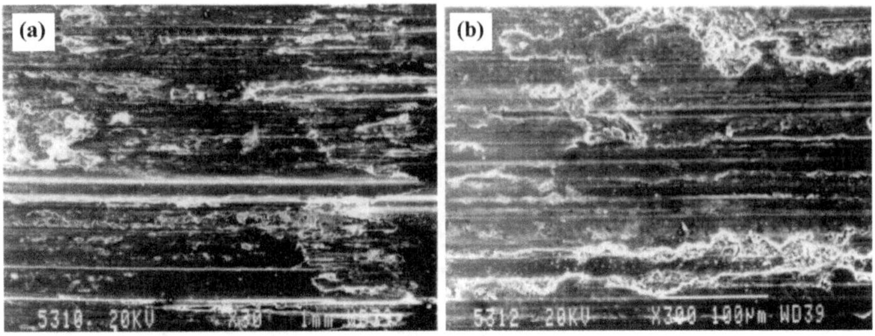

Fig. 9.14 SEM images of worn surface of 12 vol.%$(Al_3Zr + Al_2O_3)_p$/ZL101A composite (V = 0.42 m/s, L = 58.8 N, t = 90 min)

grooves (furrows) with relatively clear outlines formed by microscopic cutting of abrasive particles on the worn surface, while there are also the uplifts (parallel lines in Fig. 9.14) on both sides of furrows formed by the plastic extrusion due to the ploughing of abrasive particles.

Figure 9.15 shows the worn subsurface morphology of $(Al_3Zr + Al_2O_3)_p$/ZL101A composite. As can be seen from Fig. 9.15a, fine particles (about 2 μm) are distributed in the subsurface layer, and no cracks are found. As can be seen in Fig. 9.15b, bare particles are exposed in the worn surface due to matrix grounding away or ploughing. Figure 9.15c shows the morphology of the furrow side (longitudinal section parallel to the sliding direction).

The biggest difference between the furrow and the pit formed by peeling off in Fig. 9.12d is that there is a large plastic deformation in the front and side layers of the furrow. Pits formed by peeling or tearing have no such plastic deformation and thus result in tear scars or debris (Fig. 9.12b). Therefore, the friction and wear characteristics of $(Al_3Zr + Al_2O_3)_p$/ZL101A composite are mainly abrasive wear, which is more obvious with the increase of particle volume fraction.

Figure 9.16 shows the SEM image of the worn surface of 12 vol.%$(Al_3Zr + Al_2O_3)_p$/Al composite. It can be seen from the figure that the worn surface has both the parallel furrows formed by the relative sliding between the abrasive particles and the matrix and the features left by the adhesion transfer (the white blocky area in Fig. 9.16a). Figure 9.16b shows the high-resolution image of the white block. It can be seen that the foggy adhesive layer bears a certain amount of plastic deformation, which is obviously different from the pit structure left by the spalling of ZL101A matrix in Fig. 9.12c.

Figure 9.17 shows the SEM images of worn subsurface of 12 vol.%$(Al_3Zr + Al_2O_3)_p$/Al composite. Figure 9.17a shows the longitudinal section structure perpendicular to the sliding direction. It can be seen from the figure that the subsurface layer is subject to a large plastic deformation, some particles are exposed by extrusion, and a crack zone appears in the subsurface layer. The cracks expand and extend from the worn surface to the subsurface layer, and several cracks connect

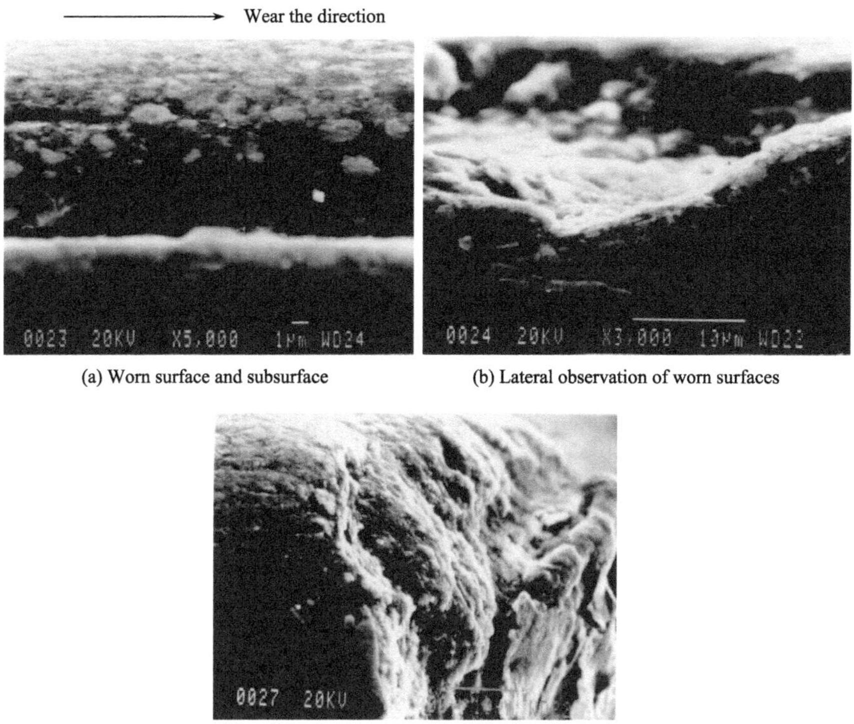

(a) Worn surface and subsurface

(b) Lateral observation of worn surfaces

(c) The sides of the furrow
(longitudinal section parallel to the sliding direction)

Fig. 9.15 Worn subsurface morphology of $(Al_3Zr + Al_2O_3)_p$/ZL101A composite ($V_f = 12$ vol.%, $V = 0.42$ m/s, $L = 58.8$ N, $t = 90$ min)

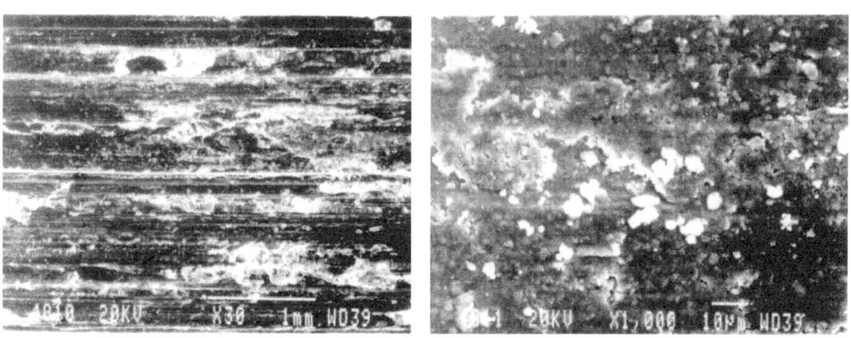

(a) Low resolution

(b) High resolution

Fig. 9.16 SEM images of worn surface of 12 vol.%$(Al_3Zr + Al_2O_3)_p$/Al composite ($V = 0.42$ m/s, $L = 58.8$ N, $t = 90$ min)

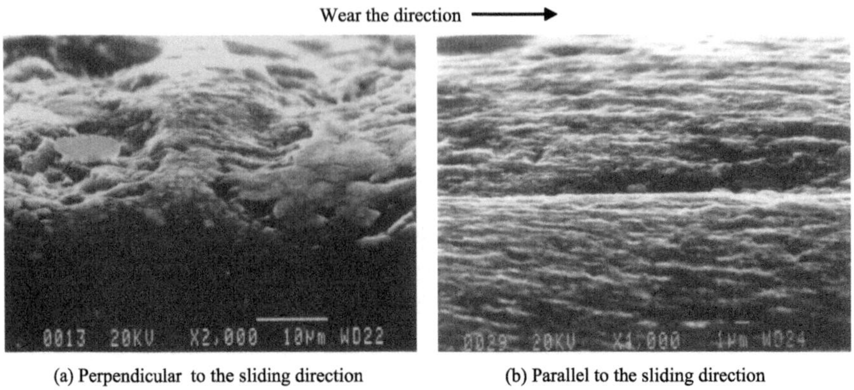

Fig. 9.17 SEM images of worn subsurface of 12 vol.%$(Al_3Zr + Al_2O_3)_p$/Al composite@@ $V = 0.42$ m/s, $L = 58.8$ N, $t = 90$ min

with each other to form a band (Fig. 9.17a). In addition, the crack layer ruptures and breaks during multiple extrusion process, followed by tearing or spalling under shear force (Fig. 9.17a). Figure 9.17b shows the longitudinal structure parallel to the sliding direction. It can be seen from the figure that there is a stream-like deformation layer structure in the worn subsurface layer, which indicates that the ploughing action of the abrasive particles causes a large plastic deformation of the subsurface layer under the composite surface. Since the matrix is pure aluminum, the microstructure characteristics of this stream-like deformation layer are more obvious. It can also be seen from the figure that the particles turn over by ploughing and the ridges formed by the plastic extrusion to both sides of the furrow due to the ploughing action of abrasive particles.

Thus, it can be conduced that the wear characteristics of 12 vol.%$(Al_3Zr + Al_2O_3)_p$/Al composite are different from the adhesive wear of ZL101A matrix alloy and the abrasive wear of $(Al_3Zr + Al_2O_3)_p$/ZL101A composite, which is a mixed type of partial abrasive wear and partial adhesive wear. With the increase of particle volume fraction, the abrasive wear characteristics become more obvious.

9.1.2.2 Wear Properties of In-Situ Aluminum Matrix Composites Prepared Under Pulsed Magnetic Field

In this section, the wear performance of in-situ $(ZrB_2 + Al_3Zr)_p$/ZL101A composites prepared by chemical reaction under the pulse magnetic field is mainly studied, and the influences of the amount of reactants, the load, and the matrix of composites on the abrasion performance are analyzed. Then, the worn surface and the subsurface microscopic morphology are observed, while the wear mechanism is analyzed and the relevant model is set up.

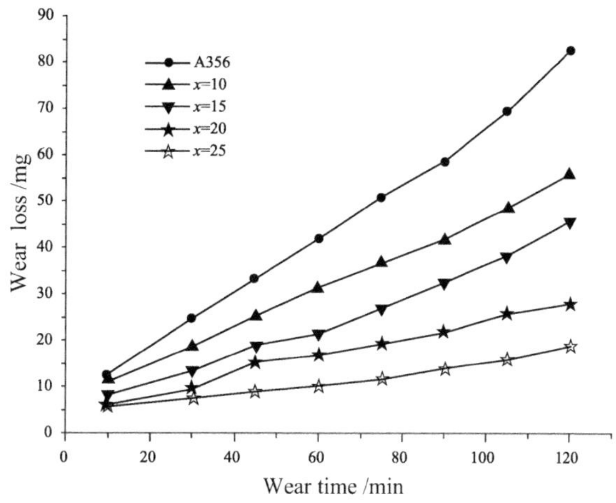

Fig. 9.18 Graphical relationship between wear amount and wear time of ZL101A matrix alloy and composites prepared from ZL101A-x%(K_2ZrF_6 + KBF_4) system under sliding velocity of 0.42 m/s and load of 58.8 N

(1) Influence of the amount of reactants on the wear performance of composites

Figure 9.18 shows the relationship curves of wear amount and wear time of ZL101A matrix alloy and (ZrB_2 + Al_3Zr)/ZL101A composite prepared from ZL101A-x%(K_2ZrF_6 + KBF_4) system under the action of pulsed magnetic field with the sliding velocity of 0.42 m/s and load of 58.8 N. It can be seen from the figure that with the increase of slip time, the wear amount of the composite increases and shows a near linear relationship with the wear time. Among those materials, the wear amount of ZL101A matrix increases greatly. At the same time, with the increase of the amount of reactants, the increase of the wear amount of composite decreases gradually. Especially in the late wear period, the wear amount of matrix alloy increases rapidly. Under the same sliding time, the wear amount of the composite is obviously less than that of the ZL101A matrix alloy. Especially, when the sliding time is 120 min and the addition amount of reactants is $x = 25$, the wear amount of the composite is only 18.97 mg, which is 23.015% of that of ZL101A matrix 82.43 mg, and its wear resistance is far better than that of ZL101A matrix alloy.

(2) Effect of reaction temperature on wear performance of composite

Figure 9.19 shows the relation curve between wear amount and wear time of (ZrB_2 + Al_3Zr)/ZL101A composite prepared from ZL101A-20%(K_2ZrF_6 + KBF_4) system under the effect of pulsed magnetic field at different melt temperatures with the sliding speed of 0.42 m/s and the load of 100 N. For the convenience of analysis and explanation, the composite samples prepared at 850 °C, 900 °C, 950 °C, and 1000 °C are numbered S1, S2, S3, and S4. It can be seen from Fig. 9.19 that, at the beginning of the wear, the wear amount of composite does not change significantly when the wear time is 15 min. With the increase of wear time, the wear amounts of S1, S2, S3, and S4 also increase. When the sliding time is 120 min, the wear amount of S3 is 81.57 mg, 1.39 times that of S1. After the same sliding time, the wear amount of the sample changes in the order of S1 < S4 < S2 < S3. Comparing

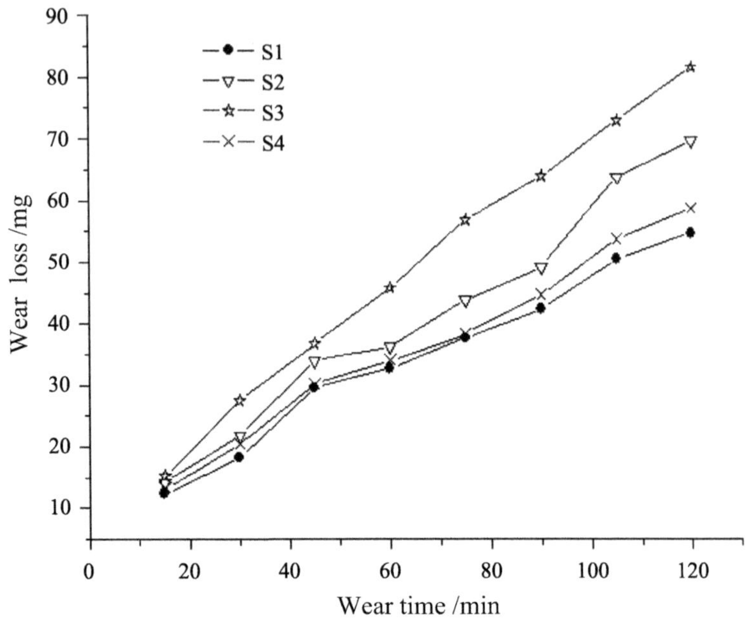

Fig. 9.19 Variation of wear weight loss of S1, S2, S3, and S4 with sliding time

the wear amounts of S1 and S4, it can be seen that the difference of wear amount between S1 and S4 is slight during all the sliding time.

(3) Influence of load on wear performance of composites

Figure 9.20 shows the influence of the sliding wear time and loads on the wear amount of $(ZrB_2 + Al_3Zr)$/A356 composite prepared from A356-x%$(K_2ZrF_6 + KBF_4)$ system under the pulsed magnetic field with the sliding velocity of 0.42 m/s. It can be clearly seen from Fig. 9.20 that at different wear stages, the wear amount of the composite and the matrix increases with the increase of the applied load, but the increase of the wear amount of the composite with the increase of the load is much smaller than that of the matrix. Under the condition of a certain load, the wear amount of composite is less than that of A356 matrix alloy at different wear stages. After the same wear time, the wear amount of composite under a certain load decreases with increasing reactant amount. When the sliding time is 120 min, the load is 100 N, the addition amount of reactants is 25%, and the wear amount of prepared composite is 33.65 mg, which is only 24.54% of that of A356 matrix alloy (i.e., 137.11 mg). This is mainly because the presence of endogenous reinforcement particles increases the transient load, delays the transition from microwear to severe wear, and reduces the effective contact area between the composite and the dual, which effectively weakens the adhesive wear. The hard particles in the composite have bearing capacity. When the soft matrix is worn under the action of external forces, the particles will stand out on the surface of the sample and bear most of the load, which is conducive to

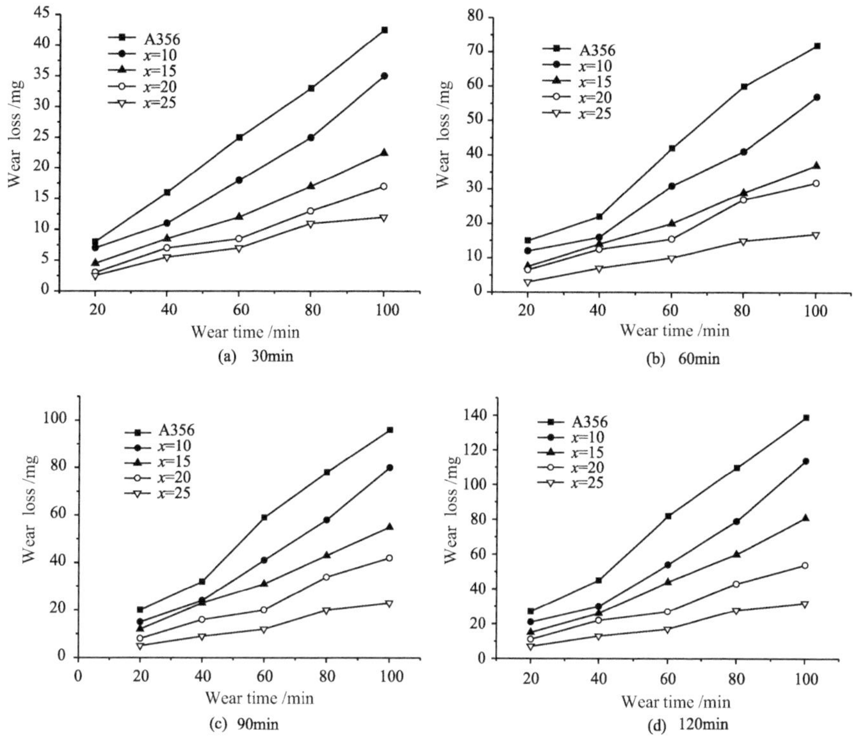

Fig. 9.20 Influence of sliding time and load on wear weight loss of ZL101A matrix alloy and $(ZrB_2 + Al_3Zr)_p$/ZL101A composite when the sliding velocity is 0.42 m/s

preventing the soft matrix from directly participating in the friction process. At the same time, the interface between the reinforcement particles synthesized by in-situ melt reaction under the physical field is strongly bonded with the matrix, which eliminates the tendency of the wear due to falling off from the matrix, enhances the resistance to adhesion and deformation of the composites, and thus improves the wear resistance of the composites [6].

According to the literature [5–7], the wear rate *k* of composite is defined as the wear amount per unit time, and the anti-wear property *R* of composite is the reciprocal of the wear rate. The wear period from 0 to 30 min is defined as the initial stage of the wear process, 30 min to 90 min is the intermediate stage, and 90 min to 120 min is the late stage. Figure 9.21 shows the variation curve of the wear resistance of the composite with the load during different wear periods. It can be seen from the figure that, at different wear stages, the wear resistance *R* of the composite prepared with the same amount of reactants *x* decreases with the increase of the load. At different wear stages and under the same load, the wear resistance *R* of the composite increases with the increase of the added amount of reactants. At the same wear stage and under the same load condition, the wear resistance *R* of composite also increases

Fig. 9.21 Variation of wear resistance of composites with load during different wear periods

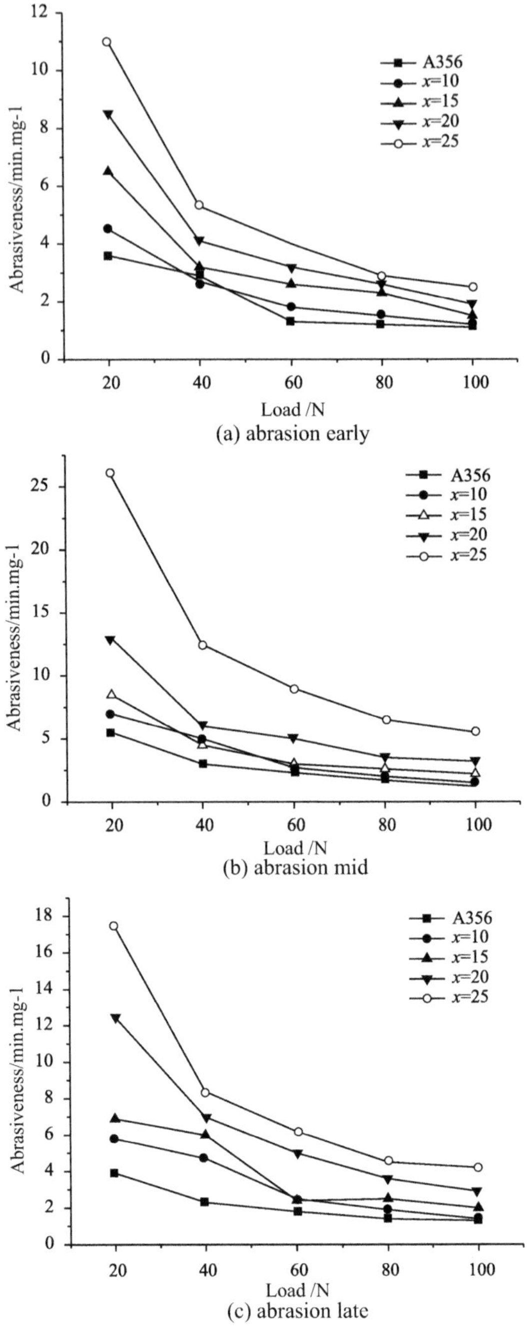

with the increase of the reactant amount. Under the same loading condition, the wear resistance of the composite with certain reactant amount of x changes with the law of increasing first and then decreasing at the three wear stages. At the middle stage of wear, when the applied load is 100 N and the added reactant amount is $x = 25$, the wear resistance of the composite is $R = 8.114$, which is 4.722 times that of ZL101A matrix alloy.

(4) Wear surface and subsurface structure of composites

Friction and wear of the composite is a systematic dynamic process, which is kept in developing and changing. Each instantaneous change of the process is very complex and unable to dynamicly track. However, as material wear is a surface failure phenomenon, the wear surface morphology is the ultimate direct recording of friction and wear process, especially the wear subsurface has become the important basis for judging the wear process. Therefore, according to the analysis of the wear surface and subsurface morphology of the composites at different wear stages, the wear mechanisms of composite can be determined, and the related model can then be established.

1. Wear surface and subsurface of ZL101A matrix alloy.

Figure 9.22 shows the SEM images of wear surface of the ZL101A alloy at a sliding speed of 0.42 m/s and a load of 100 N after different sliding times. It can be seen from Fig. 9.22a that when ZL101A alloy has been worn for 30 min, obvious pits appear on the wear surface of the alloy, accompanied by deformation and migration layer caused by extrusion. During the friction and wear process, this is the low hardness of ZL101A matrix alloy and the softening and deformation of the surface layer as the heat generated that makes the surface layer temperature of ZL101A alloy rise, which is easily to occur when the load is large and the sliding speed is high. With continuous friction and wear process, the migration layer has obvious accumulation under the action of shearing force, as shown in Fig. 9.22b. The accumulation layers form raised adhesive points with ploughing effect on the friction surface. With the progress of sliding, the adhesive point gradually increases, the peeling of ZL101A alloy is increased under the action of shearing force, and then, large-area surface avulsion and transfer occur due to the grinding, resulting in large-area pits. At the same time, due to the incompatibility of the Al phase and Si phase in the matrix during the deformation process, the interface is separated, cracks are generated and spread to the subsurface layer, and then, shearing tear or spalling occurs under the action of shearing force, as shown in Fig. 9.22c. The wear characteristic is mainly manifested as adhesive wear and peeling wear.

Figure 9.23 shows the SEM images of wear subsurface of ZL101A matrix alloy at a sliding speed of 0.42 m/s and a load of 100 N after different slip times. It can be seen from Fig. 9.23a that at the early stage of the wear process, the plastic deformation accumulation of the worn surface of the composite is serious and extends to the subsurface layer. It can be seen from Fig. 9.23b that there is a crack zone in the subsurface of ZL101A alloy matrix, and the cracks mainly appear in the worn surface and extend to the subsurface layer, which connect with the cracks generated

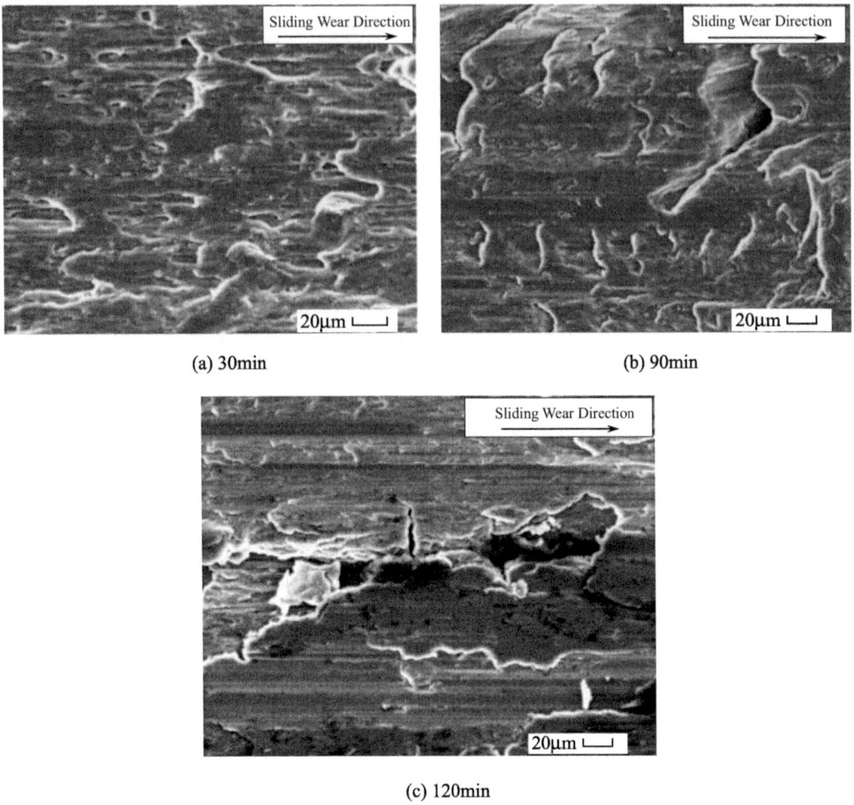

Fig. 9.22 SEM images of wear surface of ZL101A matrix alloy at sliding speed of 0.42 m/s and load of 100 N at different sliding times

in another part. As shown in Fig. 9.23c, at the late stage of wear, there is a large tearing pit in ZL101A matrix alloy after wearing for 120 min.

2. Wear surface and subsurface of $(ZrB_2 + Al_3Zr)_p$/ZL101A composite.

Figure 9.24 shows the low- and high-resolution SEM images of the wear surface of $(ZrB_2 + Al_3Zr)_p$/ZL101A composite prepared under the action of pulsed magnetic field and with different reactant addition amount x with the sliding velocity of 0.42 m/s and the load of 100 N.

It can be seen from Fig. 9.24a–d that there is an obvious plastic deformation accumulation in the worn surface. Along the sliding direction, grooves with clear contours are formed by the ploughing action of abrasive particles. The ridges formed by the ploughing action of the abrasive particles are observed to be plastically extruded to both sides of the grooves. The plastic deformation is of a large proportion, and the ploughing action is relatively weak. At this time, the wear mechanism of composites is mainly adhesive wear plus abrasive wear. With the increase of the amount of reactant addition, the volume fraction of in-situ ZrB_2 and Al_3Zr particles increases

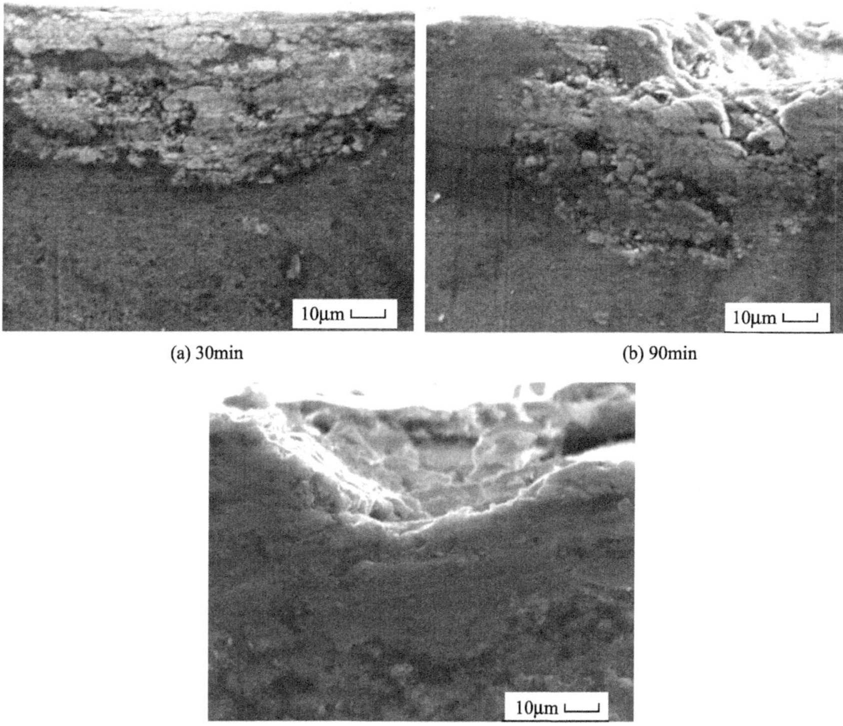

(a) 30min (b) 90min

(c) 120min

Fig. 9.23 SEM images of wear subsurface of ZL101A matrix alloy at sliding speed of 0.42 m/s and load of 100 N after different sliding times

gradually, the ploughing effect becomes more and more obvious, and the plastic deformation characteristics are gradually weakened. As shown in Fig. 9.24c–h, the wear of composites at this time is dominated by abrasive wear. In particular, when the addition amount of reactants is $x = 25$, the typical abrasive wear characteristics appear in the composite wear surface, as shown in Fig. 9.24i, j. This proves further that as the reactant amount continues to increase, the volume fraction of particles synthesized by the reaction increases. The more particles there are, the greater the role of supporting and anti-wear will be during the wear process, and the more wear-resistant the composite will be, which is consistent with the test results of wear performance.

Figure 9.25 shows low- and high-resolution SEM images of the wear subsurface (perpendicular to the sliding direction) of 25wt.%$(ZrB_2+Al_3Zr)_p$/ZL101A composite with the sliding velocity of 0.42 m/s and the load of 100 N. After sliding for 120 min, it can be seen from Fig. 9.25a that fine particles are scattered in the wear subsurface layer and the worn surface is exposed to bare particles caused by the removal of the matrix or the ploughing effect when the reactant addition amount is 25%. The furrows are deep, and there are granular grinding chips, so the ploughing

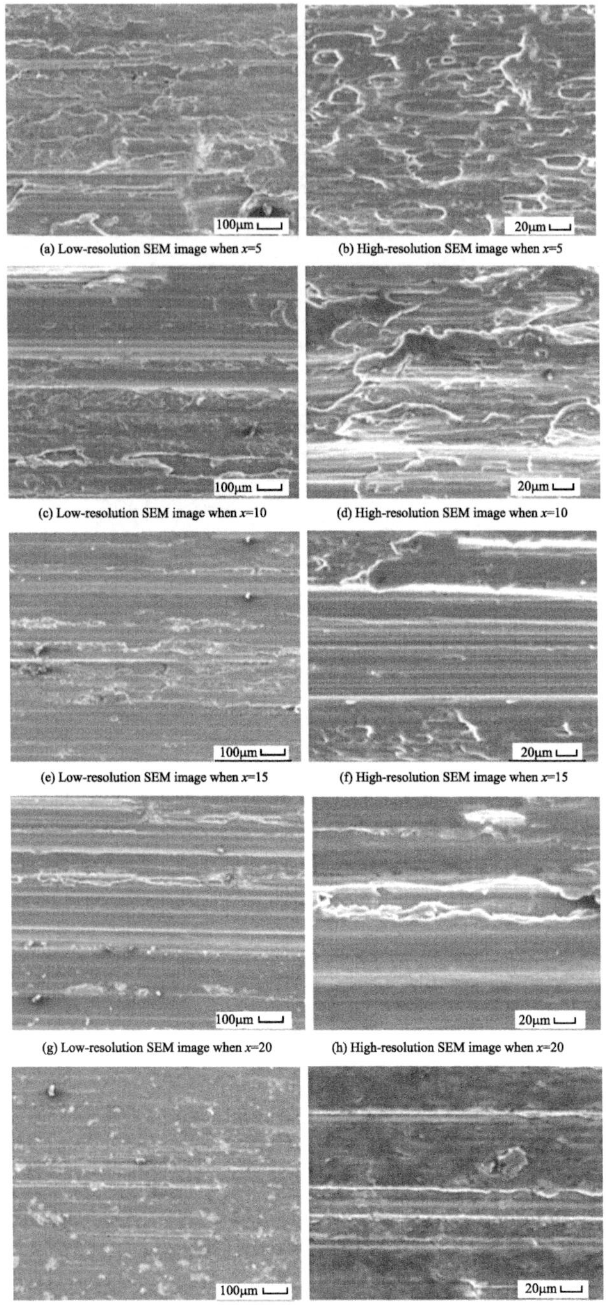

(a) Low-resolution SEM image when x=5 (b) High-resolution SEM image when x=5

(c) Low-resolution SEM image when x=10 (d) High-resolution SEM image when x=10

(e) Low-resolution SEM image when x=15 (f) High-resolution SEM image when x=15

(g) Low-resolution SEM image when x=20 (h) High-resolution SEM image when x=20

(i) Low-resolution SEM image when x=28 (j) High-resolution SEM image when x=28

Fig. 9.24 SEM images of wear surfaces at low and high power after 120 min of sliding wear at the sliding speed of 0.42 m/s and the load of 100 N of $(ZrB_2 + Al_3Zr)_p$/ZL101A composite prepared with different addition levels

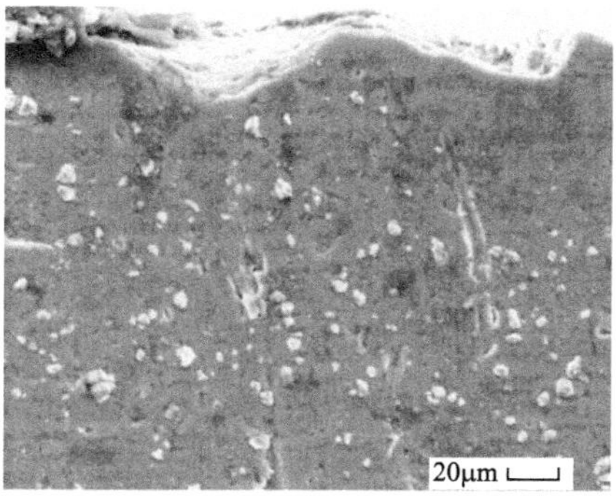

(a) Low-resolution SEM images

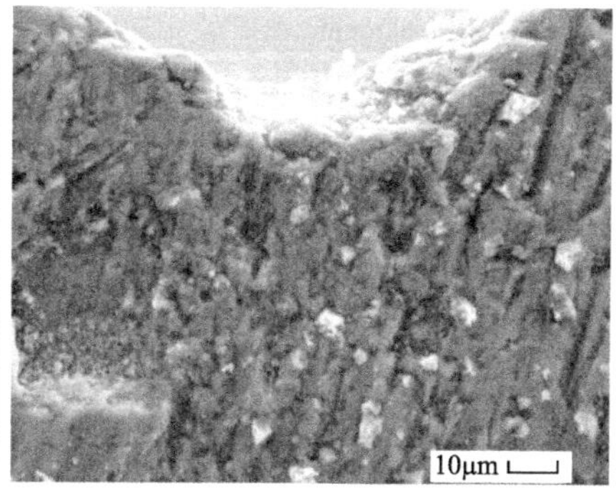

(b) High-resolution SEM images

Fig. 9.25 SEM images of the wear subsurface (perpendicular to the sliding direction) of ($ZrB_2 + Al_3Zr$)/ZL101A composite when the slip velocity is 0.42 m/s and the load is 100 N after 120 min of sliding with addition of reactants $x = 25$wt%

effect is very obvious. It can be seen from the groove morphology in Fig. 9.25b that the grooves are deep and there are granular abrasive particles, which is caused by the ploughing effect caused by the microcutting of the abrasive particles and formed by the plastic extrusion pushed to both sides of the grooves. The deep scratches indicate that embedding of hard particles into the soft matrix causes abrasive wear.

9.1.2.3 Wear Properties of Composites Prepared Under High-Energy Ultrasonic Field

(1) ($ZrB_2 + Al_3Zr)_p$/ZL101A system composite

Figure 9.26 shows the variation curve of wear amount of ($ZrB_2 + Al_3Zr)_p$/ZL101A composite prepared by ZL101A-20%($K_2ZrF_6 + KBF_4$) system with different ultrasonic power with sliding time under the condition of a sliding velocity of 0.42 m/s and load of 100 N. It can be seen from Fig. 9.26 that the wear amount of composite increases with the increasing wear time. With the introduction of ultrasonic field, the increase of wear amount of composite significantly decreases, especially the composite prepared by applying 1.2 kW high-energy ultrasonic. For the same wear time, with the increase of ultrasonic power, the wear amount of composite decreases significantly, especially during the later stage of wear. When the wear time is 120 min, the wear amount of composite prepared by applying 1.2 kW of high-energy ultrasonic field is 38.67 mg, which is 44.7% of the wear amount of composite (79.84 mg) prepared without high-energy ultrasonic field. This is mainly due to the introduction of high-energy ultrasound, which leads to the change of composite microstructure and the significant decrease of wear amount.

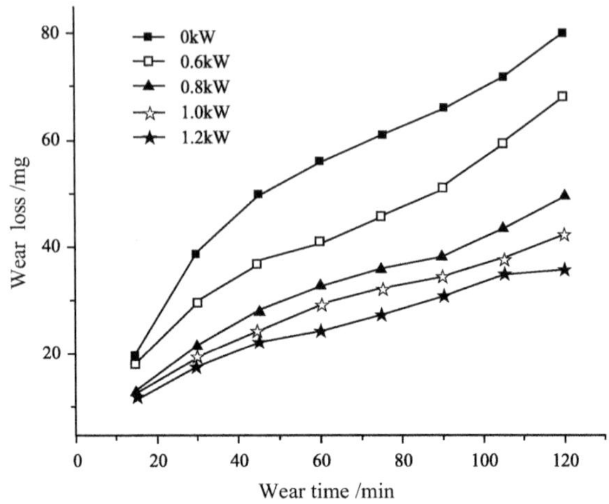

Fig. 9.26 Wear amount variation of $(ZrB_2 + Al_3Zr)_p$/ZL101A composite with sliding time under sliding velocity of 0.42 m/s and the load of 100 N

(2) Al_2O_{3p}/Al system composite

Figure 9.27 shows the weight loss variation of Al_2O_3/Al composite prepared by Al-x% $Ce_2(CO_3)_3$ ($x = 5, 10, 15, 20, 25$) with sliding time under the condition of a sliding velocity of 0.42 m/s, the load of 100 N, and the ultrasonic power of 1.2 kW. It can be seen from Fig. 9.27 that with the extension of wear time, the wear amount of the composite increases, which is less than that of the pure aluminum matrix. At the early stage of wear, for the same wear time, the wear volume change is not obvious with the increase of the added reactant amount of composite. However, at the late stage of wear, especially when the sliding time is 120 min, the wear amount of composite with an added reactant amount of 25% is 148.94 mg, which is 79.2% of that of composite with an added reactant amount of 5% (184.29 mg), and 66.5%

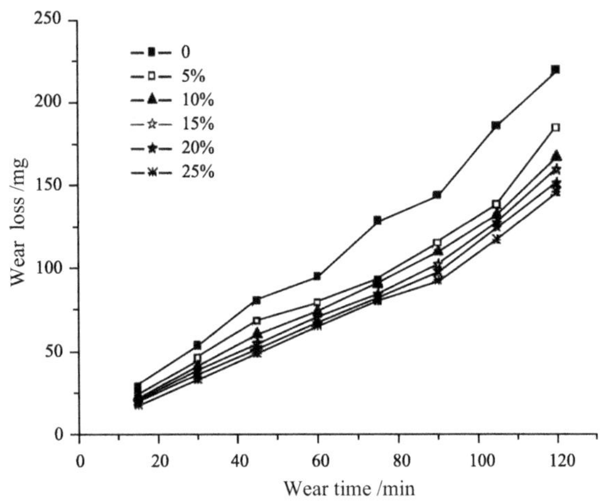

Fig. 9.27 Relationship between the wear weight loss and wear time of composite prepared by Al-x%$Ce_2(CO_3)_3$ system under the high-energy ultrasonic field

of that of the pure aluminum matrix (219.45 mg). The wear resistance of composite with an added reactant amount of 25% is far better than that of the pure aluminum matrix.

9.2 Wear Performance of In-Situ Aluminum Matrix Composites at High Temperature

9.2.1 Test Conditions of High-Temperature Friction and Wear

The high-temperature friction and wear test were carried out on MG-2000 high-temperature and high speed friction and wear testing machine using pin-disc dry sliding mode. The test device is shown in Fig. 9.28, and its working principle is shown in Fig. 9.29. The test sample is pin shaped with the size of ϕ 6 mm × 12 mm. The grinding material is 40Cr steel (HRC40) with the size of ϕ 70 mm × 10 mm. The wear was measured by the weight difference of sample before and after the wear. MA110 electronic analytical balance with an accuracy of 0.0001 g was used for weighing.

The experimental temperatures are 100 °C, 200 °C, and 250 °C, respectively. The loads are 50 N, 100 N, 150 N, and 200 N, while the revolution is 2000r, the sliding distance is 376 m, and the rotation speed is 200r/min. The sample is pre-ground for 3 min before the test to ensure that the sample can be synchronized with the grinding and be in complete contact with the grinding parts. The experimental

Fig. 9.28 MG-2000 high-temperature and high speed friction and wear testing device

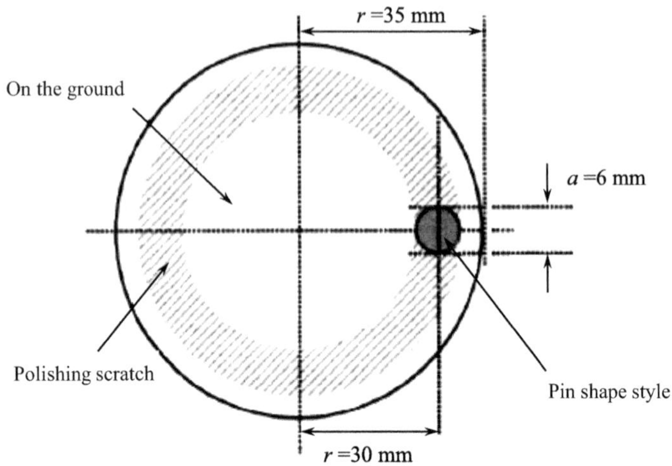

Fig. 9.29 Schematic diagram of high-temperature friction and wear test

materials are Al-22Si matrix alloy, 5.7 vol.%$(Al_2O_3 + Al_3Zr)_p$/Al-22Si composite, 8.4 vol.%$(Al_2O_3 + Al_3Zr)_p$/Al-22Si composite, and 11.2 vol.% $(Al_2O_3 + Al_3Zr)_p$/Al-22Si composite. Before each test, the sample and the grinding block shall be cleaned with acetone, and after each test, the grinding block shall be sanded to No. 1200 before the next test, so as to ensure that the surface roughness of the grinding block is constant for all tests.

9.2.2 *High-Temperature Friction and Wear Performance of High Silicon Aluminum Alloy and Its Composites*

9.2.2.1 Influence of Load on the Friction and Wear Performance of Matrix Alloy and Its Composites

Figure 9.30 shows the variation of friction coefficient (average value of steady-state friction coefficient) and wear weight loss of Al-22Si matrix alloy, 5.7 vol.%$(Al_2O_3 + Al_3Zr)_p$/Al-22Si composite, 8.4 vol.%$(Al_2O_3 + Al3Zr)_p$/Al-22Si composite, and 11.2 vol.%$(Al_2O_3 + Al_3Zr)_p$/Al-22Si composite with load at 100 °C. It can be seen from Fig. 9.30a that all the friction coefficients of the four materials decrease with the increase of load. When the load is less than 100 N, the friction coefficients of 5.7 vol.%$(Al_2O_3 + Al_3Zr)_p$/Al-22Si composite, 8.4 vol.% $(Al_2O_3 + Al_3Zr)_p$/Al-22Si composite, and 11.2 vol.%$(Al_2O_3 + Al_3Zr)_p$/Al-22Si composite are lower than that of Al-22Si matrix alloy, and the friction coefficient decreases with the increase of in-situ particle volume fraction. When the load exceeds 150 N, the friction coefficient of composite and the matrix alloy is almost the same.

Fig. 9.30 Influence of load on friction coefficient and wear amount of samples

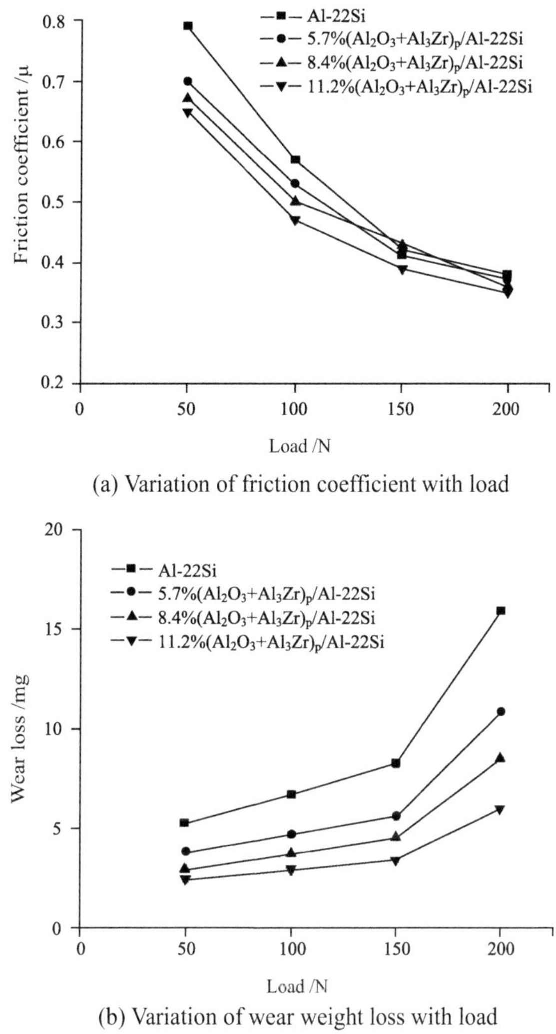

In Fig. 9.30b, all the wear amounts of the composite and the matrix alloy increase with the increase of the load, but the wear amount of the composite is always smaller than that of matrix alloy. The wear amount of composite decreases with the increasing volume fraction of in-situ particle. When the load is less than 150 N, the wear amounts of matrix alloy and its composites increase linearly. However, when the load is greater than 150 N, the wear amount increases rapidly, which indicates that both the matrix alloy and the composites show a state of rapid wear although the increased amplitude of composite is smaller than that of matrix alloy. The larger the particle volume fraction is, the smaller the increase of wear amount is. This is mainly due to the introduction of fine Al_2O_3 and Al_3Zr particles, which increases the anti-adhesion wear performance of the composite.

9.2.2.2 Effect of Temperature on Friction and Wear Performance of Matrix Alloy and Its Composites

Figure 9.31 shows the variation of friction coefficient and wear weight loss of Al-22Si matrix alloy, 5.7 vol.%$(Al_2O_3 + Al_3Zr)_p$/Al-22Si composite, 8.4

Fig. 9.31 Effect of temperature on the friction coefficient and wear amount of $(Al_2O_3 + Al_3Zr)_p$/Al-22Si complex material

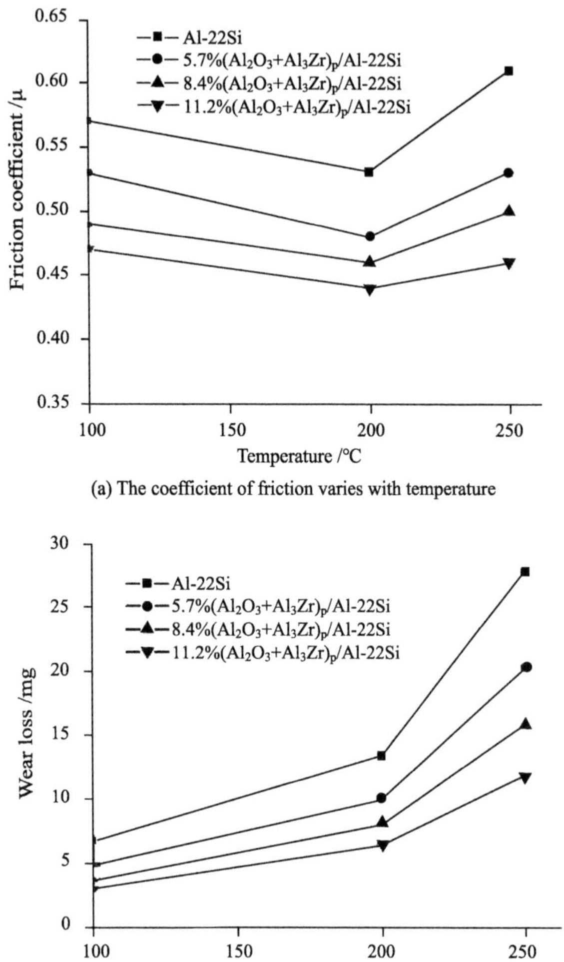

(a) The coefficient of friction varies with temperature

(b) The amount of wear varies with temperature

vol.%(Al_2O_3 + Al_3Zr)$_p$/Al-22Si composite, and 11.2 vol.%(Al_2O_3 + Al_3Zr)$_p$/Al-22Si composite with environment temperature under a load of 100 N. It can be seen from Fig. 9.31a that as the wear environment temperature increases from 100 °C to 200 °C, the friction coefficients of matrix alloy and its composites decrease gradually. When the temperature is 250 °C, the friction coefficients of the matrix alloy and its composites increase sharply, but the friction coefficient of composite is always lower than that of matrix alloy. The larger the particle volume fraction, the smaller the friction coefficient. From Fig. 9.31b, it can be seen that as the temperature increases from 100 °C to 200 °C, and the wear loss of matrix alloy and its composites increases linearly. When the temperature is 250 °C, the wear loss of the matrix alloy and its composites increases rapidly. At this time, the four kinds of materials have serious adhesive wear, but the wear loss of composite is always less than that of matrix alloy, which indicates that the adhesion resistance of the composite is higher than that of the matrix alloy. At the same time, the larger the volume fraction of particles, the smaller the wear loss and the smaller the increased range of wear loss. This is mainly due to the introduction of in-situ reinforcement particles which result in better high-temperature wear resistance of composites.

9.2.2.3 High-Temperature Worn Surface Morphology of Matrix Alloy and Its Composites

(1) Worn surface morphology of matrix alloy and its composites at 100 °C

Figure 9.32 shows the worn surface morphology of matrix alloy and its composites at 100 °C under a load of 150 N. It can be seen from Fig. 9.32a that under such conditions, the matrix alloy has undergone severe adhesive wear, resulting in large plastic flow in the surface and obvious wavy lines. As shown in Fig. 9.32b, the worn surface of 5.7 vol.%(Al_2O_3 + Al_3Zr)$_p$/Al-22Si composite also shows signs of adhesive wear and relatively large plastic flow. The grooves are relatively straight, while flake spallation is obvious, and they are deep with plastic deformation at the edges. Local plastic flow can be seen from Fig. 9.32c, but the degree of deformation has decreased, the peeling pit has also decreased significantly, and the plastic deformation at the edge of the furrow is still relatively serious. Figure 9.32d shows the worn surface of 11.2 vol.%(Al_2O_3 + Al_3Zr)$_p$/Al-22Si composite. It can be seen from the figure that the worn surface is obviously a relatively shallow abrasive wear groove. By comparing Fig. 9.32b–d, it can be seen that with increasing volume fraction of Al_2O_3 and Al_3Zr particles, the wear mechanism changes from adhesive wear and abrasive wear to abrasive wear.

(2) Worn surface morphology of matrix alloy and its composites at 250 °C

Figure 9.33 shows the worn surface topography of the matrix alloy and its composite at 250 °C under 100 N load. It can be seen that both the matrix alloy and its composite have serious adhesive wear at 250 °C. In the wear surface of matrix alloy (Fig. 9.33a), it can be found that the plastic flow and adhesive coating on the worn surface are

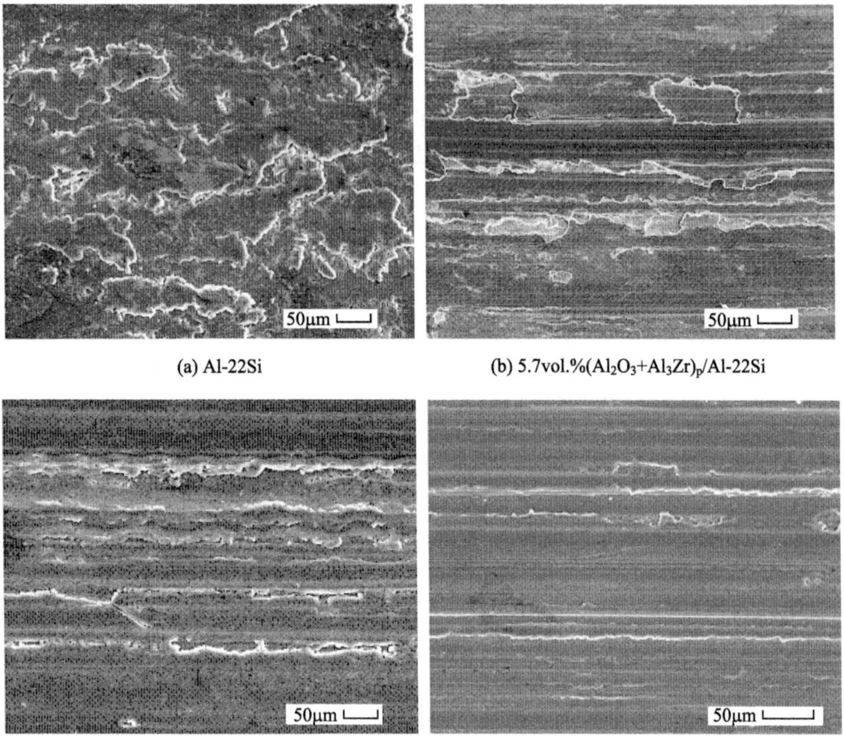

(a) Al-22Si

(b) 5.7vol.%$(Al_2O_3+Al_3Zr)_p$/Al-22Si

(c) 8.4vol.%$(Al_2O_3+Al_3Zr)_p$/Al-22Si

(d) 11.2vol.%$(Al_2O_3+Al_3Zr)_p$/Al-22Si

Fig. 9.32 SEM images showing surface wear of matrix alloy and $(Al_2O_{3+}Al_3Zr)_p$/Al-22Si composite

quite serious, and there are even pits formed by adhesive shedding (Fig. 9.33b). There are clear primary Si particles, indicating that the softening of the matrix is quite serious. Figure 9.33c shows that in 11.2 vol.%$(Al_2O_{3+}Al_3Zr)_p$/Al-22Si composite the adhesive surface wear phenomenon is not as that serious as in the matrix alloy, there are a laminated peeling phenomenon and deep grooves on the edges of plastic deformation (Fig. 9.33d). At this temperature, the wear mechanism of composite is mainly adhesive wear.

Singh and Alpas [8] have shown that for aluminum alloy and its composite/steel friction system, iron-rich oxide layer is generated on the friction surface due to the action of friction heat, which plays an in-situ self-lubricating role and leads to the reduction of friction coefficient and wear rate. At the same time, with the increase of ambient temperature, this effect is gradually enhanced. However, with the increase of environment temperature, the matrix softens gradually, and the adhesion between the matrix and the grinding block is strengthened, while the matrix material gradually transfers to the grinding block. Therefore, there is a critical value of the influence of

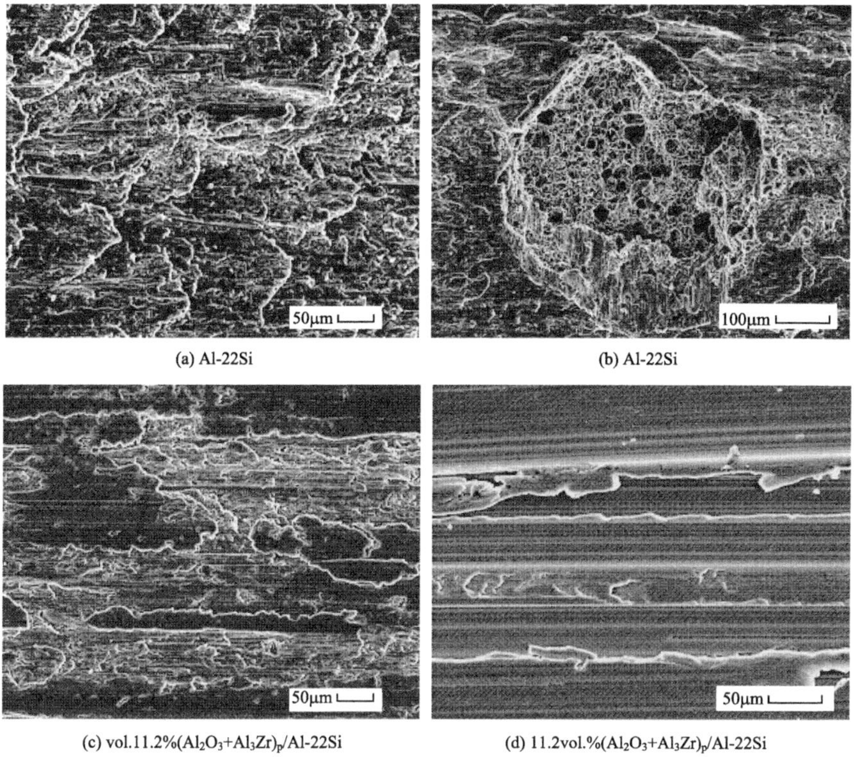

Fig. 9.33 SEM images showing wear surface of matrix alloy and $(Al_2O_3 + Al_3Zr)_p$/Al-22Si composite at 250 °C

temperature on the friction coefficient and the wear rate of the friction system, and the cowes pording influence can be minimum at this critical value.

During the wear process, there is a repeated action of stress and tangential friction in surfaces of composites and the matrix alloy. With the rising temperature of wear environment, the matrix alloy performance is declined, and the resistance ability to plastic deformation decreases gradually. In the range of wear environment temperature, the performance of reinforcement particles has no significant change, which is the hard phase and can strengthen the matrix. The hardness and the strength of composites are improved obviously, while the bearing capacity and thermal stability of matrix are enhanced. During the wear process, they can effectively play the role to block the plastic deformation of matrix and thus reduce the material transfer to the mill pair. There are only massive primary Si particles in the matrix alloy and a relatively large amount of voids at grain boundaries. Therefore, the Al-22Si substrate alloy suffers serious adhesive wear when the wear environment temperature is 100 °C. Meanwhile, the adhesion resistance of in-situ composites could be improved significantly.

It can be seen that Al_2O_3 and Al_3Zr particles generated by the chemical reaction can improve the thermal stability of the matrix and the resistance ability of the matrix to plastic deformation, but reduce the plastic flow caused by the temperature increase of the matrix alloy and effectively prevent the generation of adhesive wear. By improving the high-temperature performance of the matrix and reducing the friction force, the plastic flow and "hot adhesion" caused by thermal softening can be effectively prevented, and the critical transition temperature of the composite from slight wear to severe wear can be greatly increased as compared with the matrix alloy. However, when the temperature reaches 250 °C, severe adhesive wear will also appear in the composite.

Zhang [9] believes that when the wear of ceramic particle-reinforced metal matrix composite is relatively slight, its friction factor is caused by the action of three parts: adhesion, groove action, and external force required for the fracture of reinforcement. The addition of the ceramic reinforcement reduces the adhesion between the matrix and the pair, which is beneficial to the reduction of friction factor. However, with the increase of load, the wear-resistant reinforcement, as a cutting tool, will scratch the grinding block and form a groove in the surface, which will in turn act on the composite wear surface to enhance the furrow effect and offset the resistance to reduce the adhesion effect. Therefore, with the increase of in-situ particle volume fraction, the friction coefficient of composites and matrix alloys tends to be close as the load increases. However, with the increase of the test temperature, the adhesion of the matrix is enhanced due to "thermal softening", and then, the friction factor and wear rate of the composite increase with the increase of the temperature. In addition, from the influence of volume fraction at different temperatures on the wear rate of composites, it can be seen that the wear resistance of composite reaches the best when the volume fraction is 11.2%, which is consistent with the results of previous studies on normal temperature wear. This is due to the following factors: First, the matrix structure and performance are improved due to the existence of fine in-situ particles, and in-situ particles have a certain effect on the wear transfer of matrix, so the wear amount of the composite decreases with the increase of volume fraction. Second, after the addition of in-situ particles, the microstructure of matrix alloy is optimized, the size of primary Si particles is reduced, while the eutectic silicon is refined.

9.3 Wear Mechanism of In-Situ Aluminum Matrix Composites

9.3.1 *Analysis of the Worn Surface Morphology of Composite*

9.3.1.1 Worn Surface Morphology Analysis of Al-22Si Alloy

Figure 9.34 shows the SEM image of the worn surface of Al-22Si matrix alloy after sliding for 120 min when the sliding speed is 0.523 m/s and the load is 60 N. According to Fig. 9.34a, the worn surface morphology of Al-22Si matrix alloy along the sliding direction is relatively rough, and there are large peeling pits and some furrows. The deformation migration layer caused by deformation and extrusion is obvious and dense. The matrix is in direct contact with the steel grinding surface, and there is a large amount of plastic rheological behavior, leading to the formation of friction layer. From the observation of high-resolution white areas in Fig. 9.34a, b, it can be seen that the white area is obvious tearing pits, which are deep and wide. As the hardness of Al-22Si matrix alloy is low, the temperature of matrix surface layer

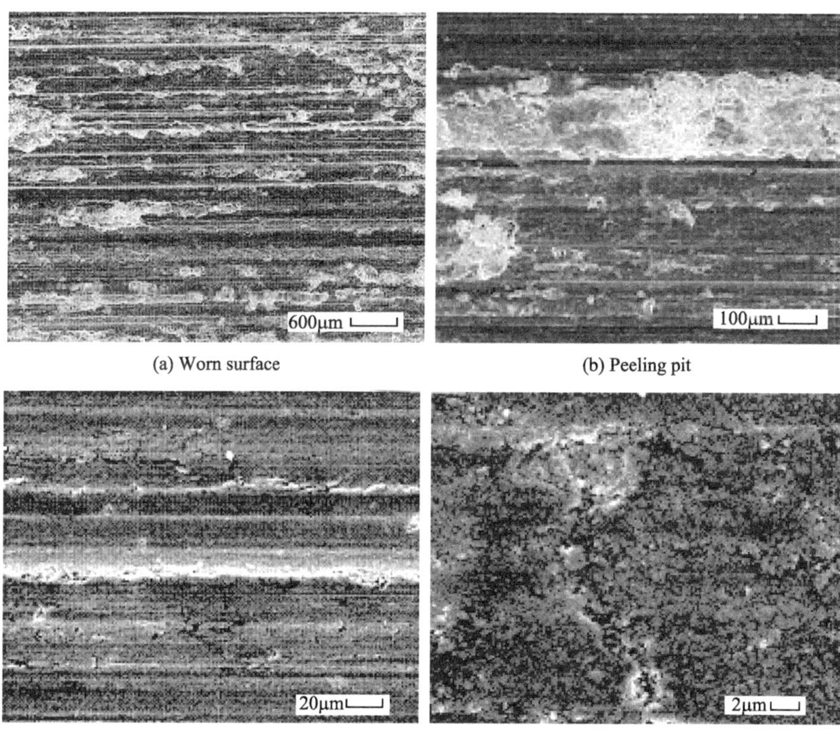

(a) Worn surface (b) Peeling pit

(c) Plastically deforming area (d) Surface crack

Fig. 9.34 SEM image of abrasion surface of Al-Si22 alloy increases during the process of friction, which leads to the softening and deformation of the surface layer. Then adhesion points with the role of ploughing forms in the worn surface, and increase gradually with the ongoing friction and wear process. Under the action of shear force, the spalling of aluminum alloy matrix increases, and there are severe deploring and mass transfer from the surface due to grinding, resulting in pits with large area [10]. Figure 9.34c shows the wear marks and furrow morphology in the worn surface of Al-22Si matrix alloy. The wear marks and furrows are wide, and the surface has no particle attachment, showing a large plastic deformation in the worn surface and a large number of tear grains perpendicular to the wear direction, which are mainly manifested as adhesive wear and abrasive wear. Figure 9.34d shows an enlarged image of a crack in a worn surface. The crack occurs mainly in the surface and extends toward the subsurface layer, connecting with a crack generated from another site.

9.3.1.2 Worn Surface Morphology Analysis of $(Al_2O_3 + Al_3Zr)_p$/Al-22Si Composites

Figures 9.35, 9.36, and 9.37, respectively, show the SEM image of the worn surface of 5.7 vol.%, 8.4 vol.%, and 11.2 vol.%$(Al_2O_3 + Al_3Zr)_p$/Al-22Si composites after sliding for 120 min at a sliding velocity of 0.523 m/s and a load of 60 N. It can be seen from Fig. 9.35 that the worn surface of 5.7 vol.%$(Al_2O_3 + Al_3Zr)_p$/Al-22Si composite has obvious plastic deformation and ploughing action characteristics, and the proportion of plastic deformation is still large, but the degree of plastic deformation is obviously lower than that of matrix alloy. Figure 9.35b shows the high-resolution microstructure of the white pit in Fig. 9.35a. At this time, the pits caused by tearing of the worn surface due to grinding are significantly smaller than those in the Al-22Si matrix alloy (Fig. 9.34b). This is because of the matrix strengthening by the fine reaction-generated reinforcement particles and resistance to the microcutting damage of grinding material to the composite surface, and thus, the resistance to shear

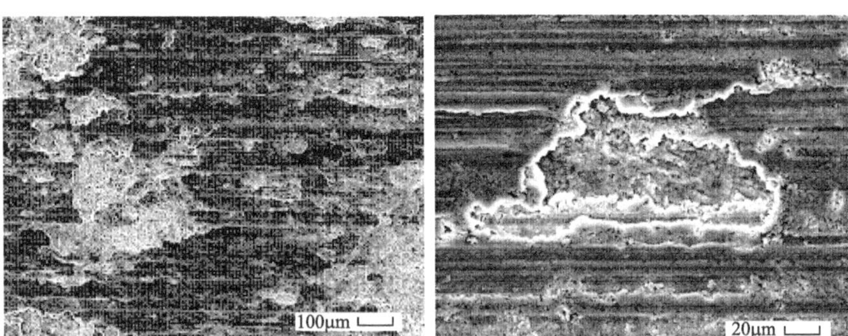

(a) Worn surface (b) Peeling pit

Fig. 9.35 SEM images of worn surface of 5.7 vol.%$(Al_2O_3 + Al_3Zr)_p$/Al-22Si composite

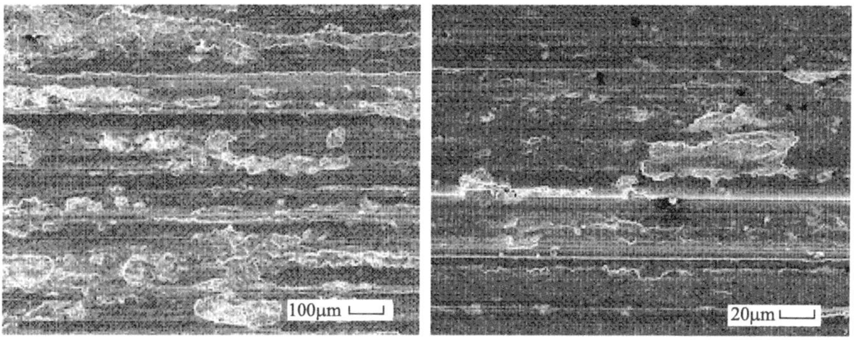

(a)Worn surface (b) Spalling pits and grooves

Fig. 9.36 SEM images of worn surface of 8.4 vol.%(Al_2O_3 + Al_3Zr)$_p$/Al-22Si composite

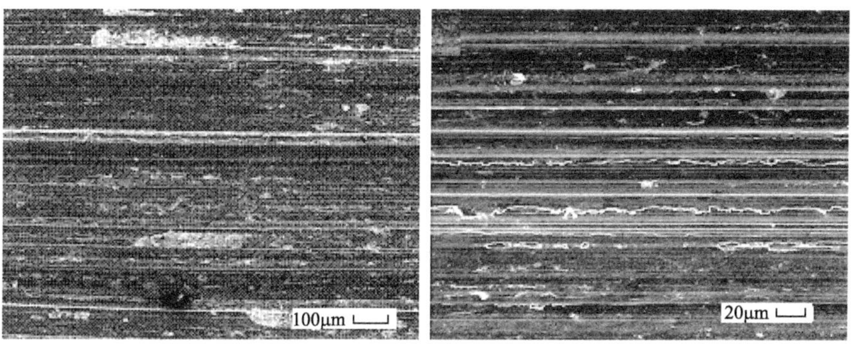

(a) Worn surface (b) Grooves

Fig. 9.37 SEM images showing worn surface of 11.2 vol.%(Al_2O_3 + Al_3Zr)$_p$/Al-22Si composite

strain of matrix alloy is improved. It is conducive to the formation of friction transfer film in the friction surface, which reduces the area bonded to the friction surface of Al-22Si matrix alloy, and then, the size and the number of adhesive avulsions in the worn surface are reduced [11].

Figure 9.36 shows the worn surface morphology of 8.4 vol.%(Al_2O_3 + Al_3Zr)$_p$/Al-22Si composite along the sliding direction. The surface is relatively smooth, and the pits are small, showing good wear resistance. The primary Si, Al_3Zr, and Al_2O_3 particles play the role of bearing the load together, and the ploughing effect is more obvious. Figure 9.37 shows the worn surface morphology of 11.2 vol.%(Al_2O_3 + Al_3Zr)$_p$/Al-22Si composite along the sliding direction, and it can be seen that the surface is smooth, the characteristics of plastic deformation reduce greatly, and the plough cutting characteristics are significantly increased. As can be seen from the figure that there are clear furrows formed by the abrasive cutting and the uplift (white lines) caused by the plastic deformation due to plough cutting. Compared with the matrix alloy, the strip lines formed in the won surface are fine and shallow, and the groove morphology is the main characteristic of abrasive wear [12]. Figure 9.37b is an enlarged image of furrows. It can be seen that the furrows are relatively shallow with small plastic deformation, and no large number of cracks caused by plastic deformation is found, while particle adhesion can be seen in the surface layer of grooves.

In a summary, the sliding friction wear of $(Al_2O_3 + Al_3Zr)_p$/Al-22Si composite is mainly abrasive wear, and the characteristics of abrasive wear become more obvious with the increasing particle volume fraction.

9.3.1.3 Wear Morphology of Al-22Si Matrix Alloy and Its Composites After Alloying

Figure 9.38 shows the worn surface morphology of alloyed Al-22Si matrix alloy after sliding for 120 min when the sliding velocity is 0.523 m/s and applied load is 60 N. It can be seen from Fig. 9.38 that compared with the worn surface of Al-22Si binary alloy, the worn surface of Al-22Si after alloying is flatter with only slight scratches and furrows. The plastic deformation of the furrow edge is very small. The spalling pits in Fig. 9.38b are also shallow and small, without a large area of adhesion. This is because the addition of alloying elements improves the hardness and the strength of matrix alloy, which increases the bonding strength of the interface between the primary Si particles and the matrix and improves the shear resistance of the material. In addition, the heat resistance of the matrix alloy can also be improved due to the addition of alloying elements. Under the same load, the friction heat generated is not easy to soften the matrix and produce plastic deformation, which is the premise of adhesive wear. The reason for the spallation is that the vacancy at the interface between the large primary Si particles and Al matrix provided a site for the crack nucleation.

Figure 9.39 shows the worn surface morphology of alloyed 11.2 vol.%$(Al_2O_3 + Al_3Zr)_p$/Al-22 Si composite after sliding for 120 min when the

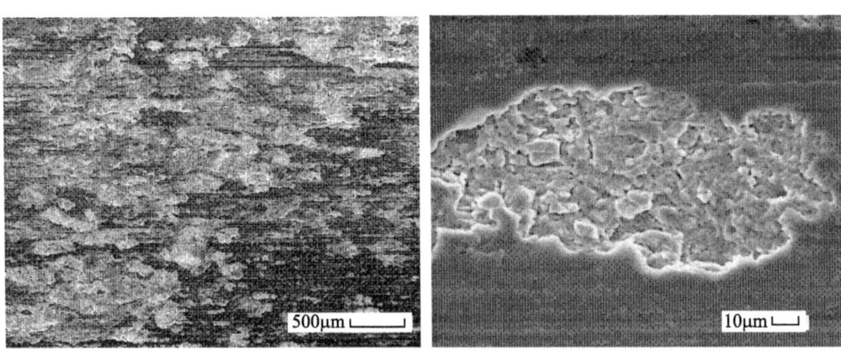

(a) Worn surface (b) Peeling pit

Fig. 9.38 SEM images of worn surface of Al-22Si matrix alloy after alloying

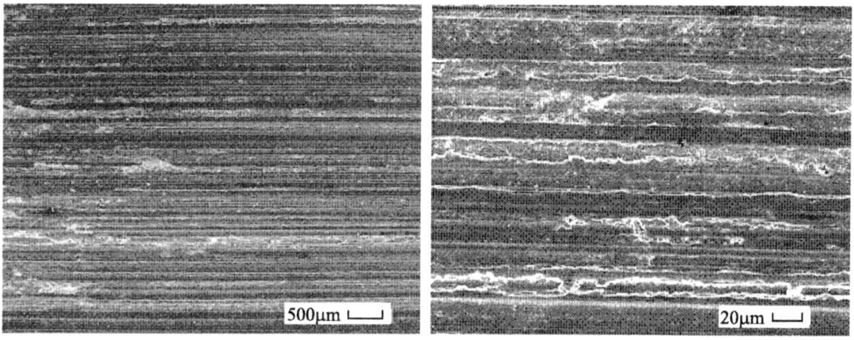

(a) Worn surface (b) Grooves

Fig. 9.39 SEM image of worn surface of 11.2 vol%$(Al_2O_3+Al_3Zr)_p$/Al-22Si composite

sliding velocity is 0.523 m/s and the load is 60 N. It can be seen from Fig. 9.39a that the worn surface of 11.2 vol.%$(Al_2O_3+Al_3Zr)_p$/Al-22Si composite is flat with few peeling layers, and there are relatively fine furrows and wear marks. This is because the energy condition of crack nucleation cannot be satisfied when the particle size is less than 2.5 μm [13], but the particles synthesized by reaction in $(Al_2O_3+Al_3Zr)_p$/Al-22Si composite are fine, usually only 0.5 ~ 2 μm, which is not easy to satisfy the condition of crack nucleation. Figure 9.39b is a high-resolution image of grooves. It can be seen that the grooves are very shallow and the plastic deformation at the edge is very small. Therefore, the wear mode of the composite is abrasive wear.

9.3.1.4 Grinding Chip Analysis of Al-22Si Matrix Alloy and Its Composites

Figure 9.40 shows the SEM images of grinding chips of Al-22Si matrix and its 11.2 vol.% $(Al_2O_3+Al_3Zr)_p$/Al-22Si composite. As can be seen from Fig. 9.40a, the grinding chips of Al-22Si matrix alloy are large, and their surface is with obvious plastic deformation and flaccid tearing grain, which is covered with bean dreg-like particles. However, the grinding chips of 11.2 vol.%$(Al_2O_3+Al_3Zr)_p$/Al-22Si composite material are a mixture of small pieces and bean dreg-like particles. The chip shape is affected by the plasticity and toughness of material and that of material with good plasticity is mainly flake, while that of material with poor plasticity is mainly small pieces and bean dreg-like. Although the matrix alloy can be enhanced by the primary Si particles, it is relatively soft and easy to flake off. However, with the addition of fine Al_2O_3 and Al_3Zr particles, the matrix is further strengthened, its plasticity is reduced, and the grinding chips are mainly particles and small pieces.

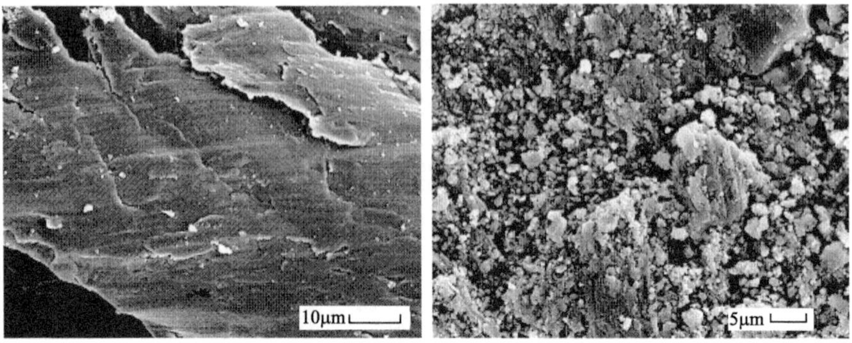

(a) Al-22Si (b) 11.2%$(Al_2O_3+Al_3Zr)_p$/Al-22Si

Fig. 9.40 SEM images of grinding chips from 11.2 vol.%$(Al_2O_3 + Al_3Zr)_p$/Al-22Si composite and Al-22Si matrix alloy

9.3.1.5 Wear Subsurface Morphology Analysis of Al-22Si Matrix Alloy and Its Composites

Composites and matrix alloys are subjected to the repeated action of positive pressure and tangential friction during wear, and their subsurface will undergo plastic deformation. The degree of plastic deformation is closely related to the load, the sliding speed, and the wear distance. It is generally believed that the larger the load, the higher the sliding speed, and the longer the wear distance, the greater the deformation [14]. As the subsurface deformation plays an important role in determining the wear mechanism, it is necessary to study the wear subsurface morphology.

Figure 9.41 shows the SEM images of the wear subsurface of Al-22Si matrix alloy under a load of 20 N and a wear time of 60 min. The subsurface in the vertical direction of wear is flat and smooth, with no plastic deformation zone. There are saw tooth wear marks and the longitudinal profile of the furrow. The subsurface in parallel wear direction is also relatively flat, but no cracks are found. In other words, under

Fig. 9.41 SEM images of subsurface worn of Al-22Si matrix alloy under 20 N load

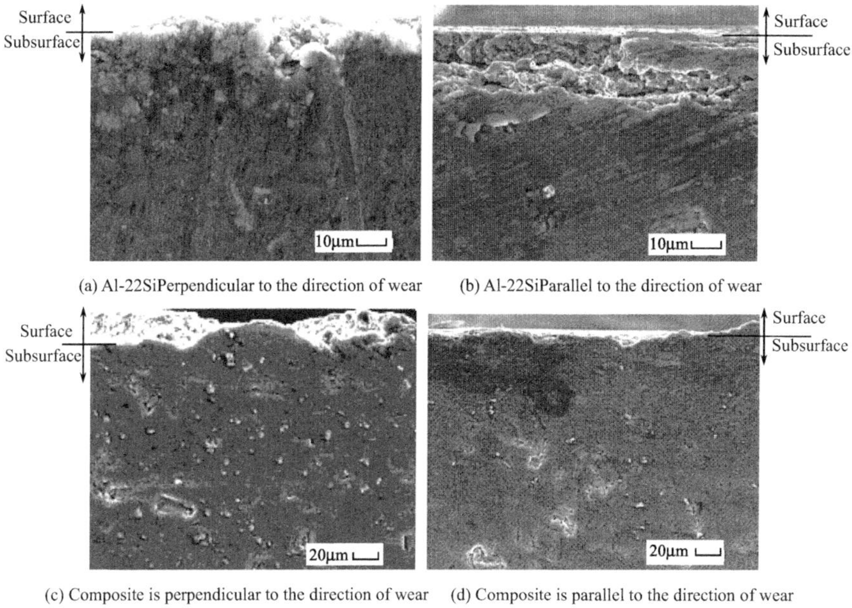

Fig. 9.42 SEM images of worn subsurface of Al-22Si matrix alloy and 11.2 vol.%$(Al_2O_3 + Al_3Zr)_p$/Al-22Si composites under 80 N load

20 N load, there is only slight wear and no plastic deformation in the subsurface area of the matrix alloy.

SEM images of matrix alloy and its composites are shown in Fig. 9.42 when the load is increased to 80 N. It can be seen from Fig. 9.42a that there is a large plastic deformation zone (the white part in Fig. 9.42a) in the wear subsurface of matrix alloy (perpendicular to the sliding wear direction). It can be seen from Fig. 9.42b that along the wear direction, there is a certain plastic deformation and flow, and there is an obvious crack zone, whose trend is parallel to the wear direction. It can be seen that the matrix alloy has undergone serious adhesive wear under 80 N load. This is due to the increase of wear time, which leads to the increase of wear interface temperature, which in a certain range also improves the capacity of matrix Al-22Si to accommodate hard brittle polyhedral primary silicon crystals. When the temperature exceeds a certain critical value, thermal softening of materials will occur in the subsurface area, which will lead to a wide range of plastic deformation under a certain load during sliding [15]. On the other hand, in the worn surface layer, the discoordination between the Al phase and Si phase during the deformation process leads to interface separation [16], leading to cracks in the worn surface of Al-22Si alloy which will expand to the subsurface. Under the action of shear force, tearing or peeling appears, resulting in tearing bands or peeling pits.

Above the crack area, the area close to the worn surface is the unshed layer, which is often referred to as the mechanically mixed layer in the literature [17].

Under the unloosed detachment layer, there is the plastic deformation zone, which is also the area of the subsurface with crack formation, where tiny but isolated cracks can be observed. As the wear processes, the upper unloosed layer will be worn off gradually, the deformation degree of plastic deformation area will increase gradually, while cracks will extend and connect to each other along the sliding direction, and the new stripping layer will be formed. The area below the crack formation zone is not affected by plastic deformation and is called the unaffected zone.

However, for 11.2 vol.%($Al_2O_3 + Al_3Zr$)$_p$/Al-22Si composite, the subsurface along the vertical wear direction is smooth and flat, the plastic deformation zone is small, there are some white structures only close to the worn surface (Fig. 9.42c), and there is also a serrated groove formed by ploughing. It can be seen from Fig. 9.42d that in the subsurface parallel to the wear direction, there are fine cracks in the area near the worn surface, the exfoliated layer formed is very thin, and the worn surface is smooth with almost no peeling pits. This is mainly due to the high bonding strength between the fine in-situ particles and the matrix, which prevents the development of composite toward adhesive wear. Moreover, the finer the particles are, the less the nucleation condition of interface crack between the substrate and particles can be satisfied, and the more effective the peeling wear can be prevented [18].

When the load is increased to 100 N, it can be seen from Fig. 9.43a that a white plastic deformation zone exists in the composite wear subsurface (perpendicular to the sliding wear direction). It can be seen from Fig. 9.43b that the composite wear subsurface (parallel to the sliding wear direction) has a stripping layer that is about to be peeled off, thus it can be seen that the worn surface is seriously damaged. This is because with the further increase of the load, the composite will gradually soften due to the effect of friction heat, and the subsurface crack area will gradually expand and finally expose the surface so that the peeling layer will be damaged and flake off, losing its protective effect, leading to a sharp increase in wear rate and serious adhesive wear, resulting in wear failure.

To sum up, the wear mechanism of Al-22Si alloy under different load conditions is different. Under 80 N and the Al-22Si matrix alloy sticks and begins to wear rapidly, while the ($Al_2O_3 + Al_3Zr$)$_p$/Al-22Si composite is abrasive wear. When the load is

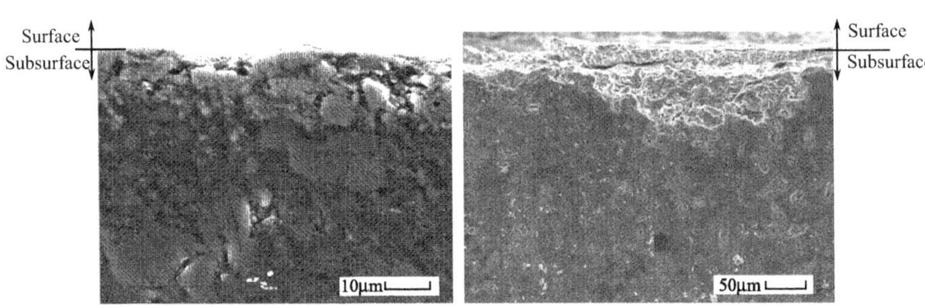

(a) Perpendicular to the direction of wear (b) Parallel to the direction of wear

Fig. 9.43 SEM images of worn subsurface of 11.2 vol.%($Al_2O_3 + Al_3Zr$)$_p$/Al-22Si composite under a load of 100 N increased to 100 N, 11.2 vol.%(Al_2O_3 + Al_3Zr)$_p$/Al-22Si composite begins to wear rapidly, and the wear mechanism is that the adhesive wear plays a major role, which is consistent with the previous analysis of the load effect on the composite wear.

9.3.2 *Analysis of Dry Sliding Wear Mechanism of Composites*

Based on SEM the observation and analysis of friction and wear surface, it can be concluded that the dry sliding wear forms of Al_3Zr and Al_2O_3 particle-reinforced composites by the melt reaction are mainly divided into two types: one is the mixed wear of adhesive wear and abrasive wear of composites, the other is the abrasive wear of composites. The two wear processes and mechanisms are analyzed as follows.

9.3.2.1 Adhesion and Abrasive Wear Mechanism of Composites

By observing the subsurface layer of worn composite samples, it is found that there is a layer of white structure in the subsurface layer of samples, as shown in Fig. 9.43a. The white tissue is a common wear product of adhesive wear, whose formation is related to the temperature rising of friction surface. Due to the welding of local contact points of friction pair material under the action of normal load, the bonding points are subjected to shear stress during the sliding process. On the other hand, the local temperature of worn surface increases during the wear testing, and then, the adhesive points in the material surface are prone to plastic deformation under the action of shear stress. The white layer observed under the scanning electron microscope is the deformed layer.

As shown in Fig. 9.43b, cracks are found in the transition zone. According to the delamination theory of adhesive wear [19], under the action of normal load and tangential load, the dislocation in the sample generates motion, as shown in Fig. 9.44. At the early stage of wear, Al_3Zr, Al_2O_3, and primary Si particles are directly immersed into the matrix under the normal applied load. The particles are wrapped by the Al matrix and cannot play a supporting role, and the Al matrix is in direct contact with the 40Cr grinding pair. As the hardness of the 40Cr grinding pair is much higher than that of Al matrix, the surface and subsurface of (Al_2O_3 + Al_3Zr)$_p$/Al-22Si composite have a large plastic deformation and flow under the action of shear force, and a large number of dislocations appear. Due to the presence of reinforcement particles in the composite, the dislocations are packed, as shown in Fig. 9.44a. In addition, due to the large difference between the thermal expansion coefficient of the particles and the matrix in the composite, the temperature of the sample will increase under the action of friction heat, resulting in a large thermal mismatch stress at the interface between the particles and the matrix. When the tensile stress and thermal mismatch stress on the particle and the interface are greater than the strength of the interface between the particle phase and the matrix, the interface between the particle and the matrix will be separated, and microcracks will form, as shown in Fig. 9.44b.

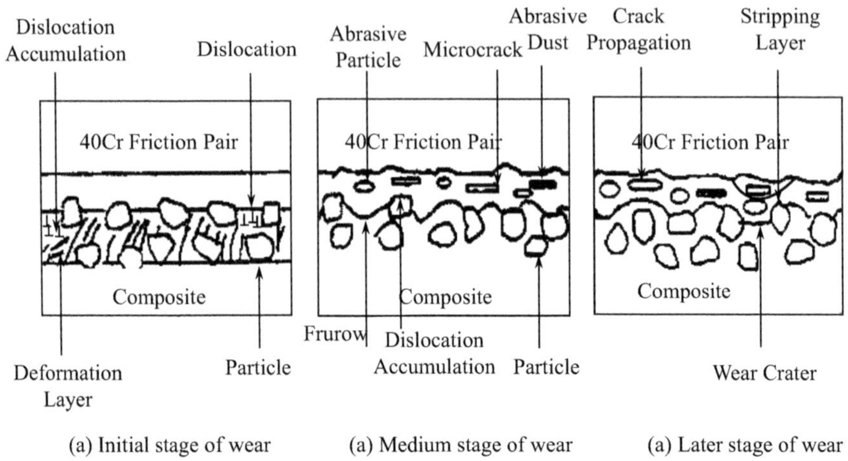

Fig. 9.44 Mixed wear mechanism model of adhesive wear and abrasive wear

The depth and size of the plastic deformation layer and crack nucleation zone increase with the increase of load, the crack expands along the sliding direction under the action of orthogonal shear stress, and some cracks extend to the material surface. Adjacent cracks may be connected to form a longer crack. When the critical length is reached, the material above the crack will peel off, and the lamellar grinding chip will be formed, as shown in Fig. 9.44c. Due to the good plasticity and fluidity of aluminum matrix, the surface of composite is quickly ground away, resulting in the shedding of particles. There are obvious traces of abrasive wear in the surface of composite: The shed particles and foliated chips become grinding particles and are pressed into the surface of composite under large load, while they slide with the movement of grinding block in the friction direction, resulting in the ploughing effects on composite and furrows in the worn surface of composite. At the same time, the ridge formed by the accumulation of a large number of particles and grinding chips is characterized as abrasive wear because the particles result in the ploughing effects and the plastic squeeze on both sides of the trench. As the friction and wear continue, the abrasive wear is enhanced, the number of abrasive chips and shedding particles increases, the cracks in the deformation layer of matrix surface increase, the grooves deepen, and the deformed ridges break, resulting in tearing of large chunks from the composite surface and large grinding pits, which finally leads to the wear failure.

9.3.2.2 Abrasive Wear Mechanism of Composites

It can be seen from the wear surface morphology of the matrix alloy and its composites that the furrow on the worn surface becomes more and more obvious with the increase

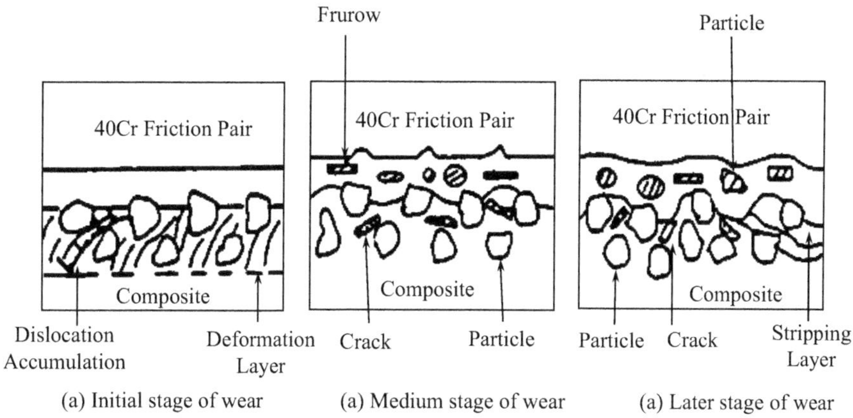

Fig. 9.45 Model of abrasive wear mechanism

of in-situ particle volume fraction. Grooves are the main feature of abrasive wear [20].

With the increase of the volume fraction of Al_3Zr and Al_2O_3 particles, the bearing particle spacing decreases, the hardness of composite increases further, the effective resistance of the microconvex points on the friction pair surface is enhanced, and thus, the wear resistance of the composite material increases with the increase of the volume fraction of particles. With the further development of the wear process, the grinding surface of the composite will be heated up due to the heat generated by friction, and the surface layer will be softened. The matrix of the composite is constantly ground away, resulting in the exposed particles, and the abrasive particles will be formed on the grinding surface of the composite. Its characteristic is abrasive wear, and the corresponding model is shown in Fig. 9.45.

At the normal wear stage, particles dropping from the grinding surface of the composite become abrasive particles. On the one hand, the ploughing action between composites and the 40Cr grinding pair result in grooves with a clear outline. On the other hand, due to the plastic extrusion of Al_3Zr and Al_2O_3 particles and the primary Si particles on the composite, grooves form on both sides of the ridge. With the development of wear, there are more and more grooves with the increasing depth, and the ridge will fracture and drop to form grinding chips. In addition, the accumulation of plastic deformation in the matrix will lead to cracks in the deformation layer, which will result in the formation of small chip shedding under the action of shear force and microcutting of abrasive particles, and thus, the wear loss of samples will be enhanced. With the increase of abrasive particles and grinding chips, the composite and the surface of the grinding pair will be damaged seriously, and the stage of rapid wear will be entered into. At the interface of particles, due to periodic tensile and compressive stresses, fatigue cracks are generated in the matrix around the particles, and the support to the reinforcement particles is weakened. When it is insufficient to resist the effect of friction on the reinforced particles, the particles will fall off from the matrix and form pits, and thus, the composite will eventually fail due to wear loss.

In conclusion, with the increase of the volume fraction of in-situ Al_3Zr and Al_2O_3 particles, the composite material gradually experiences from the mixed wear of adhesive wear and abrasive wear to abrasive wear when the load, the sliding speed, and the sliding time are same. As the load changes, the wear mechanism of composite changes from the abrasive wear to the mixed type of adhesive wear and abrasive wear. When the applied load exceeds a certain limit, adhesive wear will play a major role.

References

1. Huo X, Zhao Y, Chen G, et al. Wear properties and mechanism of silumin and its in-situ composites in dry sliding wear. Foundry. 2007;56(4):375–9 (in Chinese).
2. Zhang S, Dong X, Zhao Y, et al. Preparation and wear properties of TiB_2/Al-30Si composites via in situ melt reactions under high-energy ultrasonic field. Trans Nonferrous Met Soc Chin. 2014;24(12):3894–900.
3. Chen D, Zhao Y, Zhu H, et al. Microstructures and dry sliding wear properties of ZrB_2/A356 composites synthesized by magneto-chemistry in situ reaction. J Wuhan Univ Technol (Mater Sci Ed). 2013;28(2):384–8.
4. Jiao L, Zhao Y, Zhang Z, et al. Preparation and wear properties of (Al_3Zr+ZrB_2)P/2124 composites material with in-situ process. Advan Mater Res. 2013;616–618:1736–40.
5. Dwivedi DK. Wear behaviour of cast hypereutectic aluminum silicon alloys. Mater Des. 2006;27:610–6.
6. Erarslan Y. Wear performance of in situ aluminum matrix composite after micro-arc oxidation. Trans Nonferrous Met Soc Chin. 2013;23(2):347–52.
7. Kumar S, Chakraborty M, Subramanya Sarma V, et al. Tensile and wear behaviour of in situ Al-7Si/TiB_2 particulate composites. Wear. 2008;265(1–2):134–42.
8. Singh J, Alpas AT. High-temperature wear and deformation processes in metal matrix composites. Metall Mater Trans A. 1996;27(10):3135–48.
9. Zhang GD. Friction and wear behavior of aluminum allays reinforced with SiC. Tribol Lett. 2004;17:91–8.
10. Gautam G, Mohan A. Effect of ZrB_2 particles on the microstructure and mechanical properties of hybrid (ZrB_2+Al_3Zr)/AA5052 in situ composites. J Alloy Compd. 2015;649:174–83.
11. Onat A. Mechanical and dry sliding wear properties of silicon carbide particulate reinforced aluminium–copper alloy matrix composites produced by direct squeeze casting method. J Alloy Compd. 2010;489(1):119–24.
12. Toptan F, Kerti I, Rocha LA. Reciprocal dry sliding wear behaviour of B_4C_p reinforced aluminium alloy matrix composites. Wear. 2012;290–291:74–85.
13. Mazahery A, Shabani MO. Microstructural and abrasive wear properties of SiC reinforced aluminum-based composite produced by compocasting. Trans Nonferrous Met Soc Chin. 2013;23(7):1905–14.
14. Singh M, Mondal DP, Modi OP, et al. Two-body abrasive wear behaviour of aluminium alloy-sillimanite particle reinforced composite. Wear. 2002;253(3–4):357–68.
15. Zhu H, Jia C, Li J, et al. Microstructure and high temperature wear of the aluminum matrix composites fabricated by reaction from Al-ZrO_2-B elemental powders. Powder Technol. 2012;217:401–8.
16. Zhao Y. Microstructure and tensile property of in-situ composite synthesized by the direct melt reaction in the Al-Zr-O system. Foundry. 2002;51(3):153–6 (in Chinese).

17. How HC, Baker TN. Characterization of sliding friction-induced subsurface deformation of Saffil-reinforced A16061 composites. Wear. 1999;232:106–15.
18. Diler EA, Ipek R. Main and interaction effects of matrix particle size, reinforcement particle size and volume fraction on wear characteristics of Al-SiC_p composites using central composite design. Compos B Eng. 2013;50:371–80.
19. Kumar S, Panwar RS, Pandey OP. Effect of dual reinforced ceramic particles on high temperature tribological properties of aluminum composites. Ceram Int. 2013;39(6):6333–42.
20. Shockley JM, Descartes S, Vo P, et al. The influence of Al_2O_3 particle morphology on the coating formation and dry sliding wear behavior of cold sprayed Al-Al_2O_3 composites. Surf Coat Technol. 2015;270:324–33.

Printed by Printforce, the Netherlands